Principles of
Mineral Behaviour

GEOSCIENCE TEXTS

SERIES EDITOR

A. HALLAM

Lapworth Professor of Geology
University of Birmingham

1 Principles of Mineral Behaviour
A. PUTNIS and J. D. C. MCCONNELL

2 An Introduction to Ore Geology
A. M. EVANS

3 Sedimentary Petrology: an introduction
MAURICE E. TUCKER

GEOSCIENCE TEXTS VOLUME 1

Principles of Mineral Behaviour

A. PUTNIS
BSc, PhD
Senior Research Associate,
University of Cambridge

J.D.C. McCONNELL
MA, PhD
Reader in Mineralogy,
University of Cambridge

BLACKWELL SCIENTIFIC PUBLICATIONS
OXFORD LONDON EDINBURGH BOSTON MELBOURNE

Editorial Offices:
Osney Mead, Oxford, OX2 0EL
8 John Street, London, WC1N 2ES
9 Forrest Road, Edinburgh, EH1 2QH
52 Beacon Street, Boston, Mass., USA
214 Berkeley Street, Carlton
Victoria 3053, Australia

First published 1980

Sole distributors in U.S.A. and Canada:
Elsevier North-Holland Inc
52 Vanderbilt Avenue, New York,
New York 10017

Typeset by Enset Ltd
Midsomer Norton, Bath
printed by Billing & Sons Ltd
London, Guildford and Worcester
and bound by
Kemp Hall Bindery, Oxford

British Library
Cataloguing in Publication Data

Putnis, A
Principles of mineral behaviour.—
(Geoscience texts; vol. 1).
1. Mineralogy
I. Title II. McConnell JDC III. Series
553'.1 QE363.2

ISBN 0-632-00045-3
ISBN 0-632-00583-1 Pbk

Contents

Preface

One of the most important objectives in the study of minerals is an understanding of the processes which they undergo as a function of their geological history, whether these depend simply on changes in temperature and pressure or involve changes in chemical environment. The task of studying the behaviour of minerals is, in many cases, difficult as it is often quite impossible to carry out experiments on a laboratory time scale. Thus it is necessary to deduce the detailed characteristics of reaction mechanisms and their rates from the evidence of the microstructures in the natural mineral. This cannot be done successfully without a thorough appreciation of both the nature of the atomic mechanisms which operate in minerals and their likely kinetic parameters.

The study of mineral behaviour will have increasingly important implications in related disciplines within the Earth Sciences. Such an understanding is fundamental to the analysis of large-scale phenomena which are ultimately due to processes taking place within the minerals. One example is the interpretation of palaeomagnetism in rocks. Similarly a real appreciation of the behaviour of olivine during subduction depends on understanding its possible transformation mechanisms and their related kinetics. As we learn more about the special characteristics of mineral processes it is likely that a great deal of information will become available which is directly useful in determining both the time scale and the environment of geological events.

In the last decade the Earth Sciences have developed rapidly on many fronts, particularly in the field of large-scale tectonic processes. A result of this development seems to be that the study of mineralogy has become seriously undervalued in many universities. The extent to which major advances in the theory of processes in minerals have failed to penetrate the undergraduate textbook literature suggests the need for a broad survey at a level suitable for both undergraduates and lecturers assembling courses in mineralogy.

This book is an attempt at introducing the basic concepts of mineral behaviour in the simplest possible way. Very little previous knowledge of crystallography or thermodynamics is assumed. We have taken an entirely atomistic and microstructural approach and avoided a formal mathematical treatment of processes. As the book is intended for undergraduates the treatment is deliberately descriptive and attempts to show the broad picture, the wood rather than the trees.

The sequence of presentation is intended to parallel the way in which mineral problems are approached. Initially we must look at the structural and chemical make-up of minerals and the flexibility in structure and composition likely at higher temperatures. This is done in Chapters 2 and 3. Under a different set of physical or chemical conditions a different state may become more stable, and we next ask the question: what is the ideal behaviour of the mineral under equilibrium conditions? The various possibilities are described in terms of the thermodynamics of simple binary systems in Chapter 4. Real behaviour however involves the actual mechanisms by which transformations take place, and an appreciation of the rates of different mechanisms. These are described in Chapters 5 and 6. This approach enables us to estimate the extent to which equilibrium will be maintained and the likelihood of non-equilibrium, metastable behaviour.

In the second half of the book this philosophy is applied to a number of mineral systems which illustrate certain processes or aspects of mineral behaviour. The minerals chosen are some of those which have been studied in this way and the choice reflects the authors' own current interests. No attempt has been made to be comprehensive, although it is hoped that similar methods of approach could be applied to any mineral system. One of the principal techniques used is transmission electron microscopy which has made an important contribution to our understanding of the way minerals behave. Electron micrographs are used as illustrations where possible although the electron diffraction patterns which provide the key to the interpretation of microstructure are, for the sake of simplicity, omitted.

ACKNOWLEDGEMENTS

We are pleased to acknowledge the support of the Natural Environment Research Council which has funded research in this area by Studentships, Fellowships and Research Grants to members of the mineral sciences group at Cambridge. It is this original research which has stimulated the ideas for this book. We are grateful to the mineralogists who have provided copies of their micrographs for publication and to our colleagues in Cambridge for helping us formulate the ideas described in this book. Dr E.J.W. Whittaker of Oxford University has read the entire manuscript and we appreciate his valuable suggestions, although any remaining errors or infelicities are of course ours.

A. Putnis
J.D.C. McConnell

I

Mineral Behaviour—An Introduction

1.1 Mineral transformations

From the time a mineral first crystallizes from a melt or a hydrothermal solution, it it subjected to changes in both its physical and chemical environment. One of the remarkable features of minerals is the degree to which they respond to such changes by adapting their structures and compositions to this new environment. These adaptations may be subtle changes in bond lengths or major structural transformations; they may involve chemical changes on an atomic scale or reactions to form new species. All of these processes, however, have a common goal—to decrease the free energy of the mineral or mineral assemblage under the new set of conditions.

This book deals entirely with the ways in which minerals change or transform to decrease their free energy. We use the term transformation in a very general way to include any structural or chemical change which occurs in response to a change in the environment. We will be mainly concerned, however, with changes as a function of temperature, and to some extent pressure, and will not discuss the broader aspects of transformations such as chemical weathering and reactions between minerals during metamorphism. The general principles of solid state processes do, however, still apply. If we confine ourselves, in the first instance, to dealing with mineral transformations which depend on temperature, i.e. thermally activated processes, we will include most of the transformations of interest to mineralogists.

1.1.1 POLYMORPHISM

The simplest transformation is the inversion from one structural form to another as the temperature changes. Many elements as well as minerals have different structural modifications under different temperatures and pressures. This phenomenon of polymorphism is very common and may be illustrated by considering the various stable forms of silica, SiO_2, which are known to exist. In Fig. 1.1 the stability fields of the silica polymorphs are plotted against pressure and temperature. Each phase is stable only within the specific region of pressure and temperature and at room temperature and pressure there is only one stable phase, i.e. α-quartz. However, tridymite, cristobalite and coesite are all found in rocks exposed at the earth's surface; in other words, the transformations from these phases to α-quartz ought to occur, but the mechanism of the transformations is often too sluggish to be observed at low temperatures. This problem of kinetics will feature very strongly in most of the mineral transformations we will discuss, especially those which are geologically the most important. Polymorphism is a very important phenomenon in mineralogy because the different crystal structures assumed by one compound are controlled by the temperature and pressure of its crystallization. Once the phase diagram and

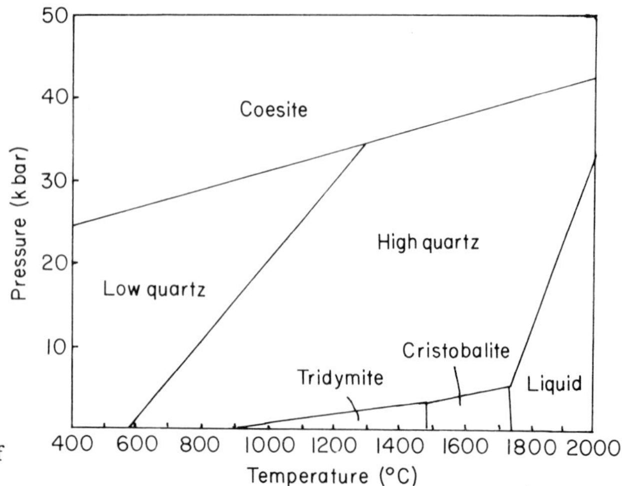

Fig. 1.1 Stability fields of some of the polymorphs of silica.

the transformation behaviour between the phases is properly understood, the conditions under which the mineral formed may be defined.

1.1.2 ORDER–DISORDER TRANSFORMATIONS

A more subtle transformation within a single compound is that of order–disorder. The simplest example of this phenomenon occurs in a simple alloy of two elements. Consider an alloy AB of the elements A and B, and assume that at some temperature it is completely disordered. Fig. 1.2(a) represents this situation. If at a lower temperature the two elements tend to surround themselves with unlike neighbours a state of complete order may be reached, as shown in Fig. 1.2(b). At some inter-

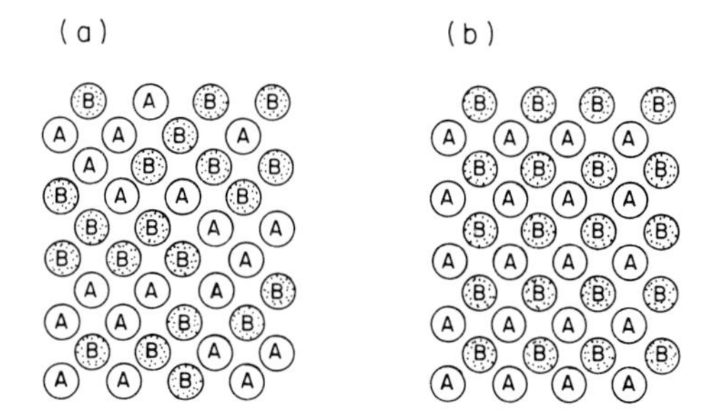

Fig. 1.2 (a) Disordered distribution of elements A and B and (b) an ordered distribution of A and B in an alloy AB.

mediate temperature a transformation from the state shown in 1.2(a) to that shown in 1.2(b) must occur. This type of ordering transformation has been widely studied by metallurgists in alloys, and although the structures of most minerals are far more complex, similar transformations are found. These transformations will be particularly important, mainly because of their extreme sluggishness, in the most common minerals, such as feldspars and pyroxenes. Thus there is an extended time scale over which to study the progress of the transformations and an opportunity to study the stages between complete disorder and complete order. This will provide the basis for the main geological applications of phase transformations—in defining the thermal history of a mineral from a study of the processes which have taken place as it has cooled.

1.1.3 EXSOLUTION

We have mentioned briefly two types of transformations in which the atomic arrangement has changed without any appreciable transport of the atoms themselves. The local chemistry has not changed. There are many cases, however, where a homogeneous mineral phase under one set of conditions, breaks up into regions of different composition under a new set of conditions. Consider Fig. 1.2(a) again. If at a lower temperature the elements A and B had tended to surround themselves with like neighbours, a situation like that shown in Fig. 1.3 would eventually develop, with the minimum number of atoms having unlike neighbours. This is

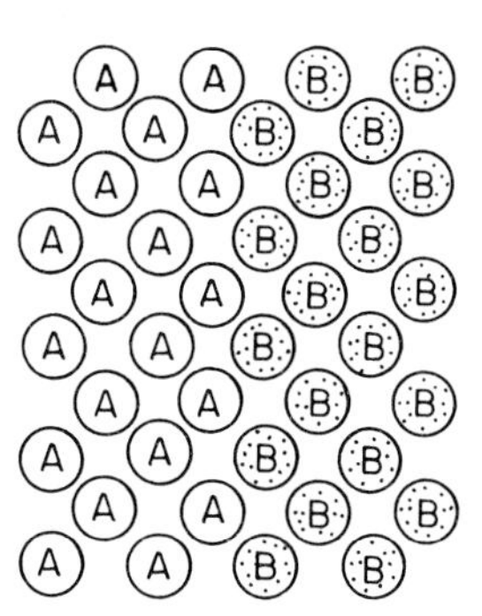

Fig. 1.3 Separation of elements A and B—exsolution.

again an ordered state, but the system is no longer chemically homogeneous. In minerals this type of process is usually termed exsolution or precipitation, and again is a very widespread phenomenon. Such a process clearly involves the migration, or diffusion, of atoms through the crystal structure, and more often than not this is an extremely slow process in most minerals. Therefore we again have the opportunity of studying the progress of a transformation over an extended time scale, and the many intermediate stages which will take place in a complex mineral structure undergoing exsolution.

The processes of exsolution and ordering should not be regarded as being mutually exclusive, and there are cases where both occur and at times even compete with one another creating a situation which, while being structurally very complicated, can yield a substantial amount of information about the thermal history of the mineral.

1.2 Electron microscopy and diffraction

In order to study the types of transformations that have been mentioned an instrument which can provide the magnifications required to observe the progress of the transformation on a very fine scale (ideally down to the unit-cell scale) and structural information from the regions being observed is essential. The transmission electron microscope is ideal for such work and its use in mineralogy over the last decade has made many important contributions to our understanding of processes in minerals.

For transmission electron microscopy the mineral specimen must be thin enough to be transparent to the electrons. In a 100 kV microscope the specimens must be thinned down to thicknesses of the order of 0.05–0.5 μm, depending on the density of the mineral. The usual procedure is to thin flakes of the mineral or parts of a standard optically thin section by ion bombardment. Alternatively simple crushing of separated mineral fragments will often result in enough grains with suitably thin edges. On passing through the specimen the transmitted electrons are focussed by a suitable optical system to form an image on a fluorescent screen or photographic plate. A schematic view of the column of a transmission electron microscope is shown in Fig. 1.4.

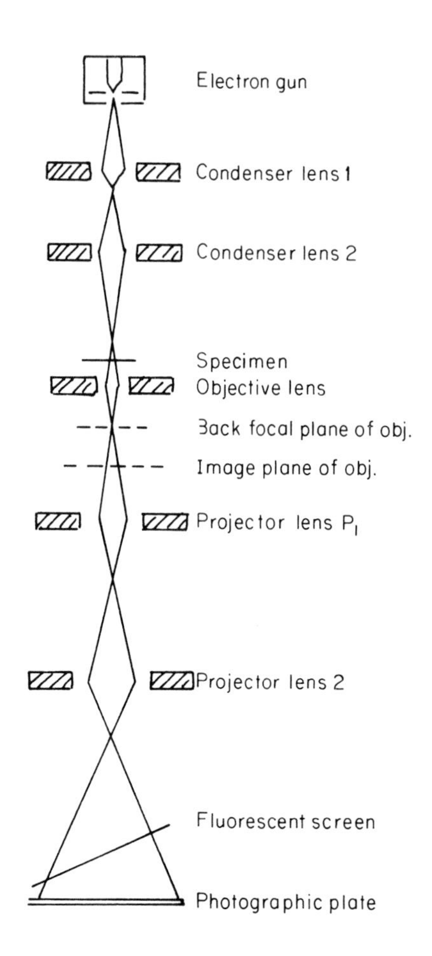

Fig. 1.4 Schematic view of the column of a transmission electron microscope.

Although we are not concerned with the details of the electron microscope technique here, many of the plates in the latter part of the book will be electron micrographs, and in order to be able to interpret even the simplest of these a few words should be said about the origin of the contrast we observe. In a crystalline specimen, the electron beam may be diffracted as it passes through the thin foil. Fig. 1.5 illustrates how such diffraction produces contrast in the image. The rays that are diffracted are blocked out of the transmitted image by an objective aperture. Consequently those parts of a crystal foil in a diffracting position, i.e. satisfying the Bragg condition for diffraction, will appear darker in the image due to the loss of the diffracted rays from the image. This effect is illustrated by the electron micrograph of fine scale polysynthetic twinning in cubanite (Fig. 1.6). One set of twins is in a

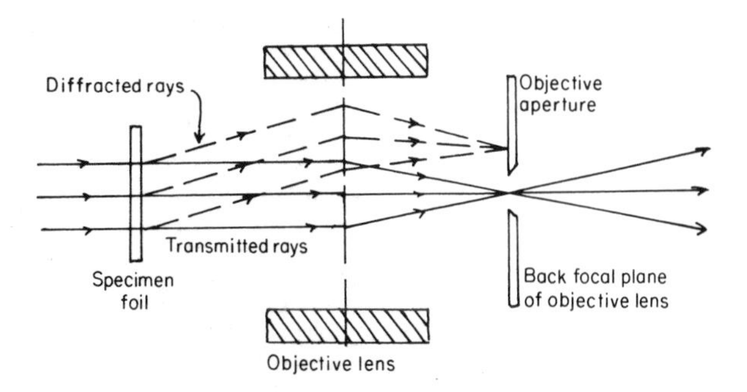

Fig. 1.5 Ray diagram illustrating the origin of diffraction contrast.

Fig. 1.6 Transmission electron micrograph illustrating the diffraction contrast from fine-scale twins in cubanite, $CuFe_2S_3$. The scale bar is 0.2 μm.

more strongly diffracting position and appears as darker lamellae. This is a bright-field image as it is formed by using only the undeviated beam. It is also possible to produce an image using any of the diffracted beams. This is called a dark-field image. Under these conditions only those parts of the crystal which are diffracting into the selected beam will appear bright. This technique provides valuable information on very fine intergrowths of phases, as well as on defects in the structure.

A thin foil is not generally perfectly planar, so that while some parts are diffracting strongly, other parts of the same crystal are not. This leads to broad bands of contrast in the image, termed bend contours. Again, the dark areas define those regions which are diffracting the electron beam out of the objective aperture. Another type of contrast arises from differences in thickness of a foil. This gives rise to thickness fringes in much the same way as in the optical diffraction from films. Fig. 1.7 shows such fringes in a thin specimen of rutile.

The presence of any structural or chemical inhomogeneities in the crystal will modify the contrast observed in the image. Specific examples will be described as they occur in the text. The contrast theory appropriate to such cases is in itself a separate topic on which research is by no means concluded. Some references to this may be found in the reading list at the end of the chapter.

Structural information from the specimen is contained in the diffraction pattern. The diffracted rays are all focussed in the back focal plane of the objective lens as shown in Fig. 1.8. Each diffracted beam arises from one set of lattice planes in the specimen which satisfies the Bragg condition, and the resulting diffraction pattern

Fig. 1.7 Thickness extinction contours at the edges of a thick crystal of rutile.

can be used to determine the crystallographic orientations and spacings of the lattice planes of the mineral. To observe this diffraction pattern on the screen it is only necessary to change the projector lens P_1 by increasing its focal length until it focusses on the back focal plane of the objective. A small aperture is inserted into the image plane of the objective lens to restrict the area of the specimen from which diffraction is observed, as illustrated in the ray diagram of Fig. 1.8.

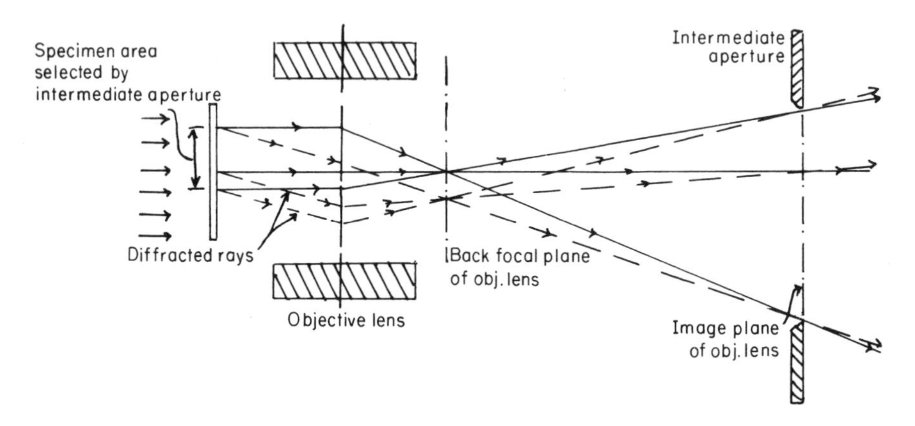

Fig. 1.8 Ray diagram illustrating selected area diffraction by an intermediate aperture.

The structural changes which may occur in a mineral during a transformation will produce a change in its diffraction pattern, and the ability to observe the course of such transformations in the image, as well as in the diffraction patterns from selected areas of the image, is one of the great advantages of electron microscopy in the study of mineral behaviour. In order to obtain precise structural information on the positions of individual atoms other diffraction techniques, which are more amenable to quantitative investigations such as X-ray or neutron diffraction, may be used. Each has its advantages in specific cases, as do the many other physical measurements which can be made on a mineral, but as transmission electron microscopy has had the greatest impact on recent research in mineral behaviour this technique will almost exclusively be referred to here.

1.3 Thermodynamics

As already stated, the mineral processes which we will discuss all tend to reduce the free energy of the system. In other words, the driving forces for change will be found by considering the thermodynamics of the system. These concepts can be treated with rigour and long equations, but this path will not be followed here. Throughout this book the aim is to establish a qualitative understanding of the processes and the relative rates at which competing processes proceed. The level of thermodynamic discussion will be extremely elementary and in no way attempts to introduce a thermodynamic approach to the study of transformations. A thermodynamic approach would attempt to define the parameters needed to calculate stabilities of the various physical and chemical configurations of the elements present. These parameters could then ideally be used to show how the stable configuration of a system will vary with the environmental conditions. Experimental determinations of such parameters are at present providing important data on minerals and on their ideal thermodynamic behaviour. The approach here, however, is directly the opposite. Our interest lies in what actually happens to a mineral as it cools, and simple thermodynamic arguments only will be applied in cases where they aid in the interpretation of this observed behaviour.

Thermodynamics deals with the beginning and the end state of a mineral process, and is not concerned with the mechanism of transforming from one state to another. In this sense, thermodynamics describes the ideal behaviour, divorced from the realities of the process, and assumes that the minerals are in equilibrium with their environment. There is no guarantee, however, that a mineral was ever totally in equilibrium with its surroundings. The rates of the processes involved are very often sluggish and if, for example, the cooling rate of a mineral was faster than its ability to adapt to the changing temperature, large departures from equilibrium would occur. Under conditions of non-equilibrium the behaviour of a mineral can be quite different from the stable equilibrium behaviour, and we will find many examples of this important assertion in the examples we discuss. Thus the real behaviour is determined primarily by the rates of the processes, and to understand this behaviour a discussion of the mechanism by which the processes take place is necessary.

1.4 Mechanisms of transformations

If we return to our schematic example of the disordered alloy AB and assume that on cooling it will begin to order at some temperature, T, ultimately to form a distribution such as shown in Fig. 1.2(b), we must ask ourselves the question: how does this change come about? Do the atoms cooperatively change places so that the number of atoms with unlike nearest neighbours continuously increases until a state of perfect order is reached? Or does a small region of perfect order form within the disordered alloy, such as shown in Fig. 1.9? This is one aspect of the problem of how the transformation starts, i.e. what is the mechanism by which the new phase forms? The first suggestion represents some type of continuous transformation mechanism, whereas in the second a discontinuity occurs with a distinct boundary between the parent and the product phases, in other words a nucleation event.

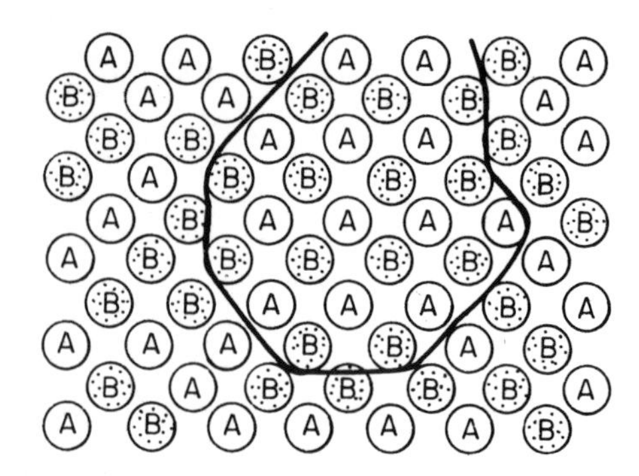

Fig. 1.9 Region of perfect order within the disordered matrix in an AB alloy.

An analogous situation exists if the AB alloy tends to ultimately break down to separate regions of A and B such as in Fig. 1.3. Under some conditions a continuous mechanism may lead to a distribution of some regions richer in A and others richer in B, whereas under different cooling conditions distinct domains of nearly pure A or pure B may form within the disordered structure. A study of the fine micro-structure which forms during the transformation may reveal which mechanism is operating. This distinction between the two types of mechanisms will be explored in some detail later, and will be found to be particularly relevant in mineral behaviour.

The study of mechanisms, in the broad definition of the term, is concerned with the relationships between the parent and product phases: their spatial dimensions, distribution and crystallographic relationships. Most importantly we want to know why some transformations fail to occur, or do so extremely sluggishly. To understand such constraints on transformations we must combine our knowledge of the thermodynamics of the process with an investigation of the structures of the phases involved. The relationship between parent and product structures will reveal points of similarity which will often give a clue to the way in which one changes to another. If the structures are very similar it might be possible to carry out some transformations with relatively few chemical bonds being broken. We would expect this to be an easier transformation than one in which all of the bonds have to be broken and rearranged in a different configuration. At this point we will relate the kinetics of the transformations to the structural relationships between the phases. It is reasonable, from the outset, to assume that transformations between phases with similar structures will be faster and so less likely to involve large departures from equilibrium.

Where the beginning and the end state of a process involve markedly different structures, we may find that the mineral finds it expedient to go through a number of transitional stages, each a little more like the ultimate end structure. While it may manage such a stepwise process, in that each step can occur at a reasonable rate, the direct transformation from the beginning to the end state may be kinetically negligible. Thus it can be seen that the kinetics will depend strongly in the structures of the mineral phases involved, and these will be discussed in some detail when specific transformations are described.

Similarly, if one considers the nature of the boundary between the parent and product phases, the relationship between the structures will determine the degree of misfit across the boundary. This misfit gives rise to a surface energy term associated with the presence of the boundary, and if it is remembered that the whole aim of the transformation is to reduce the free energy, the presence of additional energy terms will need some evaluation with regard to the net energy reduction. The types of boundaries formed during a transformation, their distribution and their energy contributions will therefore also require our attention.

Some of the ramifications of our approach to mineral behaviour should now be clear, but there is one further aspect of the transformation process which, although implied, has not yet been mentioned.

1.5 Diffusion

In the example of the disordered AB alloy we assumed that the ultimate configuration at low temperatures might be like that shown in Fig. 1.2(b), i.e. with unlike neighbours, or like that in Fig. 1.3 with like neighbours. In both cases the essential process involves atomic migration through the crystal. Any description of the rate of a transformation must be concerned with the way in which this atomic diffusion takes place: how does the atom move from one place in the structure to another, and what controls the rate at which this will occur? The first problem we encountered in considering transformation mechanisms was how does the new phase first appear. Now we want to know how it will grow, and in almost all of the processes of geological interest this growth is diffusion controlled.

There are two ways of looking at diffusion. The first may be called an *atomistic approach*. If during any process an atom is to move from one site in a structure to another, there must be an underlying thermodynamic reason for this to take place, i.e. the atom has a lower free energy in one site than the other, and so its movement will reduce the overall free energy of the crystal. The mode of migration of the atom through the structure defines the diffusion mechanism. Specific mechanisms for diffusion will be discussed later, but a simple example might be the straightforward interchange of atom A with atom B. A series of such exchanges would change the distribution of atoms of A and B.

In most cases any movement of this type will involve some breaking of bonds in the structure, and this will require some energy input. An increase in temperature, i.e. in the kinetic energy, of the atoms will increase the chances that any particular atom will be able to break away from its neighbours and jump to a new position, and therefore the diffusion rate will be strongly dependent on temperature. We cannot ignore the structure however. Minerals are often anisotropic—their structural arrangement is such that their physical properties will not be the same in every direction. Thus there will be easier diffusion in some directions than in others, Diffusion through regions of 'good' crystal structure is termed volume diffusion. In practice there may be a variety of other short-circuit diffusion mechanisms where the atoms move along paths of even easier diffusion. These easy diffusion paths usually involve defects in the crystal, or fast diffusion directions along free surfaces or boundaries within the crystal. Different mechanisms may predominate at different temperatures, and so we have a further point to keep in mind when discussing mineral transformations. Fig. 1.10 shows a number of these possible diffusion paths. The atomistic approach is concerned with diffusion pathways and the derivation if kinetic relations using various models for the actual movement of

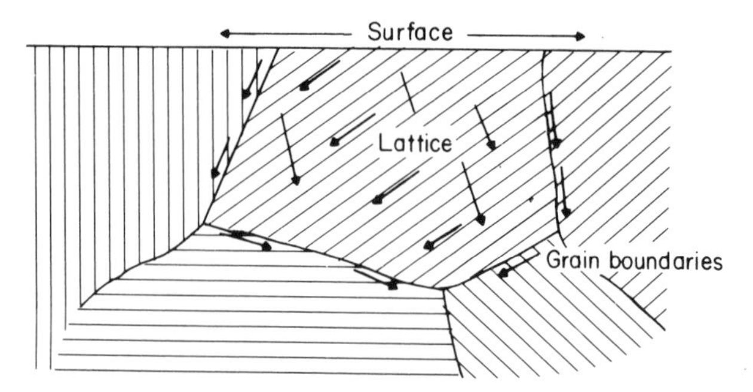

Fig. 1.10 Paths for surface diffusion, grain boundary diffusion and volume (lattice) diffusion.

individual atoms. When experimentally measuring diffusion rates, however, there is no direct way of relating our experimental results to the atomic mechanism operating, and we use what is termed a phenomenological or continuum approach.

In the *phenomenological approach* we ask the question: how can we describe, in terms of the parameters we can measure, the rate, and hence the amount, of mass transport that occurs when an element diffuses through a mineral? We consider the mineral to be a continuum without a discrete atomic structure. Most thermodynamic discussions of diffusion also adopt this approach since thermodynamics is primarily concerned with initial and final states and not directly with the atomic paths between these states.

Diffusion rates can be experimentally determined in two basically different ways. The first method can be illustrated in principle by the simple unidirectional diffusion experiment shown in Fig. 1.11. To measure, for example, the rate of diffusion of Fe in FeS, a diffusion couple is set up between pure FeS and FeS enriched in an

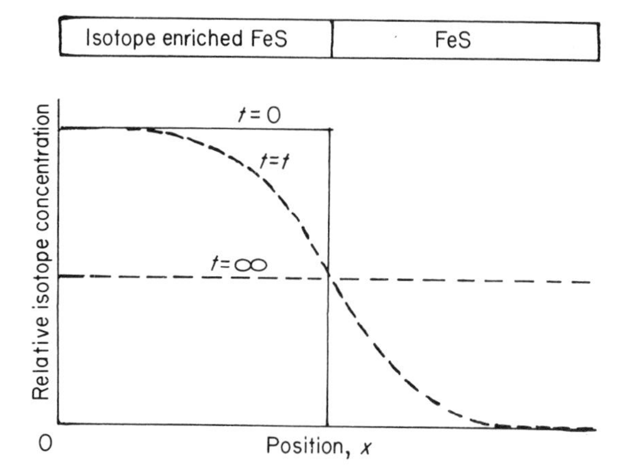

Fig. 1.11 A diffusion couple between isotope enriched FeS and standard FeS. The variation in the isotope concentration is shown as a function of position, x, and time, t.

Fe isotope which will act as a tracer. This couple is then heated to allow diffusion of the tracer isotope to occur at a significant rate. After some time at temperature, the diffusion couple is quenched to room temperature and the isotopic composition profile measured. The composition profile after time, t, might look as shown by the dashed line in Fig. 1.11. After an infinite time the composition will become constant at the average value. An experiment set up in this way enables the rate of self-diffusion of Fe in FeS to be measured at various temperatures from the shapes of the diffusion profiles. There are a number of variants of this method and complications may arise where the diffusion of one species may depend on the diffusion of another species in the opposite direction. The overall diffusion rate will then be determined by the slowest diffusing species.

The second method of determining diffusion rates is to study experimentally the progress of some transformation which depends on diffusion. For example, in an exsolution process where a single homogeneous phase breaks down into regions of different composition on cooling, the process clearly depends on the diffusion of atoms through the structure. If the exsolution process is monitored experimentally as a function of temperature and time and the distance over which diffusion has occurred measured from electron micrographs, the diffusion rates can be calculated.

In simple structures, measured diffusion rates can be compared with those predicted by the atomistic approach, and hence a likely diffusion mechanism found. The mechanism and rates of diffusion in metals are well understood and extensive data exist. Unfortunately, the same is not yet true for minerals. Very little is known about the diffusion mechanisms in complex crystal structures such as silicates, and, with a few exceptions, the fundamental rate data are virtually non-existent. The problems lie both in the sluggishness of many of the diffusion rates of geological interest, and the complexity of the structures. Experimental techniques now exist for determining reproducible diffusion data, and the increasing research efforts in this direction will hopefully improve this state of affairs.

The point has been reached where it is worthwhile to look back at what it is necessary to know about any mineral about to undergo a phase transformation. Firstly, what is the ideal behaviour of the mineral system, as determined by direct experiment or thermodynamic investigation? Although in most cases this is known, we will examine some minerals where the ideal behaviour can only be inferred. Secondly, by considering the relationship between parent and product phases, what are the possible transformation mechanisms, i.e. how is the formation of the product phase initiated? Are there likely to be severe constraints on its formation and, if so, will the transformation occur at any significant rate? If we expect large departures from the ideal equilibrium behaviour, is the non-equilibrium behaviour likely to produce a different phase (or phases) than the ideal behaviour? Assuming that the transformation does occur, what will determine the rate of the transformation, in other words, which aspect of the process will be the rate-limiting step? Where a number of different elements are involved in the diffusion, the one with the slowest diffusion rate must be identified. The kinetics of the transformation can then be calculated if the diffusion rate of this element is known. More often than not this will not be available, and the kinetics must be determined by direct experiments on the transformation. Even in extremely sluggish transformations some progress can be made in this direction by observing the course of a transformation on a very fine scale. If this point has been reached one is in a position to define the behaviour of the mineral and can begin to apply this to natural systems.

1.6 Kinetics

As we have seen, the overall rate of a transformation is determined by a number of factors. A synthesis of all of these effects and their contributions to the kinetics is a curve which will tell us the time it will take for a given fraction of the transformation to have been completed at any specified temperature. This curve is termed a Time–Temperature–Transformation (TTT) diagram. Fig. 1.12 shows the shape of such a curve. Although the underlying reasons for this C-curve shape will be discussed in some detail in a later chapter, there are a number of features which should be pointed out at this stage.

In Fig. 1.12, temperature is plotted vertically and the logarithm of the time plotted horizontally. The temperature, T_c, is the thermodynamically ideal transformation temperature, that is, the only temperature at which the product phase will be in complete equilibrium with the parent phase. The first heavy curve represents the time at the appropriate temperature at which the transformation product is first detectable, and the second heavy curve the time at which the reaction is virtually complete. The intermediate lines represent fractional amounts of transformation.

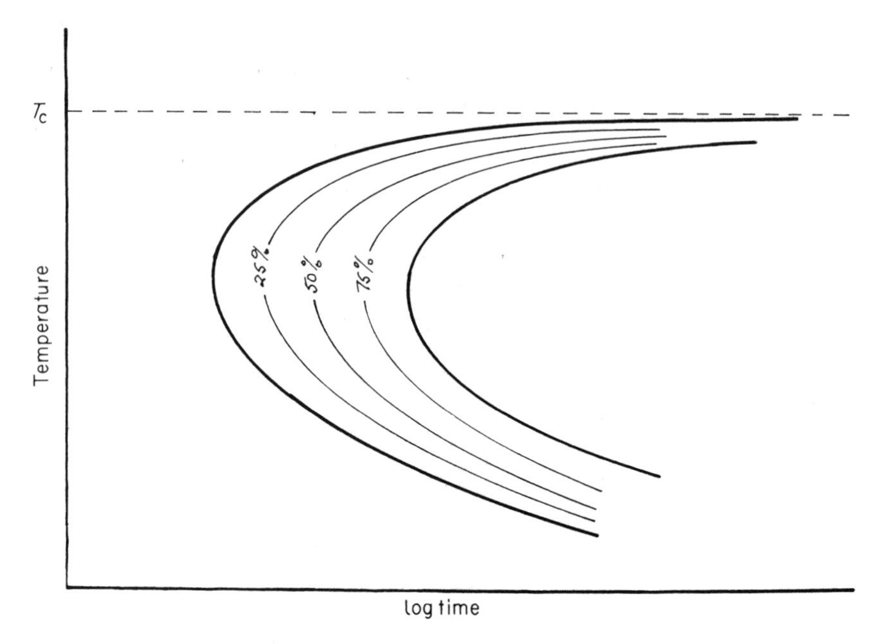

Fig. 1.12 The general shape of the Time-Temperature-Transformation (TTT) curves for a thermally activated transformation.

At temperatures near the equilibrium temperature, T_c, the rate of the transformation is vanishingly small. At lower temperatures the rate increases to a maximum, and then tails off again as the temperature is further decreased. Below a certain temperature the transformation virtually ceases due to the kinetics. This could be called the kinetic cut-off temperature.

It should be emphasized that if there were no kinetic barriers to a transformation and therefore the kinetics kept pace with the thermodynamic requirements, a single temperature would be sufficient to characterize the transformation. This is often the case in liquid–solid transformations where diffusion rates are fast. The fact that in the solid state transformations occur over a range of temperatures often extending to several hundred degrees is entirely due to their sluggishness. Because of this the importance of the TTT curve to the understanding of solid-state transformations cannot be overstated.

An equilibrium phase diagram, such as that shown in Fig. 1.1, only describes the thermodynamically ideal behaviour, and tells us nothing about the actual transformation. It is therefore absolutely imperative to consider both the TTT situation as well as the equilibrium phase diagram when considering mineral behaviour. In the past this has been neglected, so that while a large number of equilibrium phase diagrams for mineral systems exists, the TTT data are very scarce. This preoccupation with equilibrium phase diagrams has led many mineralogists to ignore the fact that the observed behaviour of many minerals cannot be characterized in terms of ideal equilibrium behaviour.

This is a theme that will be taken up again and again in the following chapters, but its importance warrants its inclusion in the introduction, even though its basis has not been fully discussed.

TTT curves can be determined experimentally by studying the progress of a transformation over a series of isothermal heating experiments for various periods of time. The mineral is quenched from the run temperature and its microstructure in the electron microscope used to determine the transformation characteristics.

1.7 Mineral transformations and thermal history

One of the aims of the study of mineral behaviour is to determine the thermal history of the parent rock. The nature and the scale of the microstructure in minerals have traditionally been used by petrologists to define, in a qualitative way, the likely cooling history of rocks. Thus the presence of tridymite (see Fig. 1.1) in igneous rock implies that it probably crystallized above 870°C and cooled sufficiently quickly to prevent the transformation of the tridymite to quartz. Tridymite is often found in volcanic rocks of rhyolite type. Alternatively, the presence of a coarse exsolution texture in a mineral is evidence that the host rock cooled sufficiently slowly for diffusion to have occurred on this scale.

If, however, the kinetics of mineral transformations are known the opportunity exists for quantifying such thermal histories. As the scale of the microstructure observed in a diffusion-controlled transformation is governed by the diffusion rates relevant for the process, a knowledge of these diffusion rates gives us direct access to the geological information contained in the mineral microscructure. An extremely fine-scaled microstructure may indicate rapid cooling if the diffusion rate is high, or it may result even after long periods at high temperatures if the diffusion process is extremely slow. Obviously, not all minerals or transformations are amenable to such applications, and for a mineral transformation to be useful in this context, the time scale for the transformation to take place must be similar to the time scale for geological processes.

In many minerals transformations are so sluggish that even after millions of years of slow cooling from high temperatures processes have not gone to completion. This gives us an extended time scale which includes most geological processes of interest. If mineral transformations were rapid relative to cooling rates their usefulness as indicators of thermal histories would be considerably lessened.

Diffusion rate data can be used to determine thermal histories without the need for a microstructure to be formed, as long as the process occurring is diffusion controlled. Consider the experimental situation in Fig. 1.11. In this experiment a knowledge of the heat treatment given to the diffusion coupled together with the measured diffusion profiles after various times, enables the diffusion coefficient to be calculated. Clearly, once this diffusion rate is known, the shape of diffusion profiles can be used to calculate thermal histories.

Before we make it seem as if the determination of quantitative thermal histories of minerals will soon be a routine procedure there are, as usual, the difficulties and even pitfalls. Under experimental conditions the number of variables will be small compared to those which may exist in natural systems. What, for example, will be the effect of trace elements and impurities on diffusion rates? Can diffusion rates measured at higher temperatures be extrapolated to lower temperatures? More seriously, how will the presence of fluid phases in a rock affect the rate of mineral transformations? One must be careful to correlate laboratory measurements with carefully analysed field situations. The transformations themselves may take place via a number of different mechanisms, often occurring together and each producing a different scale of microstructure, thereby making interpretation difficult.

These problems will be discussed when the specific transformations are described. Although we may be still some way from obtaining quantitative data on geological cooling rates, the potential of this approach to mineral behaviour is enormous. As more data become available the understanding of processes in minerals will be a necessary prerequisite to anyone interested in the evolution of rocks.

In this introductory chapter we have attempted to give a brief preview of the kinds of problems we will be discussing in the book as well as the philosophy behind our approach to the study of mineral behaviour. Each aspect of the transformation process is a major topic in itself, so that in order to present an overall view of mineral behaviour, i.e. the wood rather than the trees, the treatment is deliberately simplified and does not attempt to be rigorous.

The sequence of chapters broadly parallels the sequence of processes occurring in minerals as they cool from high temperatures. In this way the concepts introduced go from the simpler to the more complex. Few new concepts are introduced in each chapter and are reinforced by using them again and again in subsequent sections.

A convenient starting point is to consider minerals at high temperatures. Firstly, because many minerals crystallize at relatively high temperatures, and secondly, because at high temperatures their structures are the simplest and their compositions the most general. In Chapters 2 and 3 we will look at the situation minerals find themselves in at high temperatures and ask the question: what will happen as the mineral cools? This sets the stage for the processes which will follow.

A recurrent theme throughout the book is that the changes which occur in any mineral always decrease the free energy. It is appropriate, therefore, to introduce the main thermodynamic concepts which are required to discuss these changes and this is done in Chapter 4. The subsequent chapters describe the factors which dictate the real behaviour of a mineral undergoing a transformation. The actual process which occurs is governed by both the thermodynamics of the system and the kinetics of the mechanisms involved in the process, and it is this interplay between the thermodynamics and kinetics which will provide us with the more interesting aspects of mineral behaviour.

In the second half of the book the principles of transformation behaviour are applied to the behaviour of a number of mineral systems. The processes themselves form the backbone with the minerals being used as examples. The minerals dis-discussed are those which have been fairly well studied and which serve to illustrate some particular process. No attempt is made to cover a wide range of minerals; rather, fewer minerals are covered in more depth, as the principles involved can be subsequently applied to any mineral.

References and additional reading

Grundy, P.J. and Jones, G.A. (1976) *Electron microscopy in the study of materials.* Edward Arnold Ltd.

McConnell, J.D.C. (1977) Electron microscopy and electron diffraction. In *Physical Methods in Determinative Mineralogy.* 2nd Edition. (Zussman, J., Ed.) Academic Press, London.

Wenk, H.R. (Ed.) (1976) *Electron Microscopy in Mineralogy.* Springer-Verlag, Heidelberg.

2

Minerals at High Temperatures

Many of the minerals of interest to us have crystallized at relatively high temperatures. The high-temperature state of minerals is characterized by their chemical variability and their most generalized crystal structure. In other words, while the low-temperature structure of a mineral may be quite specific in terms of the nature and position of all of the constituent atoms, there is a certain amount of randomness associated with the high-temperature state. This makes it a convenient starting point for two reasons: firstly, the description of the mineral structure will be simpler, and secondly, the change from 'randomness' to 'non-randomness' is one of the important aspects of mineral transformations on cooling.

2.1 Atomic freedom and disorder

In an ideal ordered crystal structure every atom occupies a specific position and the spatial relationship between atoms leads to a regularity which is essentially infinite in extent relative to the size of the repeating unit of the structure. All unit cells in the crystal are identical in both geometry and chemical content and the structure is described completely by specifying the dimensions of the unit cell and the coordinates and nature of every atom in the cell. The geometry is described by the way in which the atoms are associated in the structure, i.e. by bond angles and bond lengths. The chemical content depends on the nature of the atoms in the unit cell.

This 'frozen' model of a crystal structure describes a perfect order which is not found in real crystals. The atoms in a real crystal oscillate about their mean positions and these thermal vibrations represent a form of disorder. The energy added to a mineral by an increase in temperature results in an overall increase in mobility of the atoms, increasing the disorder due to thermal vibrations as well as changing the mean bond distance between atoms. This is manifested in thermal expansion. Thus the size and shape of atomic arrangements may change with temperature and our description of the mineral structure at high temperatures approaches that of an averaged structure in which local geometrical differences are smoothed out.

The types of disorder found in minerals at high temperatures depend on their structures. The chemistry and the nature of the bonding will determine the degree of flexibility allowed. Some of the more important types of disorder will be described here in general terms, and will be referred to later when mineral structures are discussed in more detail.

2.1.1 POSITIONAL DISORDER

All atoms in a structure undergo thermal vibrations which can be described as disorder of position on a time basis. The amplitude of this vibration will depend on a number of factors including the nature of the bonds around the atom in question and the size of the atom relative to the size of the atomic site. In Fig. 2.1 a schematic diagram illustrates the vibrations of an atom in a relatively large site in a structure. The presence of such vibrations can be deduced from crystal structure analysis by a measurement of the uncertainty in the actual position of this atom. In some cases it is important to distinguish between two different interpretations of such positional uncertainty.

Firstly, does this apparent uncertainty arise from the anisotropic thermal motion of a centrally positioned atom, or does it represent the space average of four different positions of the atom, as illustrated in Fig. 2.1? If the latter is true, then in some parts of the structure the atom will be vibrating about mean position 1; in other parts the mean position will be 2, 3 or 4. In other words the whole crystal would then have to be a mosaic of regions or domains of these four different types. This type of disorder is termed positional disorder (on a space basis) whereas vibrational disorder implies an exactly equivalent mean position for all of the equivalent atoms in the structure.

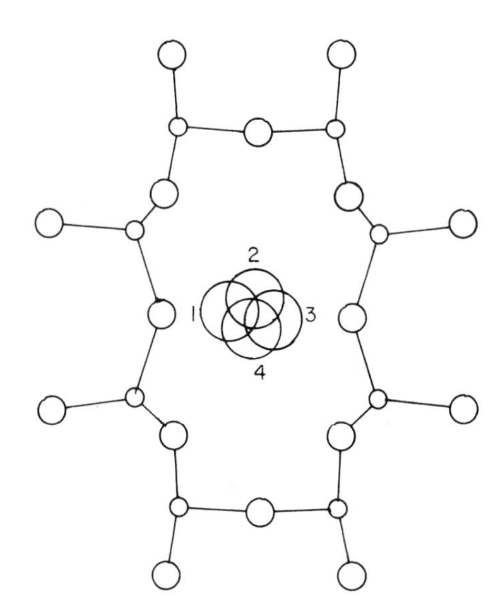

Fig. 2.1 An illustration of the possible positional disorder of an atom in a relatively large site in a structure at high temperatures.

The difference between these two interpretations is not easy to ascertain by the usual methods of crystal structure analysis. Positional disorder of a certain atom simply increases its apparent thermal motion. Careful structural analysis over a number of temperatures can help to decide which model gives the best fit with the data. Correct interpretation of this phenomenon is however very important in deducing the nature of transformations in some minerals. The position of the Na atom in albite is considered to be an example of such positional disorder, and will be discussed further in the section on albite.

2.1.2 DISTORTIONAL DISORDER

In a structure exhibiting positional disorder of an atom at high temperatures, the overall structure is described as a space average over the different possible positions of the atom within its site. In distortional disorder, the same general idea is applied to the whole structure.

The mechanism for distortional disorder is suggested in Fig. 2.2. The central diagram (b) represents the overall structure at high temperatures: square coordination groups of atoms joined by straight bonds. Suppose that in this hypothetical structure it is found that at a lower temperature a distortion takes place so that the bonds joining the square groups are now bent. Fig. 2.2(a) and (c) show the two equally likely distortions which could result. We would therefore expect that on cooling the high-temperature form, roughly equal proportions of these two alternatives should form, and this is generally found to be the case in structures of this type.

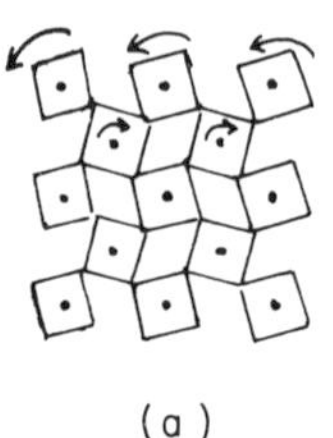

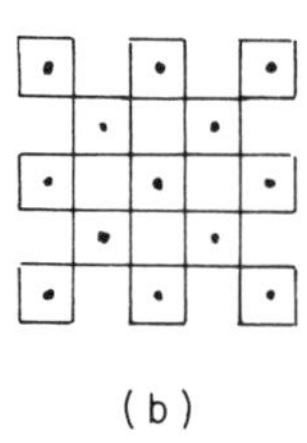

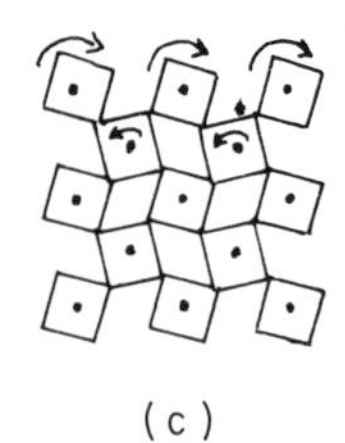

Fig. 2.2 (a) and (c) show two possible modes of distortion of the undistorted structure (b). Thermal oscillation from one distorted configuration to the the other would result in an average structure (b).

The question arises however as to the true nature of the high-temperature form. A description which involved equal proportions of these two distorted configurations intimately associated on a very fine scale would produce an average structure such as that in Fig. 2.2(b). If at high temperatures the thermal agitation of the structure produced oscillations from one distorted configuration to another, and these oscillations were correlated only over very small regions, the result would be a high-temperature structure which was a space and time average of the two low-temperature alternatives. The disorder consists in the mixture of these two structures. This is termed pure distortional disorder.

The relationship between α and β quartz is probably of this type and will be discussed in more detail in the section on the behaviour of silica.

2.1.3 SUBSTITUTIONAL DISORDER

In the two types of disorder mentioned above we have considered the effects of increased thermal agitation on the positions of atoms within a site, and secondly on the distortion of bonds between atoms or groups of atoms. These types of disorder do not involve any migration of atoms within the structure.

The next step is to consider the effect of an increase in temperature on the site occupancies in the structure. At any temperature diffusion processes will result in an interchange of atoms between sites. If there are two or more distinct atomic sites in a structure a certain equilibrium distribution of atoms will be set up between the sites. Thus we can assign average site occupancies for each site even though the atoms are very far from static. As the temperature increases, however, previously distinct sites become more similar on average, until finally the sites become indistinguishable. The interchange of cations between sites then leads to chemical disorder where the average chemical content of each site is the same. This type of substitutional disorder is characteristic of many substances at high temperatures.

The very simplest example of this type of disorder is to consider again the distribution of two elements A and B on a simple cubic lattice. In Fig. 2.3(a) we have an ordered distribution of A and B. If the interchange of A and B is allowed at random a distribution in which statistically there is an equal chance of any atomic position being occupied by A or B will eventually be produced. The site occupancy is 0.5 for A and 0.5 for B. This is shown in Fig. 2.3(b). In each figure the dotted lines outline the unit cell of the repeating unit for the structure. Note that the unit cell of the disordered structure is smaller.

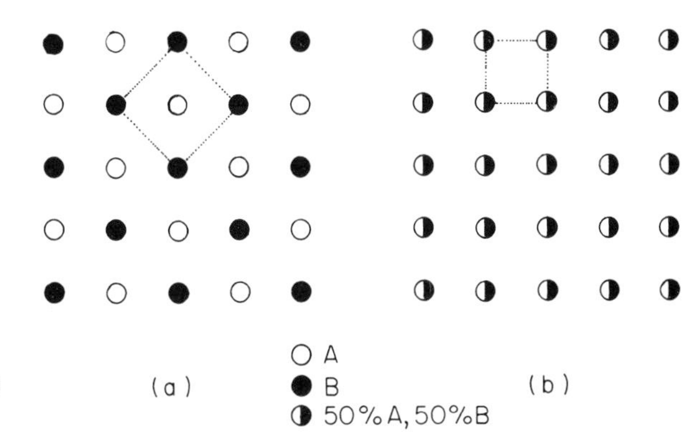

Fig. 2.3 (a) shows an ordered distribution of atoms A and B on a simple cubic lattice. In (b) the distribution is completely disordered. The unit cell is shown by the dotted lines in each case.

The structures of most minerals are considerably more complex than this simple example, but the phenomenon in complex structures can often be reduced to a very similar description. If the structure of chalcopyrite is considered [$CuFeS_2$ shown in Fig. 2.4(a)], we can see that there are two cations Cu and Fe arranged in

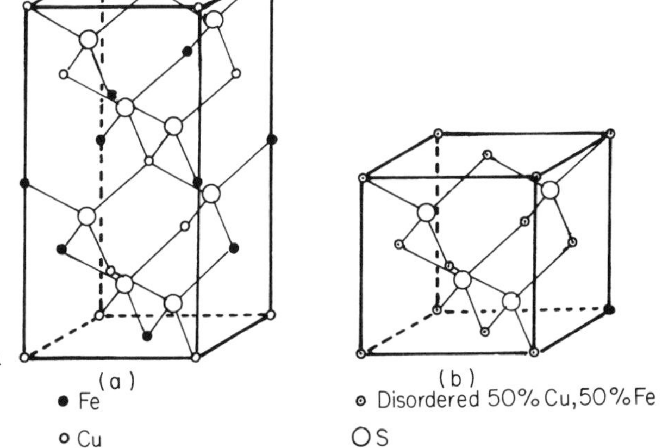

Fig. 2.4 (a) The chalcopyrite ($CuFeS_2$) structure with an ordered distribution of Cu and Fe. (b) The unit cell of the structure if Cu and Fe were disordered.

an ordered way within the framework of sulphur atoms. If at some higher temperature the distinction between a Cu site and an Fe site is lost, the distribution of Cu and Fe would become disordered. The equivalence of these two sites results in a disordered structure whose unit cell is halved in volume, as shown in Fig. 2.4(b). Throughout the disordering process the framework of the sulphur atoms can be considered to have remained essentially static.

A further consequence of the disordering of Cu and Fe in this structure is that the equivalence of the cation sites implies that in the disordered state the structure may accept a change in the proportions of Cu and Fe present. Here is the basis for an even greater degree of chemical randomness in the high-temperature state of minerals, i.e. the existence of solid solutions.

The example of $CuFeS_2$ represents a general situation in which a cation distribution may be disordered within an ordered anion framework structure. Many similar examples will be found in other sulphides as well as oxides. A similar, but more complex type of disorder is found in some silicate structures in which the framework structure, made up of (Si,Al,O), may be disordered as well as the cations within this framework. We will discuss these in more detail in the section on the crystal structures of minerals.

The net effect of an increase in temperature on a mineral is to bring about a more disordered structure both geometrically and chemically. In the description of such a structure distortions are, on average, eliminated thereby increasing the amount of symmetry. Site populations are also averaged resulting in a greater number of equivalent atomic sites and hence a smaller unit cell. At high temperatures a mineral will in general have its highest symmetry and its greatest chemical variability.

The general rule that high-temperature states will be associated with a greater degree of disorder implies that in a thermodynamic description of the change from a high-temperature state to a low-temperature state a purely statistical factor related to this degree of disorder must play an important role. This factor is termed the entropy of the system.

2.2 Disorder and entropy

When we say that a system is disordered we are describing the state of the system in much the same way that we might speak of its mass, its volume or its internal energy. The property of the system which may be related to disorder is its entropy.

The concept of disorder or randomness in a structure may be expressed statistically by the number of different ways in which atoms can arrange themselves. Thus if the disorder is due to vibrations in the lattice we are concerned with energy levels in the crystal and the different ways in which they are filled; in chemical disorder we are concerned with the number of different ways of occupying the atomic sites in the structure. The entropy is frequently separated into two parts related to these two types of distributions. These are termed vibrational entropy and configurational entropy respectively.

As entropy is a statistical function its physical nature can perhaps be best understood using simple probability theory. To illustrate this approach we will again refer to the distribution of two different elements on a simple cubic lattice, but for convenience will use equal numbers of black and white spheres. To keep the arithmetic manageable the numbers will be kept small (e.g. eight of each) and their distribution will be considered in two dimensions only. If these 16 balls were arranged at random in a square 'crystal' the kind of distribution which might result is shown in Fig. 2.5(a). The distribution is disordered and as such it is a member of a large class of distributions all of which have the common feature of being disordered. The total number of different ways of distributing the sixteen balls over the sixteen 'atomic sites' is $16!/8! \times 8!$ and of these 12 870 ways there are very few arrangements which could be described as ordered. Figs 2.5(b) and (c) show two examples of ordered distributions. Clearly, the number of distributions of the disordered kind is very much greater than those of the ordered kind, even in a small system with only 16 'atoms'. If we take a larger crystal the effect of a similar computation is to increase enormously the number of distributions of the disordered kind, but not the others.

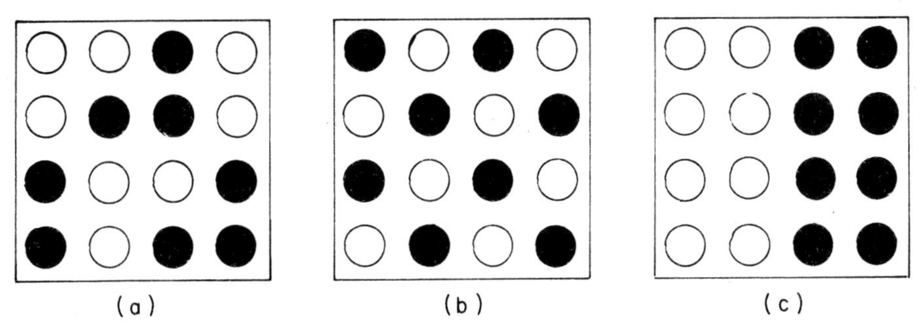

Fig. 2.5 (a) A disordered distribution of eight black and eight white spheres. (b) and (c) are two different ordered distributions.

This example describes a particular kind of disorder but the principle is general and applies to the disorder associated with atomic vibrations as well as disorder in atomic positions when bonds are broken. As a measure of the degree of disorder in a system we usually use ω, the number of distributions in the system. As the values of ω in a macroscopic crystal are very large it is more convenient to use $\log_e\omega$. The entropy, S, is then defined by the expression

$$Sl \propto og_e\omega$$
$$S = k \cdot \log_e\omega$$

where k, the proportionality constant, is Boltzmann's constant ($1.38 \times 10^{-23}\,\mathrm{JK^{-1}}$)

Thus a high-temperature disordered form of a particular phase has a higher configurational entropy and hence a higher symmetry than the lower temperature polymorph. The connection between disorder, symmetry and entropy can be made in this way.

2.3 Internal energy

The conclusion that in any system there are many more disordered distributions than ordered distributions leads us to another question. Why then do so many forms of matter exist with highly ordered distributions, especially at low temperatures? The answer lies in the fact that firstly, different distributions will have different internal energies and secondly that systems with different internal energies have different chances of turning up, i.e. different thermodynamic probabilities.

The internal energy is the sum of all the kinetic energies and energies of interaction (potential energies) of all the atoms in the system. The energies of interaction depend on the nature of the nearest neighbour atoms with which bonds are formed. In our example in Fig. 2.5 we did not consider any possible interactions between the black and the white balls. If they represent different atoms then it is most likely that there will be some interaction, e.g. there may be a preference for like or unlike neighbours. If the black and white balls preferred unlike nearest neighbours, as in Fig. 2.5(b) then the internal energy is lowest when there is a maximum number of nearest neighbour bonds between unlike atoms. Therefore the distribution in Fig. 2.5(b) has a lower energy than that of Fig. 2.5(a). If, on the other hand, like neighbours were preferred then a total physical separation of the two types of atoms would result in the lowest internal energy.

To account for our observation that ordered states are associated with low temperatures and disordered states with high temperatures, there must be some connection between the thermodynamic probability of a particular state occurring and the temperature. This connection is made in statistical thermodynamics, a field in which we will not become involved, but the general nature of the argument can perhaps be obtained by considering the following.

Any thermodynamic state of internal energy, E, will have a number of possible distributions associated with it. Let us assume that there are ω such distributions with internal energy, E. If the thermodynamic probability of such a distribution turning up at temperature, T, is $p(E,T)$, then the probability of a thermodynamic state with internal energy, E, turning up is $\omega.\ p(E,T)$. This is denoted as Z, and clearly the state with the highest Z will be the most favourable. Historically, the quantity used to define this state is not Z, but the free energy, F, where:

$$F = -kT \log_e Z$$

The free energy, F, is therefore a term which is at a minimum when Z is at a maximum. Statistical thermodynamics tells us that the probability of a distribution of internal energy, E, occurring at a temperature, T, is given by the general equation

$$p(E,T) = e^{-E/kT}$$

If there are ω distributions of internal energy, E, then

$$Z = \omega.e^{-E/kT}$$

Substituting into the above equation

$$F = -kT\,(\log_e\omega - E/kT)$$
$$F = E - kT \log_e\omega$$

Substituting for the entropy, S,

$$F = E - TS$$

This is a very important relationship and it will be explored in more detail in Chapter 4. The lowest value of F is that with the highest thermodynamic probability of occurring, i.e. the thermodynamically predicted most stable state, and this relationship shows that a state becomes more stable as its internal energy decreases and its entropy increases. A system with a low internal energy usually will also have a low entropy, as illustrated by Fig. 2.5, and therefore these two factors tend to act in opposition.

Temperature is the controlling factor. At low temperatures the TS term is sufficiently small for E to be the predominant term, and so at low temperatures states with low internal energies, i.e. ordered states will be the most stable. At high temperatures TS becomes very large for disordered states and predominates over the internal energy, E. Disordered states are therefore more stable as the temperature increases.

After our discussion on disorder at the beginning of this chapter this may seem to be an obvious statement. There is some value however in relating a physical description of disorder to the formalism of thermodynamics, and hence an outline is given at this stage. We will now return to the mainstream of this chapter and develop the thermodynamics at a later stage.

2.4 Solid solutions

Chemical disorder in a mineral implies the existence of solid solutions. If the site occupied by each atom is not completely specified, and the atoms may interchange their positions, the proportion of each atom present may also change within limits and a solid solution is formed. In most minerals such solid solutions have structures based on a definite framework with one or more cations disordered over sites within this framework. The types of sites and the ions which occupy them vary in different solid solutions and will be discussed in the section on specific mineral structures. Some mechanisms by which a continuously varying chemical composition is produced in a mineral are outlined below.

2.4.1 ATOMIC SUBSTITUTION

The factors which govern the development of substitutional solid solutions are the relative ionic radii of the atoms and their valency. These two properties determine the possibility of exchange between atoms in a particular site in the structure. Alternatively different atoms may occupy the site if their sizes and charges are such that the geometrical stability and local charge balance is maintained.

The two most common examples of such substitutions in minerals are Mg^{2+} for Fe^{2+} and Al^{3+} for Si^{4+}. An atomic site which can accept Fe^{2+} (ionic radius 0.74 Å) will also accept Mg^{2+} (ionic radius 0.66 Å) and these two ions can replace each other with great freedom in many minerals. Thus a complete solid solution may exist between the Fe end member and the Mg end member of a solid solution series. The olivines provide a simple example with the two end members having compositions Mg_2SiO_4 and Fe_2SiO_4. Almost all natural olivines have compositions intermediate between these end members, the exact composition depending on the availability of the two elements during crystallization.

Table 2.1 lists some of the common substitutions found in minerals and the ionic radii of the cations. A wide range of substitution is generally possible if the size difference between the ions is less than 15%. As the size difference increases solid solution becomes limited due to the strain produced in the structure. For example, when Na^+ substitutes for K^+ the considerable difference in ionic radius means that only a limited number of Na^+ ions can be replaced by K^+ unless the temperature is increased.

In the case of Al^{3+} substitution for Si^{4+} the need to maintain local charge balance will require a coupled substitution where some other pair of ions will also need to replace one another. A common example is the plagioclase–feldspar series in which charge balance is maintained by Ca^{2+} substitution for Na^+.

i.e. $$Na^{+} + Si^{4+} \rightleftharpoons Ca^{2+} + Al^{3+}$$

The two end-member compositions in this case are $NaAlSi_3O_8$ and $CaAl_2Si_2O_8$.

Table 2.1

Common substitutions			
Mg — Fe^{2+}		Ba — Ca	
Al — Si		Mn^{2+} — Fe^{2+}	
Na — Ca		Mn^{4+} — Fe^{3+}	
Al — Fe^{3+}		Cu^{2+} — Fe^{2+}	
Ionic radii	Å		
Ag^{+}	1.26	Fe^{3+}	0.64
Ag^{2+}	0.89	K^{+}	1.33
Al^{3+}	0.51	Li^{+}	0.68
Ba^{2+}	1.34	Mg^{2+}	0.66
Ca^{2+}	0.99	Mn^{2+}	0.80
Co^{2+}	0.72	Mn^{3+}	0.66
Co^{3+}	0.63	Na^{+}	0.97
Cr^{3+}	0.63	Ni^{2+}	0.69
Cu^{+}	0.96	Si^{4+}	0.42
Cu^{2+}	0.72	Ti^{3+}	0.76
Fe^{2+}	0.74	Ti^{4+}	0.68
Zn^{2+}	0.74		

2.4.2 OMISSIONAL SOLID SOLUTION

Less commonly, the simple removal of atoms from the structure, leaving vacancies may be the mechanism of producing appreciable compositional variations. In iron sulphide, FeS, iron atoms can be removed from the structure up to a limit of around the composition Fe_7S_8 leaving one out of every eight iron sites vacant. These defect structures with the general formula $Fe_{1-x}S$ are called the pyrrhotites. Electrostatic neutrality may be maintained if some of the Fe ions are in the ferric state.

Vacancies may also be produced in a structure to maintain charge balance during a substitution mechanism. For example, in the mullites a compositional range from at least $3Al_2O_3 \cdot 2SiO_2$ to $2Al_2O_3 \cdot SiO_2$ is achieved by replacement of Si^{4+} by Al^{3+} and the removal of oxygen atoms to maintain charge balance.

$$2Si^{4+} + O^{2-} \rightleftharpoons 2Al^{3-} + \square$$

2.4.3 INTERSTITIAL SOLID SOLUTION

Between the atoms of the structural framework of a mineral interstices exist which may not normally be used as atomic sites. In the mineral chalcopyrite $CuFeS_2$ (Fig. 2.4) extra Cu and Fe atoms may be accommodated to produce cation rich compositions of the general form $Cu_{1+x}Fe_{1+x}S_2$.

Another scheme which involves the occupation of normally unoccupied sites to balance atomic substitutions is found in the stuffed derivatives. Here a cation of lower ionic charge replaces a cation of higher ionic charge with electrostatic neutrality being maintained by stuffing another cation into the structure. The composition of tridymite, SiO_2, can be varied towards nepheline, $NaAlSiO_4$, although complete solid solution is not possible due to geometrical constraints.

The type of solid solution present in a mineral cannot simply be determined from chemical analysis which provides relative abundances. In many cases density measurements are satisfactory (interstitial solid solutions increase the density, while omission solid solutions decrease the density). In complex minerals containing a number of cations X-ray structural analysis may be necessary to determine whether a particular element resides in a regular site by substitution or whether its presence should be regarded as an interstitial impurity.

2.4.4 THE EFFECT OF TEMPERATURE

The single most important point to be made on solid solutions in the context of mineral behaviour is that the extent of a disordered solid solution is usually strongly dependent on the temperature. There is, in general, much more tolerance towards atomic substitution at high temperatures when thermal vibrations are greater and the sizes of available atomic sites less rigid. At high temperatures a fair amount of flexibility in composition of a mineral is allowed, providing suitable sites exist. While such a disordered, chemically variable structure may be the stable state at high temperatures, the situation rarely remains the same as the temperature is decreased.

As has already been seen, on lowering the temperature disorder gives way to order and within this overall scheme the individual atoms must play out their parts. The neighbours they will eventually end up with may not be entirely of their choosing—their mobility, as movement becomes increasingly difficult, may see to that. The result will often be a compromise between the means and the end, with perhaps even a change of tactics along the way. The behaviour of these solid solutions as they contend with a cooling environment will form the basis for much of what follows.

References and additional reading

Bloss, F. Donald (1971) *Crystallography and Crystal Chemistry*. Holt, Reinhart and Wilson. Ch. 8.

Broeker, W.S. and Oversby, V.M. (1971) *Chemical Equilibria in the Earth*. McGraw-Hill. Ch. 8.

Brown, P.J. and Forsyth, J.B. (1973) *The Crystal Structure of Solids*. Edward Arnold Ltd.

Fast, J.D. (1970) *Entropy*. McMillan Student Editions. Ch. 2.

McKie, D. and McKie, C. (1974) *Crystalline Solids*. Nelson. Ch. 10.

3

Mineral Structures

A SIMPLE MINERAL STRUCTURES

The behaviour of minerals as they cool from their high-temperature state is ultimately controlled by their crystal structures. In this chapter we will describe in a general way the important features of a number of mineral structure types, with special emphasis on the chemical variations possible at high temperatures. By considering only the topological symmetry of the framework structure of minerals, i.e. the maximum symmetry attainable by the framework when all distortions are eliminated and the chemical content of atomic sites ignored, the description becomes that of the high-temperature state. This then forms the basis for all the derived structures which may result from either distorting the framework or specifying the nature and distribution of the constituent atoms.

In a crystal structure the atoms are arranged such that their spatial relationship leads to a regularly repeating unit cell in three dimensions. Every unit cell is, on average, identical, and the complete structure may be described by specifying the size of the unit cell and the coordinates of the constituent atoms. Such a description is not particularly useful however if we are interested in the way the atoms are associated in the structure, and classifications have been developed which emphasize these relationships.

In considering mineral behaviour in relation to crystal structure in general one may start with the simplest structures, namely metals and alloys, and proceed through close-packed structures to those of the silicates where the formula unit is large and the structure may be quite complex. At all stages it is necessary to have a feeling for the key aspects of the structures involved and hence schematic diagrams, as a means of visualizing aspects of mineral behaviour, will be much used in what follows.

In the first part of this chapter we will consider those mineral structures which may be conveniently described in terms of a close-packed array of anions with one or more interstitial cations present which have rather rigorous co-ordination requirements. Most of the rock-forming minerals are not of this type and will be discussed in the second part of the chapter on more complex structures. However, these simpler mineral structures which include many of the common sulphides and oxides will serve as an introduction to aspects of the crystallography, crystal chemistry and the transformation behaviour of minerals. Once the necessary facility has been obtained in this it will be possible to deal with the behaviour of the more complex structures of the silicates where anion packing may be irregular and the concept of a well-defined co-ordination number is necessarily inexact.

Before discussing minerals whose structures are based on close-packed atoms, some aspects of the close-packing of identical spheres must be considered. Treating atoms as rigid inert spheres with no directional bonding requirements may seem to be an unrealistic exercise, but we will find that many mineral structures can be described quite accurately in these terms.

3.1 Close-packing of spheres

We will consider first the ways in which identical spherical atoms or spheres can be packed together in order to occupy the minimum possible volume. In two dimensions there is only one way in which such spheres can be arranged to occupy the minimum area, and this is illustrated in Fig. 3.1. The atoms in this two dimensional close-packing have their centres arranged on a hexagonal lattice, and each atom has six nearest neighbours, i.e. a co-ordination number of six. Obviously the spheres cannot fill all space completely and there are voids or interstices between them. In Fig. 3.1(a) these interstices have been divided into two groups, B and C.

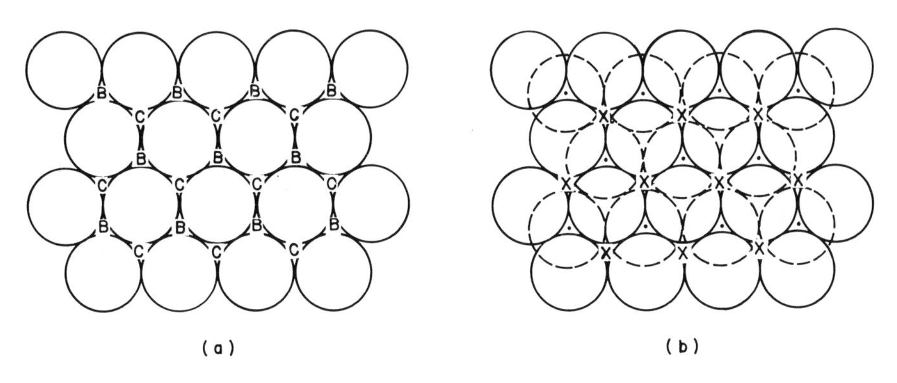

Fig. 3.1 (a) Closest packing of identical spheres in two dimensions showing two sets of hollows labelled B and C. (b) A second layer of spheres over the B positions. Two types of interstitial sites are formed: tetrahedral sites labelled · and octahedral sites labelled x.

To obtain a closest-packed three-dimensional stacking, another identical layer must be placed over this first layer such that its atoms fall over one set of interstices. If the atoms lie over the B interstices the stacking shown in Fig. 3.1(b) is obtained. If the positions of the atoms in the first layer are called A positions, then Fig. 3.1(b) is A–B stacking. Alternatively, the stacking could have been A–C, but this would be physically indistinguishable from A–B. The third layer, however, can be placed in two very different positions. If the atoms of the third layer lie over those in the first layer the stacking sequence becomes A–B–A, whereas if the atoms lie above the C insterstices the stacking is A–B–C. Continuing these stacking sequences leads to two ways in which spherical atoms can be packed together to occupy the minimum volume. It will be seen that the ABABAB . . . sequence has hexagonal symmetry and is called hexagonal close-packed, and the sequence ABCABC . . . has cubic symmetry and is hence called cubic close-packed.

The symmetry of these two structures is clearer if they are drawn as shown in Fig. 3.2. In hexagonal close-packing the layers are stacked along the c axis of the hexagonal unit cell, while in cubic close-packing the body diagonal of the cubic unit cell is normal to the layers. The unit cells of the two close-packed structures are illustrated in Fig. 3.3.

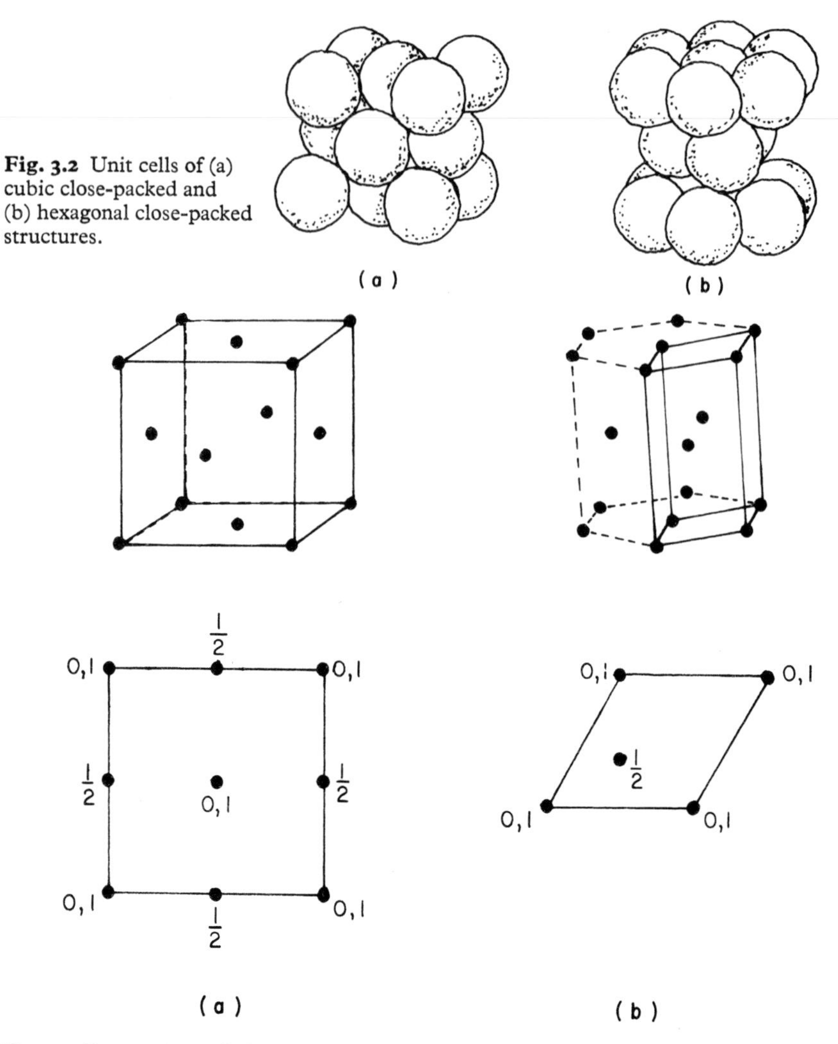

Fig. 3.2 Unit cells of (a) cubic close-packed and (b) hexagonal close-packed structures.

Fig. 3.3 Perspective and plan views of the (a) cubic close-packed and (b) hexagonal close-packed structures, with the spheres replaced by closed circles.

Although many metals adopt either the cubic close-packed or hexagonal close-packed structures, our principal interest is in relation to minerals based on an anion close-packing in which cations occupy the interstices.

3.1.1 INTERSTITIAL SITES IN CLOSE-PACKED STRUCTURES

There are two types of interstitial sites which are defined by any two closest-packed layers which are in contact. The types are therefore independent of the stacking sequence.

If Fig. 3.1(b) is examined these two types of interstitial sites can be seen. Any sphere resting in the hollow formed by three spheres in the adjacent layer forms a tetrahedral site [those marked · in Fig. 3.1(b)]. The centres of the four spheres are at the vertices of a regular tetrahedron, and any interstitial atom occupying such a site would have tetrahedral co-ordination, [4], with the nearest neighbour close-packed spheres. There are two tetrahedral sites associated with each sphere in a close-packed layer, one above and one below each the sphere. Thus there are twice as many tetrahedral sites as there are close-packed atoms in any sequence.

The interstitial sites formed between three spheres in one layer and three in the adjacent layer [those marked x in Fig. 3.1(b)] are called octahedral sites, as the six spheres lie at the corners of a regular octahedron. Thus any atom occupying this interstitial site would have octahedral co-ordination, [6]. The number of octahedral sites is equal to the number of close-packed spheres, and hence in any stacking sequence the number of tetrahedral sites is twice that of the octahedral sites.

3.1.2 SIZES OF THE INTERSTITIAL SITES

The size of an interstitial site is clearly much smaller than the size of the close-packed spheres. An estimate of the size of an interstitial site is generally made in terms of the size of a small sphere which could just fit into the site, making contact with but not distorting the close-packed spheres. If the radius of the close-packed spheres is R, and that of the interstitial site is r, then by simple geometry it can be shown that for a tetrahedral site $r/R = 0.225$, and for an octahedral site $r/R = 0.414$.

If we consider that ionic compounds can often be described in terms of close-packed arrays of anions with cations occupying these interstitial sites, then clearly the relative sizes of the two sites will place restrictions on the kinds of cations which can occupy them. The radius ratio will be one of the criteria which determine the coordination of a particular cation in a close-packed structure.

3.1.3 POSITIONS OF INTERSTITIAL SITES IN CLOSE-PACKED STRUCTURES

The unit cells of the two close-packed structures are shown in Fig. 3.3. Before structures with different atoms occupying the interstitial sites can be considered, we must locate these sites within the unit cells and determine the number of sites of each type. In Fig. 3.3(b) we can see that there are two close-packed atoms in each unit cell of the hexagonal close-packed structure (noting that the atoms at the corners are each shared by eight adjacent unit cells). Therefore in this unit cell there must be a total of four tetrahedral sites and two octahedral sites. As with all such structural details their positions can best be seen in a ball-and-spoke structure model. In Fig. 3.4 a larger section of the structure is drawn out and the positions of the interstitial sites marked.

In the cubic unit cell of a cubic close-packed structure [Fig. 3.3(a)] there are four close-packed atoms, and hence eight tetrahedral and four octahedral interstitial sites. It becomes rather cumbersome to illustrate every one of these, and so in Fig. 3.5 the position of only one of each type of site is shown.

The number of interstitial sites relative to the number of close-packed atoms in these structures is important when we are considering the cation : anion ratios of structures where the close-packed atoms are anions, and the interstitial sites are occupied by cations.

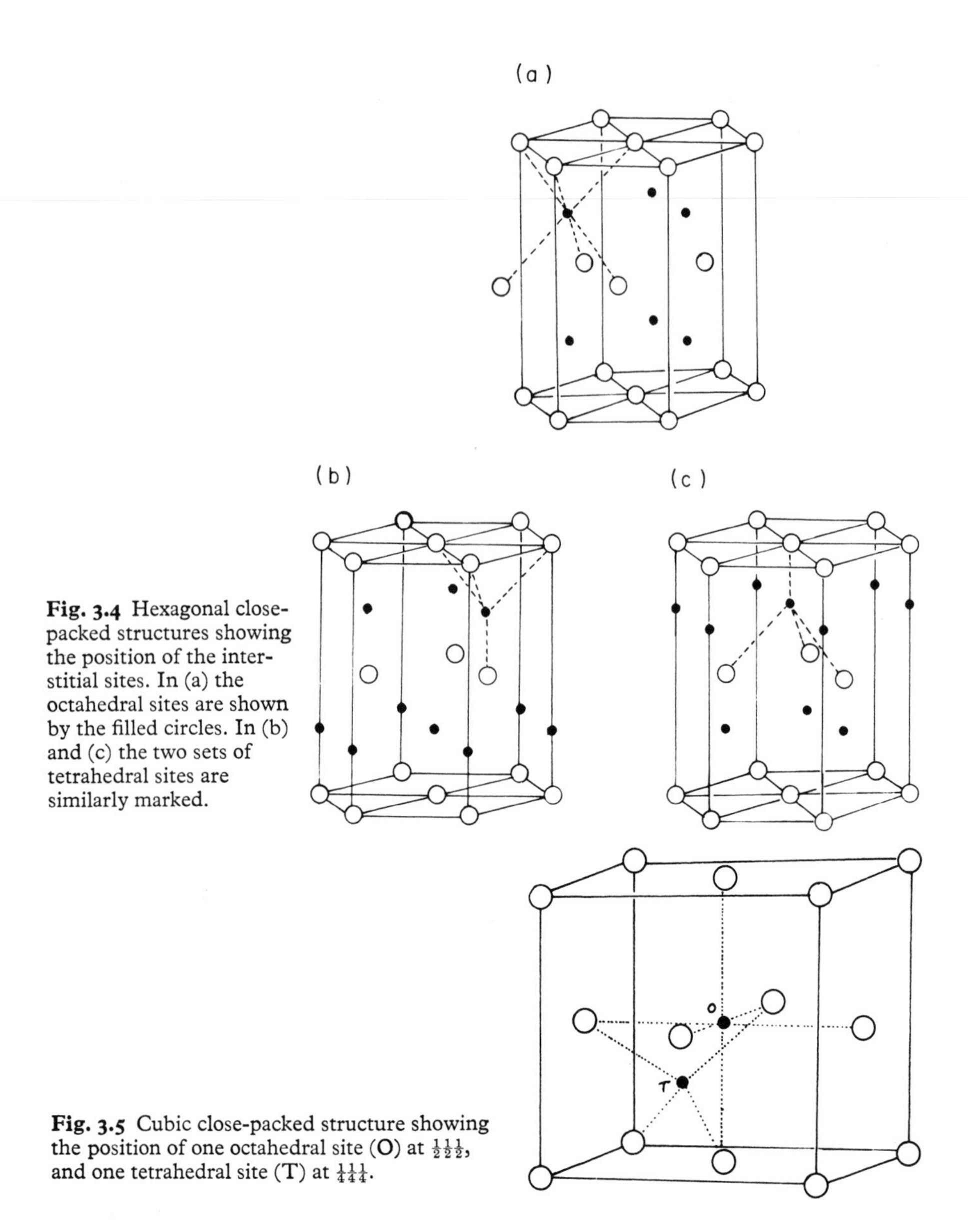

Fig. 3.4 Hexagonal close-packed structures showing the position of the interstitial sites. In (a) the octahedral sites are shown by the filled circles. In (b) and (c) the two sets of tetrahedral sites are similarly marked.

Fig. 3.5 Cubic close-packed structure showing the position of one octahedral site (O) at $\frac{1}{2}\frac{1}{2}\frac{1}{2}$, and one tetrahedral site (T) at $\frac{1}{4}\frac{1}{4}\frac{1}{4}$.

3.2 Mineral structures based on close-packing

It is evident that a large number of possible structures can be developed from close-packed arrays of atoms if we consider simply the geometrical possibilities of arranging a second group of atoms among the interstitial sites. In the description of such structures the close-packed atoms are the larger anions and the smaller cations occupy the interstitial sites. If our attention is confined to structures containing only one type of anion, the possibilities for cation distributions are as follows:

1 Tetrahedral sites may be occupied.
2 Octahedral sites may be occupied.
3 Some of the tetrahedral and some of the octahedral sites may be occupied.

In each case the number of sites occupied and the nature of the occupying cation can change, and this may result in slight distortions from ideal close-packing in the anion distribution.

We will describe these possibilities in turn, and in each case give examples of minerals which crystallize with these structures. The emphasis is on the high-temperature state of the minerals.

3.2.1 HEXAGONAL CLOSE-PACKED STRUCTURES WITH TETRAHEDRAL SITES OCCUPIED

The most important of these have half of the tetrahedral sites occupied by cations and hence the cation : anion ratio is 1 : 1. The type example is the hexagonal form of ZnS called wurtzite. The sulphur atoms are hexagonally close-packed and the zinc atoms occupy one set of the tetrahedral sites, as shown in Fig. 3.6.

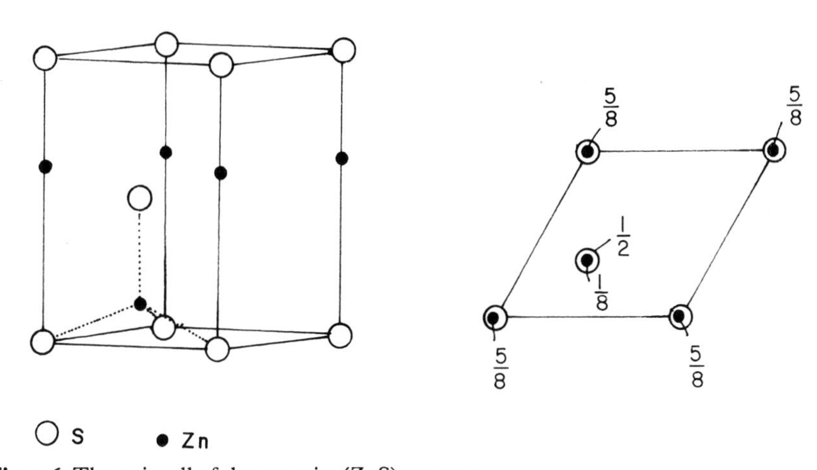

Fig. 3.6 The unit cell of the wurtzite (ZnS) structure.

A derivative of this structure is formed by replacing one third of the zinc atoms by copper and the other two thirds by iron. The formula then becomes $Cu_{\frac{1}{3}}Fe_{\frac{2}{3}}S$ or conventionally $CuFe_2S_3$. This is the mineral cubanite. If the copper and iron atoms are disordered among the set of occupied tetrahedral sites, the unit cell remains that of the wurtzite structure. If, however, the copper and iron atoms are ordered in some way the size of the unit cell must increase. This ordered unit cell is geometrically related to the wurtzite cell, and is usually termed a supercell when described in this context. Similarly, the ordered structure is termed a superstructure. The phenomenon of order and disorder of cations within a close-packed anion structure which essentially remains static is very common in these structures, especially in the sulphides, and the description of such structures will become more familiar as more examples are given.

The ordered arrangement of Cu and Fe atoms in cubanite is illustrated in Fig. 3.7. This is a rather unusual arrangement in that in one half of the cell one set of tetrahedral sites is being occupied (with co-ordination tetrahedra around the cation pointing upwards), while in the other half of the unit cell the other set of tetrahedral sites (with co-ordination tetrahedra pointing downwards) is occupied. The Cu atoms are arranged such that the nearest cation neighbours are Fe atoms. The result is an orthorhombic unit cell whose relationship to the disordered hexagonal cell is shown in Fig. 3.7. From the earlier discussion we would expect that the ordered state would exist at lower temperatures, and the disordered state at higher temperatures. This is found to be the case, although at still higher temperatures a form of cubanite with cubic close-packed sulphur atoms is formed. This is another aspect of phase transformations in close-packed structures which is not uncommon, i.e. a change in the packing sequence with temperature.

Therefore in a discussion of the behaviour of minerals based on close-packing there are two points to be considered. Firstly, the possibility of a change from one stacking sequence to another and secondly, the ordering or disordering of cations in the interstitial sites of the close-packed structure.

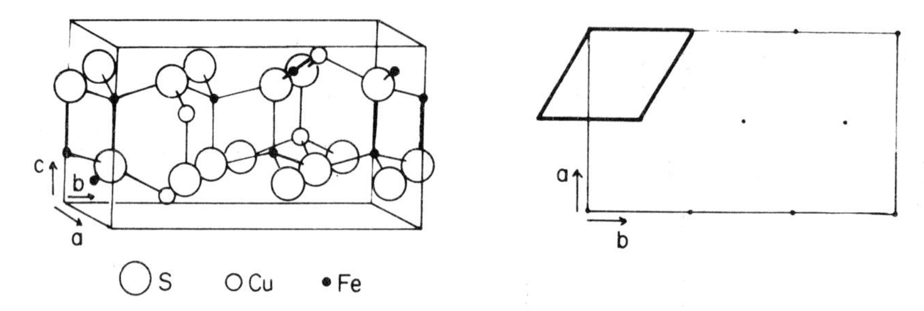

Fig. 3.7 The unit cell of the cubanite ($CuFe_2S_3$) structure showing its relationship to the wurtzite-type subcell (heavy line).

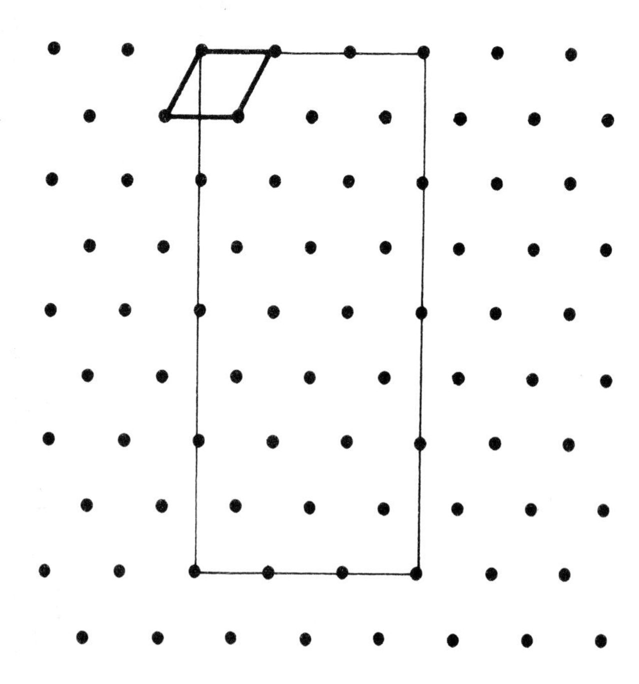

Fig. 3.8 The relationship between the unit cells of high chalcocite (disordered hexagonal) and low chalcocite (ordered pseudo-orthorhombic).

A more complex example of cation disorder within a hexagonal close-packed anion structure is in high-temperature chalcocite Cu_2S. The stoichiometry suggests that all of the tetrahedral sites must be occupied, but in fact the disordered distribution of Cu atoms is such that they statistically occupy three types of sites. Some are in tetrahedral sites, some are distorted from the centre of the tetrahedra and are in three-fold co-ordination with sulphur atoms, while the rest are even further displaced from the centre of the tetrahedral site and lie on the edge of a tetrahedraon and are therefore in two-fold co-ordination. For our purposes we will not be concerned about this actual distribution but will consider the copper atoms as simply 'disordered'. The unit cell of this high-temperature disordered state is that of the hexagonal close-packed sulphur lattice. Below about 100°C the Cu atoms lose their mobility and become fixed in the structure. The unit cell repeat must therefore become larger, again forming a superstructure of the high-temperature cell. The relationship between the unit cells of the high and low forms of chalcocite is shown in Fig. 3.8. Notice that in this case the volume of the low-temperature cell is 48 times that of the high-temperature cell, due mainly to the different types of co-ordination that Cu atoms can adopt in this structure.

A similar structure to that of chalcocite is stromeyerite, CuAgS. This is obtained by replacing one half of the Cu atoms by Ag together with some minor modifications to the distribution of cations. Again there is an ordered form at low temperatures and a disordered hexagonal form above 93°C.

Although there are many other inorganic compounds with structures based on a hexagonally close-packed anion lattice with cations in tetrahedral sites, only those of immediate mineralogical interest, i.e. those whose behaviour will be discussed in later chapters will be described here.

3.2.2 HEXAGONAL CLOSE-PACKED STRUCTURES WITH OCTAHEDRAL SITES OCCUPIED

The type example is the nickel arsenide structure, NiAs. In the stoichiometric composition (with a cation : anion ratio of 1 : 1) all of the octahedral sites are occupied by Ni atoms. The structure is illustrated in Fig. 3.9. In many compounds which adopt this structure the anions are slightly distorted from true close-packing, and the interstitial site is a distorted octahedron. The NiAs structure type is adopted by many sulphides, selenides and tellurides of the transition metals. Mineralogically the most important are FeS (troilite) and NiS (millerite).

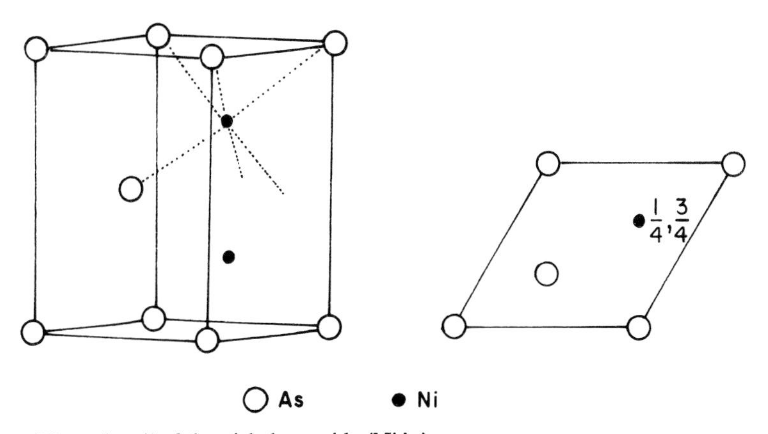

Fig. 3.9 The unit cell of the nickel arsenide (NiAs) structure.

The feature of this structure which will be of most interest here is that at higher temperatures substantial compositional variations are possible by the formation of an omission solid solution. Some of the octahedral sites remain vacant forming a range of compositions of the general form $M_{1-x}S$. In the iron sulphides $Fe_{1-x}S$ these are called the pyrrhotites and their compositional range extends to about Fe_7S_8. At high temperatures the vacancies are disordered, and from what we have already seen we would expect that at lower temperatures the vacancies should become ordered in some way. This is indeed the case, although complications will arise in trying to order vacancies in the structure when the proportion of vacancies to cations is not a simple fraction.

The octahedral sites in the NiAs structure are themselves arranged in layers, parallel to the anion layers. As our main interest in this structure will be the behaviour of the vacancies on cooling, we will consider only these cation layers and so simplify the structural description. Within the cation layers the atoms are arranged in a hexagonal array, and the layers are stacked vertically, one above the other. In the disordered defect structure $Fe_{1-x}S$ at high temperatures vacant sites are distributed randomly. At lower temperatures there is a tendency for the vacancies to achieve maximum separation and vacancy ordering takes place on this basis.

To illustrate this we can consider the ordered structure of the most cation deficient composition Fe_7S_8. The arrangement of the cation layers is shown in Fig. 3.10. Notice that the vacancies are confined to alternate layers and ordered

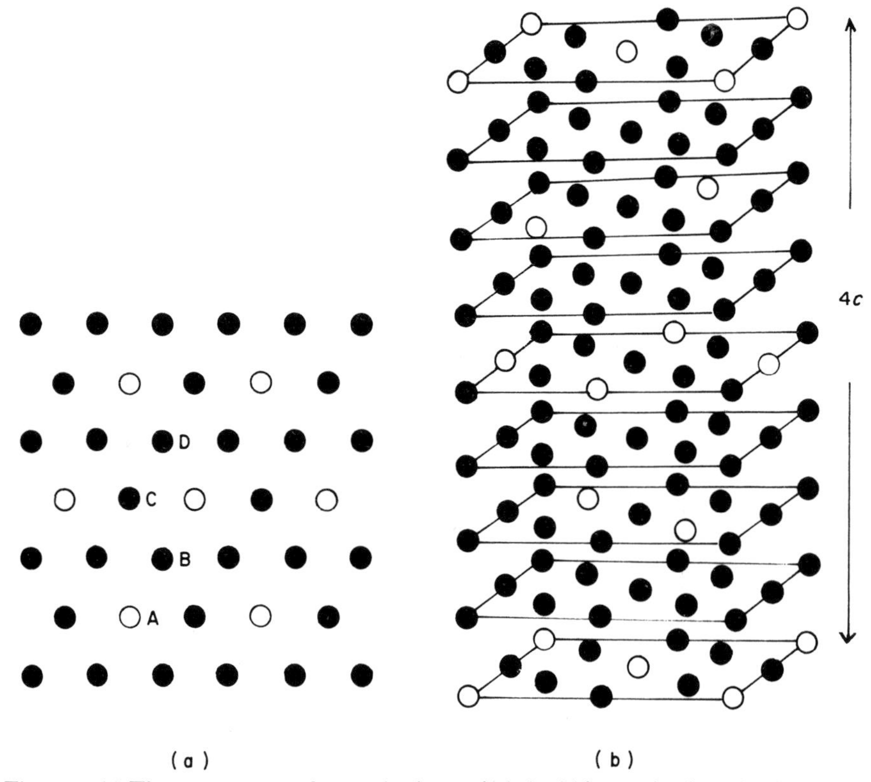

Fig. 3.10 (a) The arrangement of vacancies (open circles) within a cation layer in Fe_7S_8. (b) The arrangement of full and vacancy-containing cation layers in the ordered $4c$ structure of Fe_7S_8.

within these layers. The distribution of vacancies in such a defective layer is shown in Fig. 3.10(a). In order to achieve maximum vacancy separation between the layers, the defective layers are stacked so that they are displaced relative to one another such that an arbitrary position A in one layer occupies positions B, C, D, A in successive layers as shown in Fig. 3.10(b). The result is an eight-layer repeat, four times the repeat of the disordered state which has the NiAs structure (Fig. 3.9). As the repeat is along the *c* axis of the cell, it is referred to as the 4*c* superstructure of Fe_7S_8.

The vacancy ordered structure of Fe_7S_8 has been described briefly at this stage for a number of reasons. Firstly, it is one end member of the omission solid solution series. Secondly, it is the composition for which the criterion of maximum vacancy separation can be achieved in the simplest way, and it therefore brings us to the question of what happens in other compositions between Fe_7S_8 and FeS. A defective layer [Fig. 3.10(a)] has a vacancy : cation ratio of 1 : 4 and so the vacancy : cation ratio of 1 : 8 in Fe_7S_8 is geometrically easy to achieve. But what are the possibilities for a composition such as $Fe_{10}S_{11}$ as it cools from a disordered state? This is a very basic question and should be asked whenever the behaviour of any solid solution is investigated on cooling. In general terms, any composition $Fe_{1-x}S$ has a number of options for organizing its vacancies as the temperature is decreased. Firstly, it could try to arrange the vacancies in the most ordered way possible for the given composition, despite the size of the resultant unit cell of the superstructure. Clearly, some compositions will be able to achieve a more satisfactory (i.e. more ordered) solution than others. At the other extreme of behaviour the solid solution could perhaps break down to some regions of FeS composition and others of Fe_7S_8 structure and composition, in the appropriate proportions. A compromise solution might be the breakdown into two intermediate compositions which are able to attain a higher degree of order than the original composition of the homogeneous solid solution.

These possibilities can be summarized by considering the composition $Fe_{10}S_{11}$:

1 $Fe_{10}S_{11}$ (disordered) ——— $Fe_{10}S_{11}$ (with maximum order)
2 $Fe_{10}S_{11}$ (disordered) ——— $3FeS + Fe_7S_8$ (ordered)
3 $2Fe_{10}S_{11}$ (disordered) ——— $Fe_9S_{10} + Fe_{11}S_{12}$ (both with maximum order)

The process which operates will depend both on the thermodynamics (i.e. which end result has the lowest free energy) and the kinetics (i.e. which process is the easiest). When the behaviour of iron sulphides is considered in more detail it can be seen that the result is complex, and that a great deal of research on both synthetic and natural pyrrhotites has not yet resolved the problems. At this stage however the pyrrhotites provide an example of the questions that will be encountered in describing the behaviour of solid solutions.

Nickel sulphide, NiS, also forms an omission solid solution $Ni_{1-x}S$ at high temperatures, although on cooling the result is much simpler than for $Fe_{1-x}S$. All compositions of the form $Ni_{1-x}S$ break down at low temperatures to form NiS (millerite) + Ni_3S_4 (polydymite).

As both $Fe_{1-x}S$ and $Ni_{1-x}S$ are isostructural at high temperatures a complete solid solution exists between them with variable amounts of Fe and Ni in $(Fe, Ni)_{1-x}S$. This is often referred to as the monosulphide solid solution of the system Fe–Ni–S and an understanding of its behaviour is very important as it is generally believed to be the precursor of high temperature iron nickel sulphide ores such as those found in Sudbury, Ontario.

3.2.3 CUBIC CLOSE-PACKED STRUCTURES WITH TETRAHEDRAL SITES OCCUPIED

The mineralogically most important structures of this type are based on the sphalerite structure. Sphalerite (zinc blende) is the cubic form of ZnS and has the sulphur atoms in cubic close-packing with zinc atoms occupying one half of the tetrahedral sites (Fig. 3.11).

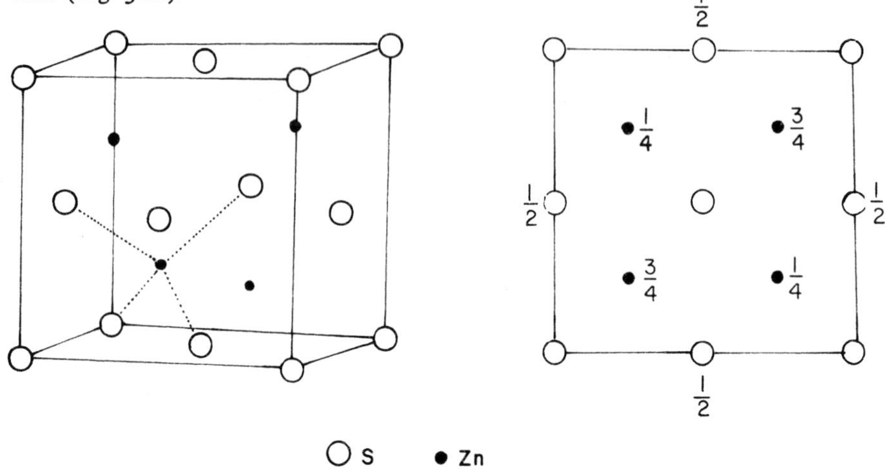

Fig. 3.11 The unit cell of the zinc blende (ZnS) structure.

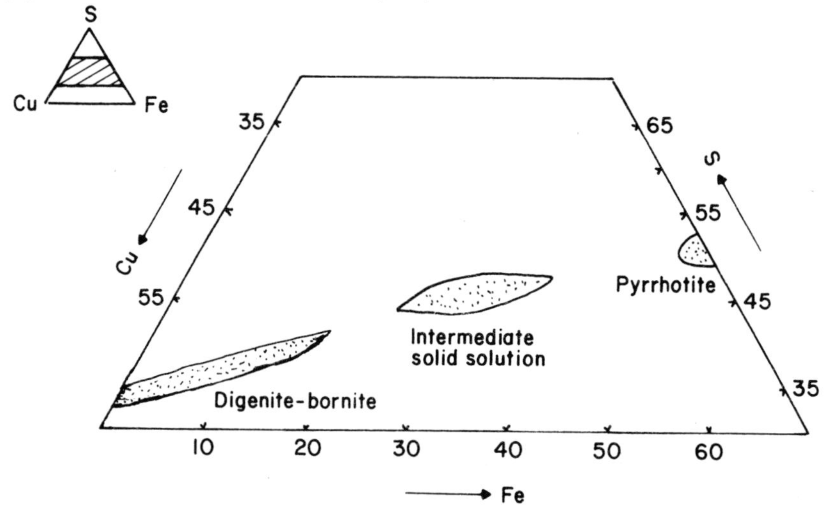

Fig. 3.12 High-temperature solid solutions in the Cu–Fe–S system at about 600°C. The digenite–bornite and the intermediate solid solution are based on cubic close-packed sulphur atoms. The pyrrhotite solid solution has sulphur atoms in hexagonal close-packing.

Derivatives of this structure are obtained by replacing the Zn atoms by other cations. For example, if one half of the zinc atoms are replaced by iron and the other half by copper the mineral chalcopyrite is formed. If the copper and iron atoms are ordered within this structure, the size of the unit cell has to be doubled. This has already been illustrated in Fig. 2.4. Stannite, Cu_2FeSnS_4 is another mineral derived from the sphalerite structure and has a cell similar to that of chalcopyrite.

Solid solutions in this structure type can be formed both by changing the proportion of each cation (for example $CuFeS_2$–$CuFe_2S_3$) without changing the cation : anion ratio, and by filling the previously unfilled tetrahedral sites by cations of either type (for example $Cu_{1+x}Fe_{1+x}S_2$). The copper–iron sulphides provide the most important examples of such solid solutions. In Fig. 3.12 the extent of these solid solutions is illustrated in the high-temperature phase diagram of the Cu–Fe–S system. The most cation-rich composition is the cubic form of digenite, Cu_2S, in which all of the tetrahedral sites are occupied.

In most of the solid solution field the disorder involves copper atoms, iron atoms and vacant sites, and during any ordering process, therefore, a number of 'degrees' of order can be envisaged. For example, the vacancies could be ordered while the cations might still be disordered amongst themselves. Once more, the extent to which the cations can order will again depend on the ratio of their compositions and for some compositions a fully ordered state will not be possible. This will raise the possibility of a number of processes operating and an evaluation of the factors which might favour one process rather than another.

3.2.4 CUBIC CLOSE-PACKED STRUCTURES WITH OCTAHEDRAL SITES OCCUPIED

The type example of this structure is sodium chloride, NaCl, which has the chlorine atoms in a cubic close-packed array with sodium atoms occupying all of the octahedral interstices. The structure is illustrated in Fig. 3.13.

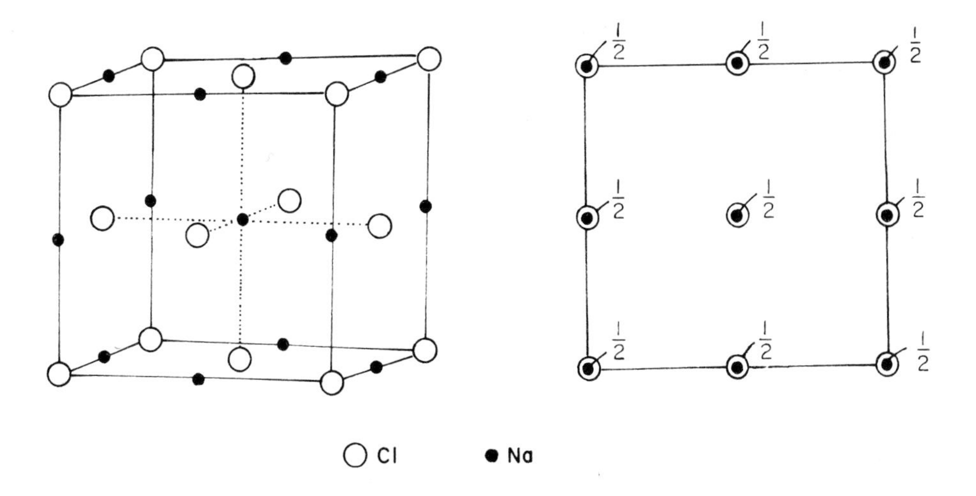

Fig. 3.13 The unit cell of the sodium chloride (NaCl) structure.

The structure is extremely common and is adopted by the majority of alkali halides, oxides and sulphides of the rare earth elements (e.g. MgO, CaS) as well as first-row transition metal oxides. The geometrical simplicity of the structure as well as the restricted environment of minerals with this structure makes it of limited interest however in the study of mineral behaviour.

3.2.5 CUBIC CLOSE-PACKED STRUCTURES WITH BOTH TETRAHEDRAL AND OCTAHEDRAL SITES PARTLY OCCUPIED

In the group of oxide structures known as the spinels, with general formula AB_2O_4 both types of interstitial site are occupied by cations. One third of the cations occupy tetrahedral sites, with the remaining two thirds in the octahedral sites. This arrangement leads to a unit cell of the spinel structure which has cell dimensions double that of the oxygen subcell which is the simple cubic close-packed cell. There are therefore eight of these oxygen subcells in the spinel structure, i.e. 32 oxygen atoms with ideally 64 tetrahedral sites and 32 octahedral sites. From the cation : anion ratios we can see that of these sites eight tetrahedral and sixteen octahedral sites will be occupied.

Due to the large number of atoms in this unit cell an illustration of the structure would not be particularly helpful, and ideally a structure model should be used to locate the sites. The spinel structure is extraordinarily flexible in the range of cations it will accept, and as well as using divalent and trivalent cations, electrical neutrality can be satisfied using combinations such as A^{4+}, B^{2+} and A^{6+}, B^{+}.

The type structure is that of the mineral spinel, $MgAl_2O_4$. In this case the divalent ions occupy the tetrahedral sites, and the trivalent ions occupy the octahedral sites. This is termed a normal (N) spinel, often written $A(B_2)O_4$. When the trivalent ions occupy the eight tetrahedral sites and the remaining eight trivalent and eight divalent cations occupy the octahedral sites, the structure is termed an inverse (I) spinel, $B(AB)O_4$. The cations are distributed at random amongst the octahedral sites, even at room temperatures.

Table 3.1

Mineral	*Formula*	*Cell dimension a* (Å)	*Structure*
Spinel	$MgAl_2O_4$	8.09	Normal
Hercynite	$FeAl_2O_4$	8.14	Normal
Gahnite	$ZnAl_2O_4$	8.08	Normal
Galaxite	$MnAl_2O_4$	8.28	Normal
Magnesioferrite	$Fe^{3+}(Mg, Fe^{3+})O_4$	8.37	Inverse
Magnetite	$Fe^{3+}(Fe^{2+}, Fe^{3+})O_4$	8.39	Inverse
Franklinite	$ZnFe_2O_4$	8.43	Normal
Jacobsite	$Fe^{3+}(Mn, Fe^{3+})O_4$	8.50	Inverse
Magnesiochromite	$MgCr_2O_4$	8.32	Normal
Chromite	$Fe^{2+}Cr_2O_4$	8.37	Normal
Ulvospinel	$Fe^{2+}(Ti, Fe^{2+})O_4$	8.54	Inverse

Table 3.1 lists the common oxide minerals which have the spinel structure, together with the composition, cell dimensions and structure type. The variation in cell dimensions is due to the differences in the ionic radii of the cations present and on this basis the spinels are subdivided into groups in which the cations have

similar radii and hence the unit cells have similar dimensions. Pure end members of these groups are rare as natural minerals, and extensive solid solutions exist between them. Many spinel solid solutions formed at high temperatures apparently remain as solid solutions down to room temperatures even over long cooling periods, but this may reflect the sluggishness of diffusion rates of the cations. Hence the low-temperature behaviour is not well understood, as any microstructures formed by a sluggish process would be on too fine a scale to be observed in an optical microscope. Very few studies on an electron-microscope scale have been carried out.

On the basis of ionic radii of substituting ions and hence of the cell dimensions of the end members we might expect that solid solutions between members in the same group might persist down to low temperatures. On the other hand solid solutions between members of different groups might be expected to be more limited, especially at lower temperatures. Therefore, although it is known that a complete solid solution exists between $MgAl_2O_4$ and $MgCr_2O_4$ at 1000°C it might be anticipated that this would not be the case at 20°C. If an intermediate composition is found to be an apparently homogeneous phase at low temperatures we should suspect that either the inhomogeneity is on a very fine scale and has not been observed, or the sluggishness of any exsolution process has prevented any perceptible result.

Very little data on such behaviour exists. Considering that the most common compositions of spinels formed in igneous rocks are of the form $(Mg, Fe)^{2+}(Fe, Cr, Al)^{3+}_2O_4$ an understanding of the processes which may occur in such complex solid solutions may be able to provide useful information on the thermal history of the igneous rock. We will return to this topic in a later chapter.

With regard to the cation distribution in spinels an important point must be made in relation to the use of ionic radii. In $MgCr_2O_4$, for example, the larger Mg^{2+} ions occupy the tetrahedral sites while the smaller Al^{3+} ions are in the octahedral sites. This is the reverse of what might be expected. The reason for this is

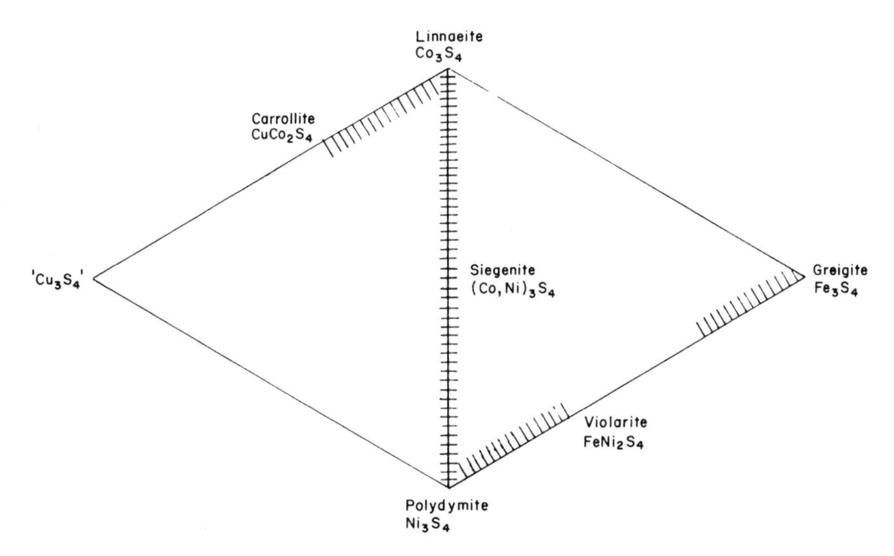

Fig. 3.14 The extent of possible solid solution in the Fe–Co–Ni–Cu–S thiospinels, shown by the hatched regions.

that the Cr^{3+} ion has a large preference for occupying octahedral sites in a structure due to crystal field effects, and this overrides the effect of the ionic radius and radius ratios. A discussion of crystal-field effects is beyond the scope of this book, and may be referred to in the book by Burns (1970).

A number of sulphide minerals also crystallize with the spinel structure, and are termed thiospinels. The compositions of most natural thiospinels fall in the range Fe–Co–Ni–Cu–S. End-member compositions include the minerals griegite, Fe_3S_4, linnaeite, Co_3S_4 and polydymite, Ni_3S_4. The compound Cu_3S_4 is unknown, as the Cu^{3+} oxidation state is very unstable. Common intermediate compositions are carrollite, $CuCo_2S_4$, violarite, $NiFe_2S_4$ and siegenite, $(Co, Ni)_3S_4$. The extent of solid solutions within this compositional field is not well known. Fig. 3.14 is a schematic diagram, based on observed natural compositions, of the possible solid solution fields in thiospinels.

B MORE COMPLEX MINERAL STRUCTURES—THE SILICATES

Compounds containing silicon and oxygen, i.e. the silicates, form the most abundant component of the earth's crust and upper mantle. Their structures consist of oxygen atoms which may sometimes be approximately close-packed, at other times less so. For most practical purposes, however, a model based on close-packing is not the most suitable for describing and classifying silicate structures. The fundamental characteristic of all silicate structures is the four-fold co-ordination of silicon by oxygen. The silicon atom and its four nearest neighbour oxygen atoms form a tightly bonded SiO_4^{4-} group, generally drawn as a simple tetrahedron (Fig. 3.15),

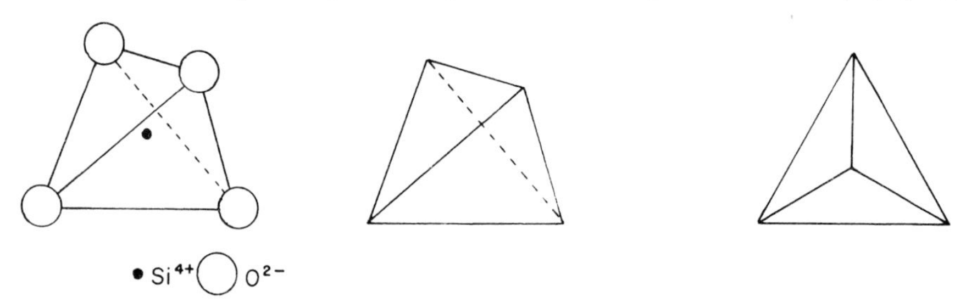

Fig. 3.15 Conventional depiction of an SiO_4^{4-} tetrahedron.

its corners at the centres of oxygen atoms with the silicon at the centre. Silicate structures are most conveniently described in terms of the arrangement of these tetrahedra.

Due to its size, Al^{3+} can replace Si^{4+} within the tetrahedron, and in many silicate structures it does so in a more or less random manner and to an indefinite extent. Therefore, the resulting tetrahedron containing Al^{3+} has the formula AlO_4^{5-} and, in general, the size of the tetrahedron is increased. Whenever such a substitution is made in a silicate structure, there must be some compensating replacement in the cation content to retain electrical neutrality (i.e. coupled substitution).

Although in some silicates isolated tetrahedra are found, the great majority have the SiO_4 groups linked together by sharing corner atoms. The silicate minerals are generally classified in terms of the way in which these tetrahedra are linked. This arrangement of linked tetrahedra forms the main structural element or skeleton which is held together by bonds to cations which lie in the interstices between the tetrahedra. The flexibility in the occupation of these cation sites depends on the arrangement of the tetrahedra, and this may lead to extensive solid solutions, especially at higher temperatures.

In dealing with the transformation behaviour in the silicate minerals a number of new principles apply. While it is no longer possible to discuss behaviour in terms of simple anion close-packing, for example, it is possible in the case of each mineral group to say something about the likely possibilities both for solid solution and polymorphism from the main aspects of the structure, and it will be helpful to think about such problems in this way. If we consider the feldspar structure, for example, we will see that the three-dimensionally linked tetrahedral framework is under a considerable constraint in adapting to different possible substituting cations. In practice it is only possible to utilize the larger cations such as Ca, Na and K. It can be said that the structure expands or contracts progressively with the substitution of cations of different size. In this case the behaviour of the structure as a whole must be homogeneous.

On the other hand, when the structure of a pyroxene is examined it will be apparent that the lack of three-dimensional linkage means that the chains of tetrahedra may adapt as units by collapse along their length, either as a function of temperature decrease or the substitution of smaller linking cations. In addition the flexibility of the structure in relation to the cross-linkage of the chains means that ions ranging from Al to Na may quite conveniently occupy the linking sites. This of course will be reflected in the immensely wide range of possible atomic substitutions observed in the pyroxene group of minerals. It is possible to make similar comments about each mineral group in turn, and these simple ideas are particularly useful in the context of mineral behaviour.

Unlike some of the examples in the first part of this chapter, order–disorder phenomena in silicates do not generally lead to the formation of superstructures in the ordered form. In the copper–iron sulphides disordering of the cations led to complete equivalence of the tetrahedral sites and hence to a state of higher symmetry with a smaller unit cell. Ordering the cations within the tetrahedral sites meant that the sites were no longer equivalent, and the result was a superstructure for the ordered form. In most silicates the cations reside in structural sites which have definite polyhedral geometries. There are generally two or more of these non-equivalent cation sites and disorder in the cations occurs whenever a particular cation occupies more than one type of site. Thus the sites retain their individual geometries regardless of their chemical occupancy, and order–disorder transformations do not change the size of the unit cell. The changes in the symmetry properties of the structure are rather more subtle and will be discussed alongside the appropriate mineral structures.

In the simplest silicate solid solutions the SiO_4 tetrahedra remain intact and compositional variations are brought about by substitutions and replacements within the cation sites. A more complex situation arises when Al replaces Si within the tetrahedra. Due to the need to maintain charge balance, cation substitutions must take place elsewhere in the structure, and the Al, Si-distribution and the cation distribution become interdependent. Order–disorder phenomena within the cations cannot in such a case be considered independently from the Al, Si-disorder within the tetrahedra, and vice versa. This will be discussed in greater detail in the section on feldspars.

We will now describe the more common silicate minerals in terms of a classification based on the arrangement of SiO_4 tetrahedra, in each case pointing out the cation sites and their possible occupation at high temperatures. Again the emphasis is on the possible behaviour of the silicate solid solutions on cooling.

3.3 Silicates with isolated SiO_4 groups

The simplest possible structural arrangement is that in which isolated SiO_4^{4-} groups are linked together only by other cations which lie between them. The minerals in this group include the olivines, the garnets, the zircon group, the aluminosilicates, sphene and topaz, but only the first two will be discussed here.

3.3.1 THE OLIVINES

The arrangement of SiO_4 tetrahedra is shown in Fig. 3.16. In this plan view projected down the axis of the unit cell the tetrahedra alternately point up and down. There are two non-equivalent sets of cation sites linking the tetrahedra, half lying between tetrahedral bases (starred in Fig. 3.16) and half between an edge and a

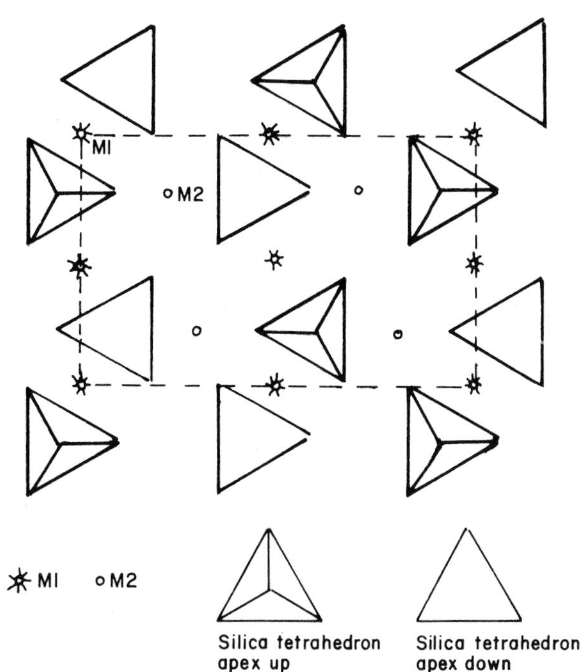

Fig. 3.16 The olivine structure showing the position of the octahedral M1 and M2 sites between the isolated SiO_4 tetrahedra.

base. The first set are termed the M1 sites, the second the M2 sites. Although in each type of site the cations are in six-fold co-ordination with neighbouring oxygen atoms, the sites are geometrically quite distinct.

As the temperature increases, the sizes of the M1 and M2 sites increase significantly but the silicate tetrahedra remain virtually identical. This probably reflects the greater bond strength of Si—O compared with that between the other cations and oxygen.

In most natural olivines the cation sites are occupied by Fe^{2+}, Mg^{2+} or Ca^{2+}. The compositions can be represented on an olivine quadrilateral (Fig. 3.17) with end members forsterite Mg_2SiO_4, fayalite Fe_2SiO_4, monticellite $CaMgSiO_4$ and kirschsteinite $CaFeSiO_4$. The separation into two solid solution series arises from the fact that the Ca^{2+} ion (ionic radius 0.99 Å) is substantially more than 15% larger than the Mg^{2+} ion (0.66 Å) or the Fe^{2+} ion (0.74 Å). Substitution of the larger Ca^{2+} ion in the structure causes strain and distorts the SiO_4 tertahedra. The structure cannot tolerate the excessive distortion produced by a random occupation of the cation sites by ions of such different sizes, and hence very little solid solution can exist between the Ca-bearing olivines and the Fe, Mg olivines even at high temperatures. In the Ca-bearing olivines, Ca^{2+} occupies the M2 site while $(Mg, Fe)^{2+}$ are randomly distributed over the M1 sites. As far as is known the Fe^{2+} and Mg^{2+} ions remain disordered down to room temperatures.

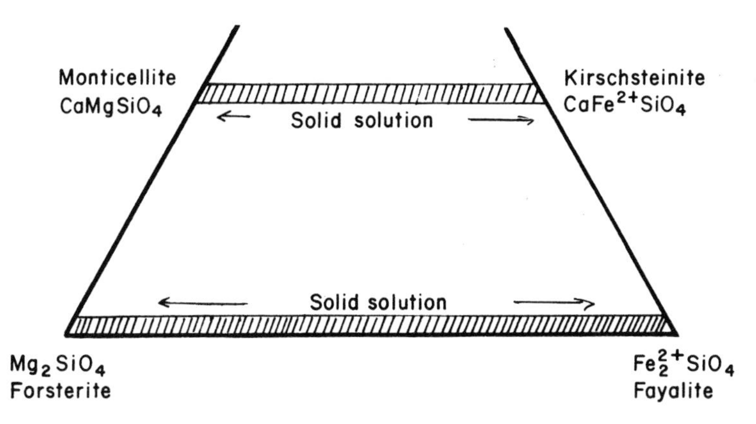

Fig. 3.17 The olivine Ca–Mg–Fe^{2+} quadrilateral showing the two solid solution series.

The solid solution between forsterite Mg_2SiO_4 and fayalite Fe_2SiO_4 is often quoted as being ideal, in the sense that there appears to be no tendency for ordering or exsolution to take place. Recently it has been found that there is a slight preference for the Fe^{2+} ions to occupy the M1 site, and so the distribution of Fe^{2+} and Mg^{2+} may not be completely random at lower temperatures. This ordering, however, is very slight.

3.3.2 THE GARNETS

The silicate garnets consist of a multicomponent solid solution, the general formula of which may be written $A_3^{2+}B_2^{3+}Si_3O_{12}$ where A^{2+} is Ca^{2+}, Mg^{2+}, Fe^{2+} or Mn^{2+} and B^{3+} is Al^{3+}, Fe^{3+} or Cr^{3+}. The larger A ions are in eight-fold co-ordination with oxygen and the smaller B ions are in six-fold co-ordination. As shown in Fig. 3.18 the SiO_4 tetrahedra are linked together by the BO_6 octahedra while the eight-co-ordinated sites are distributed among the polyhedra. The cubic unit cell is rather large and complex, but our main interest here is the extent of solid solution within the silicate garnets.

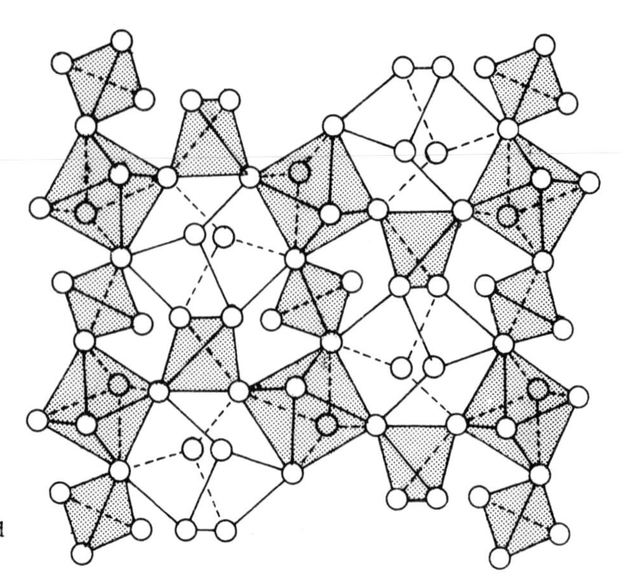

Fig. 3.18 The unit cell of the garnet structure showing the linkage between the SiO_4 tetrahedra and BO_6 octahedra. The eight-co-ordinated sites are left unshaded.

The principal species of natural garnets are commonly divided into two groups: those in which A = Ca, and those in which A ≠ Ca, B = Al. The end members and their cell dimensions are listed below.

Pyrope $Mg_3Al_2(SiO_4)_3$	11.46 Å	Uvarovite $Ca_3Cr_2(SiO_4)_3$	12.02 Å
Almandine $Fe_3^{2+}Al_2(SiO_4)_3$	11.53 Å	Grossular $Ca_3Al_2(SiO_4)_3$	11.85 Å
Spessartine $Mn_3Al_2(SiO_4)_3$	11.62 Å	Andradite $Ca_3Fe_2^{3+}(SiO_4)_3$	12.05 Å

The cell dimensions can be related linearly to the ionic radius of the A and B cations despite the complexity of the structure and the diversity of the cations. A comparison of the cell dimensions of the end members therefore gives some indication of the possibility of solid solutions between them. As a first approximation, therefore, it would be reasonable to assume that solid solutions may be extensive within each series, but limited between them.

Further clues on the ideality of possible solid solutions come from considering the nature of the cation sites in garnet. In the olivines Fe^{2+} and Mg^{2+} interchange relatively freely within the six-fold sites with little departure from ideality, in other words with little energy change associated with the substitution. In the garnets, however, Mg^{2+} is slightly too small for the eight-fold sites, a fact that is reflected in the instability of pyrope at low pressures. Consequently we may expect that the Fe–Mg substitution in garnet may be associated with a somewhat higher energy change than in olivine.

In contrast, the larger size of the eight-co-ordinated site in garnet compared to the six-co-ordinated site in olivine makes it possible for Ca^{2+} to be accommodated at high temperatures to some extent in the (Fe, Mg) garnets. Although the eight-fold site is still too small for Ca^{2+} the substitution may involve less strain than that in the olivines. These are some of the simpler criteria which we might use to infer the possible behaviour of a high-temperature garnet solid solution on cooling.

As garnets with the general composition $(Fe, Mg, Ca)_3Al_2(SiO_4)_3$ which have formed during metamorphic events at temperatures in excess of about 700°C, appear to be homogeneous under polarized light, they have generally been thought to be complete solid solutions. However, as we have already mentioned, an alternative possibility is that the sluggishness of diffusion reactions may be such that the scale on which any process has occurred on cooling may require a finer instrument such as an electron microscope to observe it.

The garnets therefore provide us with another example where at high temperatures fairly extensive solid solutions may exist. Considering the nature of the cations and the structural sites in garnet we would expect the extent of such solid solutions to be limited as the temperature falls.

3.4 Chain silicates

Two major mineral groups, the pyroxenes and the amphiboles, have silica tetrahedra linked together to form infinite chains. When each tetrahedron shares two oxygen atoms, a single chain as illustrated in Fig. 3.19(a) is formed. The formula for such a

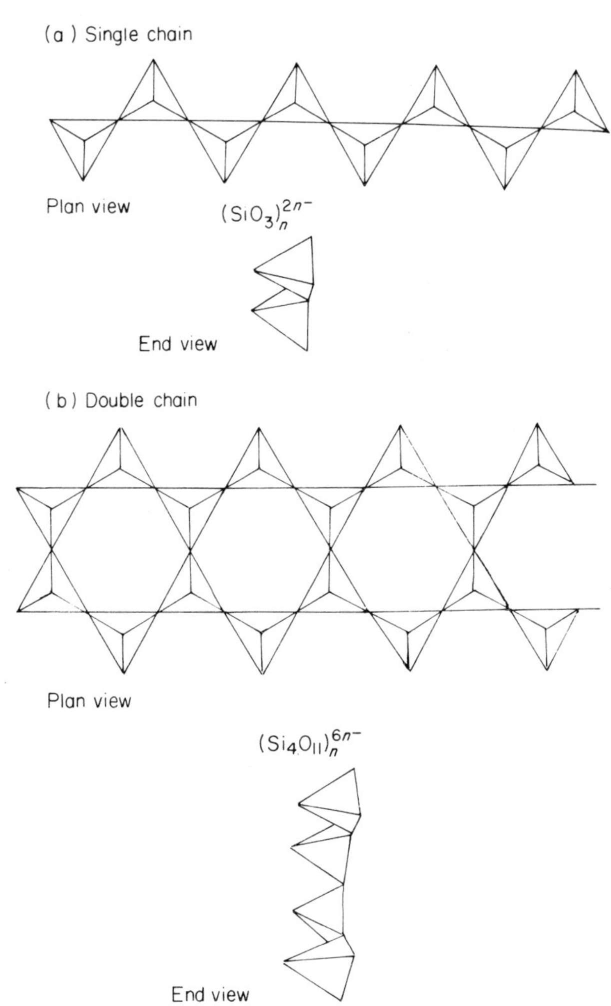

Fig. 3.19 (a) Single-chain silicate structure formed by tetrahedra sharing two corners. (b) Double-chain silicate structure formed by tetrahedra sharing two and three corners alternately.

chain is $(SiO_3)_n^{2n-}$ where n is the number of tetrahedra in the chain. The pyroxenes have structures based on single chains of silica tetrahedra.

In a double chain, the tetrahedra are linked by alternately sharing two and three oxygen atoms as illustrated in Fig. 3.19(b) and the formula becomes $(Si_4O_{11})_n^{6n-}$. Such chains are found in the amphiboles. In both the pyroxenes and the amphiboles the parallel chains are cross-linked by bonds to cations which lie between them and which give the structure electrostatic neutrality. Both mineral groups are characterized by the very extensive range of possible cation substitutions and hence solid solutions at high temperatures.

3.4.1 THE PYROXENES

The main features of the pyroxene structure are shown in Fig. 3.20 in which the silica chains are viewed from the end. The fact that each chain has a front (the apices of the tetrahedra) and a back (the bases) leads to the formation of two kinds of space between them. The cations reside in these spaces between the chains. The sites between the bases of adjacent tetrahedra are the larger and are termed the M2 sites; the smaller sites between the apices are the M1 sites.

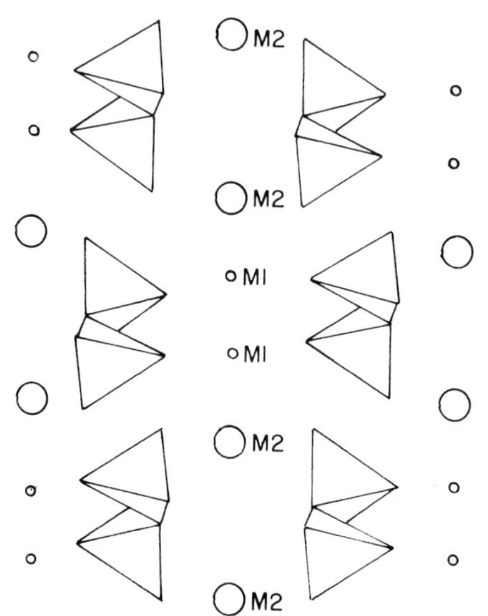

Fig. 3.20 The pyroxene structure with the single silicate chains viewed from the end. The approximate positions of the M1 and M2 sites are shown.

The chemical variations within the pyroxenes result from the variety of cations which can occupy these two cation sites. The structural variations arise from the fact that the chains themselves are able to shorten and lengthen by the rotation of the individual tetrahedra. This is illustrated in Fig. 3.21. Furthermore, adjacent chains may be displaced relative to one another along their length as well as being 'kinked' by different amounts. The chemical and structural variations are not independent however, and the nature of the cations will often determine the chain configurations which are possible.

Before discussing the cation distributions in the pyroxenes, it is instructive to outline the way in which the basic structure responds to temperature changes as this determines, to a large degree, the behaviour of the pyroxenes on cooling. As with the olivines, and probably the garnets, the size of the silica tetrahedra does not change over the whole temperature range from room temperature to around 1000°C. The M1 and M2 sites in the pyroxene structure do however expand significantly in size, the M2 site generally increasing more rapidly with temperature. Because the 'inert' tetrahedra must co-ordinate to the expanding M sites, there is a structural mismatch which is accommodated by the stretching of the silicate chains (increasing the angle θ in Fig. 3.21) or by tilting the tetrahedra out-of-plane.

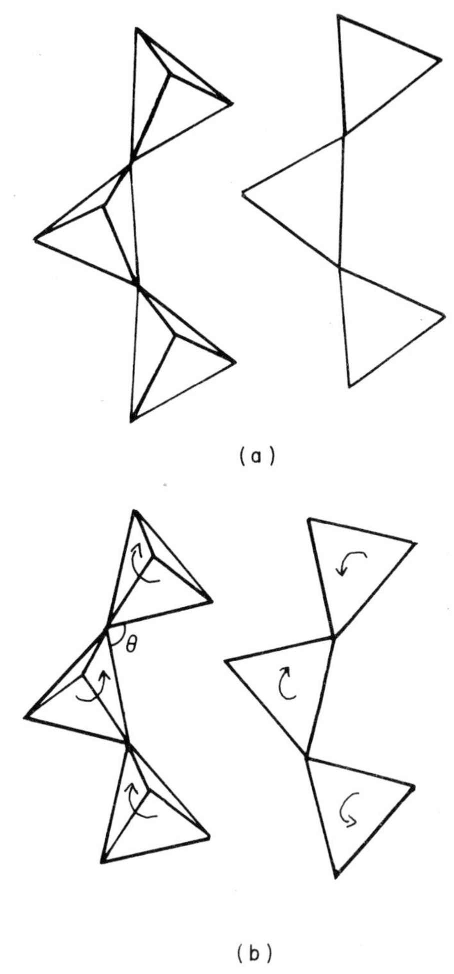

Fig. 3.21 A pair of single pyroxene chains viewed normal to their length. In (a) the chains are relatively straight as in diopside. In (b) the chains are kinked by the rotation of the individual tetrahedra.

Thus at high temperatures the chains are straighter and the cation sites larger than at low temperatures. This connection between the kinking of the chains and the size of sites imposes some restrictions on the amount of kinking possible when a relatively large cation, such as Ca^{2+} or Na^{+}, occupies the larger M2 site. The presence of Ca^{2+} effectively prevents the shortening of the chains as the temperature decreases. The effect on the M2 site of an increase in the kinking of the chains is to reduce the number of oxygen atoms which are co-ordinated to the cation occupying the site as the tetrahedral rotations change the distances between the oxygen atoms and the centre of the M2 site. Thus when Ca^{2+} occupies the M2 site it is eight-co-ordinated by oxygen atoms. If a smaller cation such as Fe^{2+} occupies the M2 site, the co-ordination is reduced to six even at high temperatures. The cation in the M1 site is always six-co-ordinated to oxygen atoms.

The structure drawn schematically in Fig. 3.20 with the chains practically straight (Fig. 3.21) may be taken as the type structure from which the other pyroxene structures may be derived by kinking or rearranging the chains. Diopside, $CaMgSi_2O_6$ has this type structure with the Ca^{2+} in the M2 site in eight-co-ordination and Mg in the M1 site in six-co-ordination. The unit cell of diopside is monoclinic and the pyroxenes derived from this structure by cation substitutions and kinking of the chains are known as clinopyroxenes. In pyroxenes which contain small cations in both the M1 and M2 sites a further, more fundamental reorganization is possible at low temperatures. The chains may be rearranged by displacements relative to one another along their lengths. The new unit cell derived in this way is orthorhombic with approximately double the *a* cell dimension of diopside and pyroxenes of this type such as hypersthene (Mg, Fe)SiO_3 are known as orthopyroxenes.

Within this general structural framework we can now discuss the variations in chemistry of the pyroxenes. The general formula may be written XYZ_2O_6 where X represents Na, Ca, Mn^{2+}, Fe^{2+}, Mg and Li in the M2 site; Y represents Mn^{2+}, Fe^{2+}, Mg, Fe^{3+}, Al, Cr^{3+} and Ti^{4+} in the M1 site; and Z represents Si and Al in the tetrahedra. Although Mn^{2+} and Li are important in some pyroxenes, these are relatively rare and will not be considered here. The compositions of some of the more important pyroxenes are generally divided into two groups : those that lie in the pyroxene 'quadrilateral' (Fig. 3.22) and the omphacitic pyroxenes (Fig. 3.23).

Most pyroxenes of igneous origin have compositions that lie approximately within the quadrilateral, and at high temperatures extensive solid solutions exist between end members with similar structures. The end members for the Ca clinopyroxenes are diopside $CaMgSi_2O_6$ and hedenbergite $CaFeSi_2O_6$, and complete solid solution is possible between them. In both cases the Ca is in eight-fold co-ordination in the M2 site, while the Mg, Fe are distributed over the M1 sites. In the augite solid solution the entry of Al^{3+} into the silica tetrahedra permits the inclusion of both divalent and trivalent cations into the M sites. A coupled substitution of the type

$$(Mg, Fe^{2+}) + Si^{4+} \longrightarrow (Al, Fe^{3+}) + Al^{3+}$$

is fairly typical and leads to a situation in which Al may be present in both the tetrahedra and in the M1 sites. The compositional variations in augites may be quite large, both in the $Fe^{2+} : Mg^{2+}$ ratios, as well as due to the presence of significant quantities of other cations such as Ti^{4+}. The pigeonite series is very similar to the augites at high temperatures, but due to the insufficient number of Ca ions in the M2 sites, the pigeonite structure at low temperatures becomes distorted by kinking of the chains. The structure however remains monoclinic.

In the orthopyroxenes the two end members enstatite $MgSiO_3$ and ferrosilite $FeSiO_3$ have Mg and Fe^{2+} in both M1 and M2 sites. A complete solid solution exists but as the temperature decreases the Mg and Fe^{2+} are partitioned between the M1 and M2 sites, with Fe^{2+} showing a strong preference for the larger M2 site.

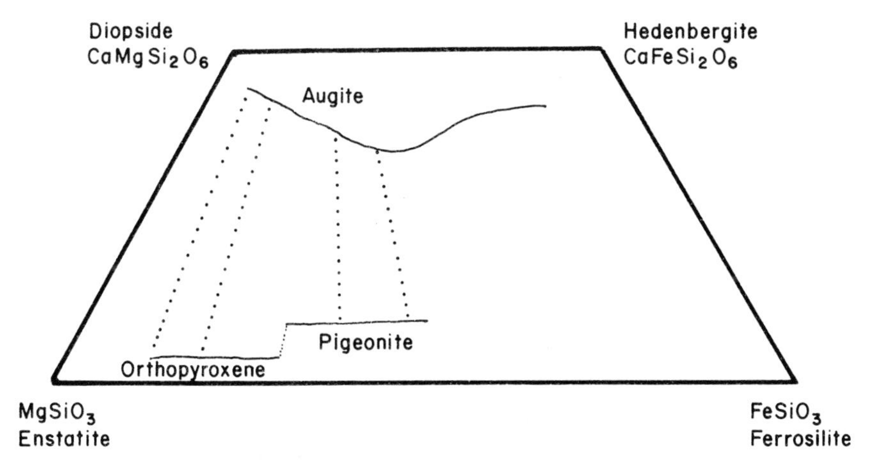

Fig. 3.22 The pyroxene quadrilateral. The solid lines show the crystallization trends of Ca-rich and Ca-poor pyroxenes from a tholeitic magma. The dotted tie lines indicate the composition of co-precipitating phases.

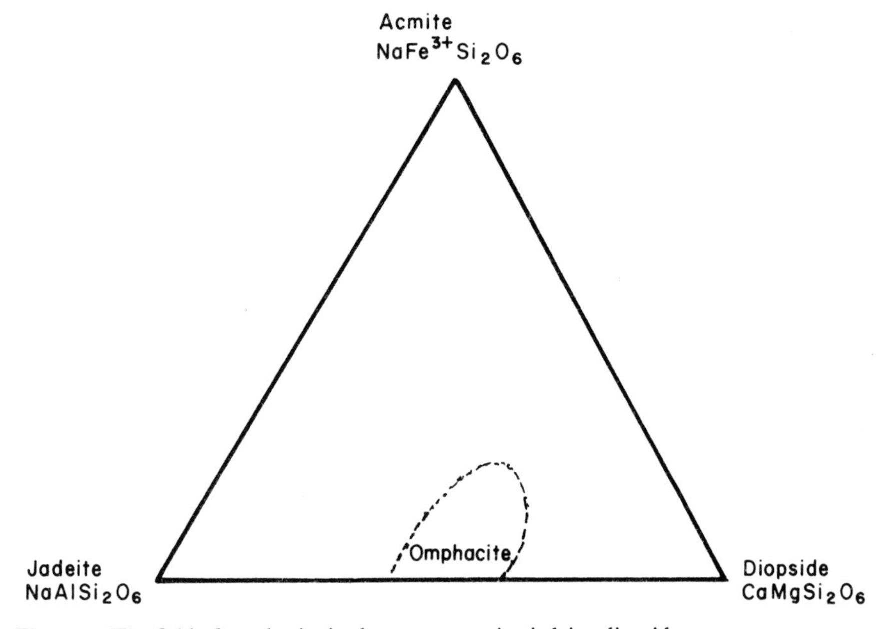

Fig. 3.23 The field of omphacite in the system acmite–jadeite–diopside.

Simply on the basis of ionic size of Ca relative to Mg and Fe^{2+} we would expect that the extent of solid solution between Ca-rich and Ca-poor pyroxenes would decrease as the temperature decreases, and this is the case, although it is further complicated by the fact that structural changes in the chains are also taking place simultaneously. As these structural changes are influenced by the cation composition, and the extent of solid solution will become even more limited as the structures between the end members become more different, a complex sequence of structural transformations and exsolution reactions will take place on cooling.

During the early crystallization history of a basaltic magma, the pyroxenes which form are commonly Mg-rich augite and orthopyroxene. As crystallization proceeds the pyroxenes become more Fe rich and pigeonites start crystallizing instead of orthopyroxenes. The tie lines in Fig. 3.22 indicate the changing composition of co-existing pyroxenes. In a later chapter we will describe in some detail the subsolidus behaviour of such pyroxenes on cooling and show that by studying the micro-structure of the low-temperature state the sequence of processes and hence the thermal history of the minerals can be understood.

In the omphacites the relevant end members are diopside $CaMgSi_2O_6$ and jadeite $NaAlSi_2O_6$ (Fig. 3.23). For intermediate compositions a coupled substitution of cations must take place and the resulting solid solution is limited to the central part of the join between the end members. In the disordered state Ca and Na ions are disordered over the M2 sites while Mg and Al are disordered over the M1 sites. On cooling, ordering of the cation pairs within the sites tends to take place, again producing microstructures which can be interpreted in terms of the metamorphic conditions prevalent at the time of the omphacite formation.

3.4.2 THE AMPHIBOLES

The structure and behaviour of the amphibole minerals bear many similarities to the pyroxenes, although the effect of the double chains of silica tetrahedra [Fig. 3.19(b)] is to increase the number of different cation sites in the structure and hence the complexity of the relationships among the different minerals. The main features of the amphibole structure are shown in Fig. 3.24, which is drawn in a similar orientation to the pyroxene structure in Fig. 3.20, i.e. viewed from the end.

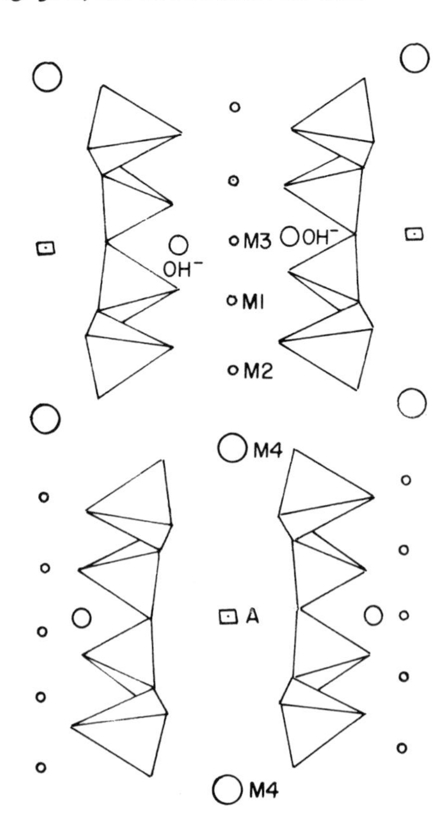

Fig. 3.24 The amphibole structure with the double silicate chains viewed from the end. The approximate positions of the cation and the OH^- sites are shown.

By analogy with the pyroxenes the cation sites in the structure are characterized by their position relative to the apices and bases of the silica tetrahedra in the double chains. The sites between tetrahedral bases of adjacent chains are termed the M4 sites (cf. the M2 sites in pyroxenes), and the smaller sites between opposed tetrahedral apices are the M1, M2, M3 sites (cf. the M1 sites in pyroxenes). The three sites arise from the fact that in the double chain the cation positions are not all crystallographically equivalent, being in different positions relative to the chains. Again by analogy with the pyroxenes, the oxygen co-ordination around the M4 site is eight when the site is occupied by a larger cation such as Ca^{2+}, or six when the site is occupied by a smaller cation such as Fe^{2+}. The M1, 2 and 3 sites are octahedral, i.e. six-co-ordinated.

The double chains lead to a third type of cation site which lies between the rings formed by opposed tetrahedral bases in the chains. Termed the A site it is a large site and may be vacant, partially filled, or fully occupied by Na and/or Ca in some amphiboles. Finally, $(OH)^-$ or F^- lies in the centre of the hexagonal rings, at the level of the tetrahedral apices.

As in the pyroxenes, chemical and structural variations within the amphiboles arise from the cation contents of the sites and the ability of the chains of tetrahedra to expand and contract under the influence of temperature changes. Again the silica tetrahedra remain inert during expansion and contraction, the volume changes taking place in the cation sites. At high temperatures, therefore, extensive solid solutions exist even with cations of dissimilar size. Contraction of the cation sites on cooling leads to a decrease in the extent of solid solution possible, leading to exsolution reactions. These are paralleled by a gradual collapse of the chains, which is limited by the size of the cations. Thus when a small cation, such as Mg^{2+}, occupies the M4 site a structural change involving a collapse of the M4 site is possible on cooling; in Ca-rich amphiboles such a collapse is not possible.

The distinction between clinoamphiboles and orthoamphiboles is analogous to that in the pyroxenes. The schematic structure drawn in Fig. 3.24 with the chains practically straight is the type structure for clinoamphiboles. Tremolite, $Ca_2Mg_5Si_8O_{22}(OH)_2$ has such a structure with the Ca ions in the M4 sites and the Mg ions distributed over the M1, 2 and 3 sites. Other clinoamphiboles are derived from this structure by cation substitutions and kinking of the chains. In the amphiboles shortening and lengthening of the double chains must preserve a mirror plane between the pair of chains as shown in Fig. 3.25. The orthoamphiboles are formed at low temperatures by a more fundamental reorganization of the double chains which becomes possible if the cations in the structure are small. In this case the chains are rearranged by displacements relative to one another along their length.

The chemistry of the amphibole minerals may be described by the general formula $W_{0-1}X_2Y_5Z_8O_{22}(OH, F)_2$. W represents the content of the large A site which may be vacant or contain varying amounts of Na/Ca. X is the content of the M4 site which may be Ca, Na, Fe^{2+} or Mg in the most common amphiboles. Y represents Fe^{2+}, Mg, Fe^{3+} or Al in the M1, M2 and M3 sites, while Z represents Si and Al in the tetrahedra. The limit of this substitution appears to be (Al_2Si_6). The variety of cation and coupled substitutions leads to a very complex chemistry and there are many possible ways of classifying the amphiboles. No attempt will be made here to describe such classifications which may be found in standard mineralogy textbooks, but the parallels between amphiboles and pyroxenes can be

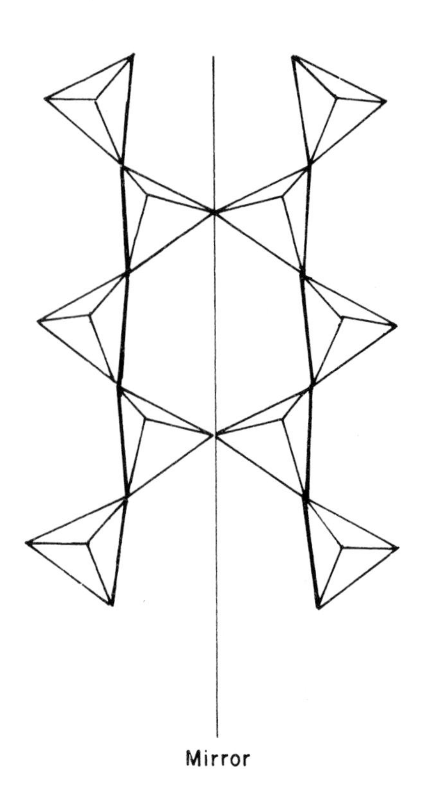

Fig. 3.25 The mirror plane between the pair of chains is preserved during kinking of the chains.

conveniently illustrated using the amphibole 'quadrilateral' (Fig. 3.26) which may be compared with Fig. 3.22 for the pyroxenes.

Complete solid solution exists between the two Ca-bearing clinoamphiboles tremolite $Ca_2Mg_5Si_8O_{22}(OH)_2$ and ferroactinolite $Ca_2Fe_5^{2+}Si_8O_{22}(OH)_2$, with Ca occupying the M4 sites and (Mg, Fe^{2+}) in the M1, 2 and 3 sites. Actinolite is a solid solution within this composition range, with a small amount of substitution of (Mg, Fe^{2+}) for Ca in the M4 sites. At high temperatures this solid solution extends towards the Ca-poor cummingtonites. As the temperature decreases however, the extent of the solid solution is reduced. At still lower temperatures the cummingtonites undergo a structural change due to collapse of the M4 site around the small cation. The larger number of Ca atoms in the actinolites prevents a similar collapse

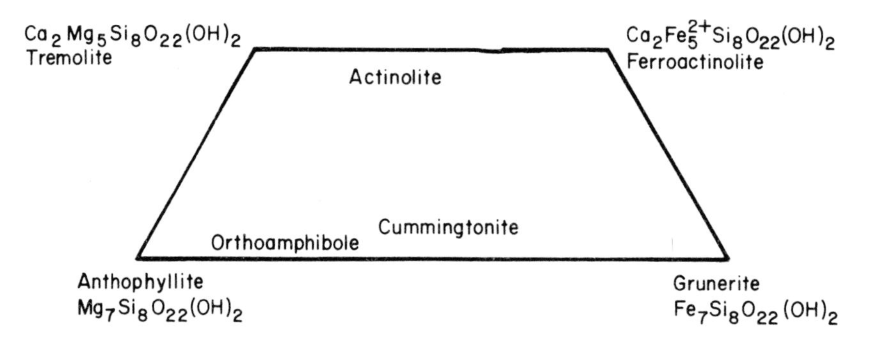

Fig. 3.26 The amphibole quadrilateral.

at these compositions. The cummingtonites may be regarded as the amphibole analogues of the pigeonitic pyroxenes.

In the Ca-free amphiboles the two end members are anthophyllite $Mg_7Si_8O_{22}(OH)_2$ and grunerite $Fe_7Si_8O_{22}(OH)_2$. In anthophyllite, which is orthorhombic, a considerable degree of substitution of Fe^{2+} for Mg is possible, with Fe^{2+} preferring the larger M2 sites at lower temperatures (see section 7.3.1).

Of the amphiboles not represented in the above quadrilateral, the most important is hornblende, which is itself a complex solid solution. Its composition may be described in terms of end members derived from tremolite by coupled substitution. Firstly, there may be entry of Na into the A sites, balanced by a substitution of Al^{3+} for Si^{4+} in the tetrahedra giving a formula of $NaCa_2Mg_5Si_7AlO_{22}(OH)_2$. Secondly, the substitution of a trivalent ion such as Al for Mg in the M1, 2 and 3 sites balanced again by Al substitution for Si in the tetrahedra leads to a formula $Ca_2Mg_4AlSi_7AlO_{22}(OH)_2$. A combination of both substitutions, i.e. $NaAl_3$ for $MgSi_2$, gives a composition $NaCa_2Mg_4AlSi_6Al_2O_{22}(OH)_2$ which may be regarded as one end member of the hornblende series. The composition of most hornblendes lies between this and tremolite, with a further substitution of Fe^{2+} for Mg.

On cooling such complex solid solutions the general principles of the behaviour will be similar to that in the pyroxenes. Structural changes in the chains will be controlled by the size of the cations in the M4 site, while the main exsolution processes will occur between Ca-rich and Ca-poor members. In other compositional series the lack of mixing at low temperatures may involve cation sites other than the M4 site. For example, the substitutions of Al for (Fe, Mg) in the M2 sites, Al for Si on the tetrahedral sites, and Na and vacancies on the A sites may all become non-ideal at low temperatures leading either to unmixing or ordering.

3.5 Layered silicates

In the layered silicates each tetrahedron shares three corners with three other tetrahedra forming a continuous two-dimensional sheet (Fig. 3.27). The formula for

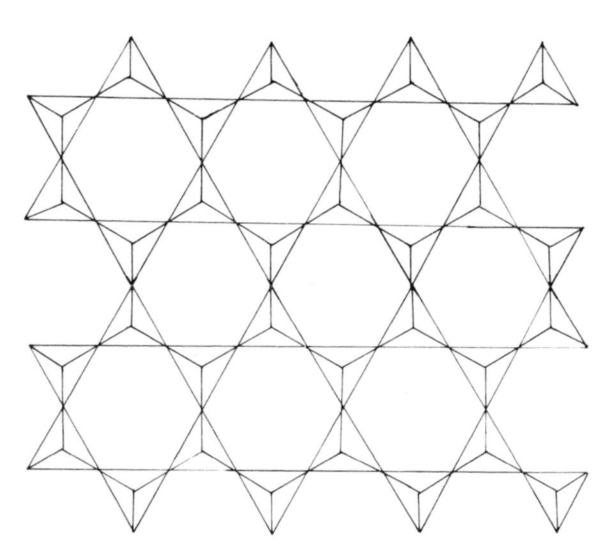

Fig. 3.27 Sheet silicate structure formed by tetrahedra sharing three corners.

the sheet is $(Si_2O_5)_n^{2n-}$. These sheets are then stacked in layers, and a variety of structures can be formed by stacking the layers in different ways and bonding them together with different cations. Although we will not discuss the layered silicates in any detail, the general nature of the structures can be illustrated by considering the mica group of minerals, taking muscovite as an example.

The muscovite structure can be regarded as a sandwich formed by two tetrahedral layers with their apices pointing inwards, cross-linked by cations which are

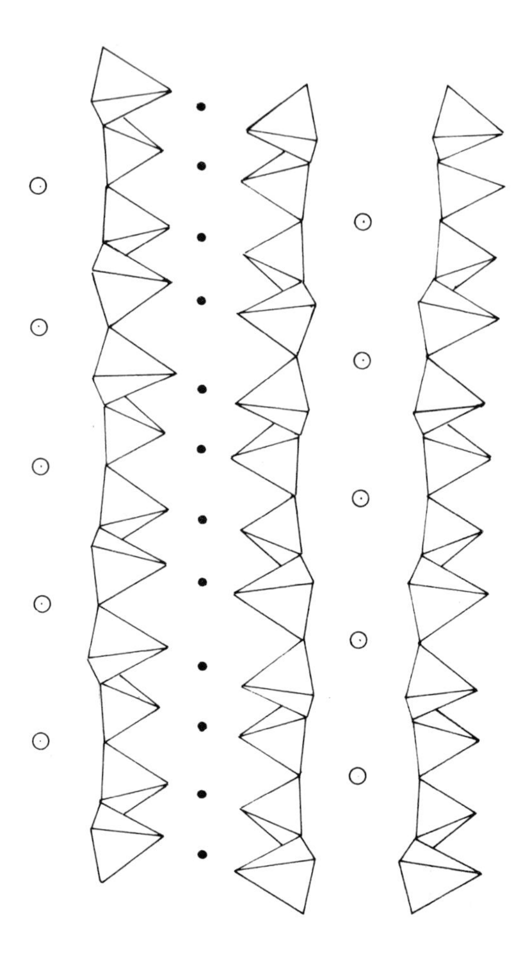

Fig. 3.28 The muscovite structure with the silicate layers viewed from the end.

octahedrally co-ordinated to the apical oxygen atoms of the tetrahedra. In order to maintain octahedral co-ordination the upper tetrahedral sheet must be staggered in the plane of the layer as shown in Fig. 3.28. The cations between the pair of tetrahedral layers form an octahedral layer and these tetrahedral–octahedral–tetrahedral units are held together by a second set of cations.

In the general formula for the micas, $W_{0-1}Y_{2-3}(Z_4O_{10})(OH,F)_2$ the W cations are those between the tetrahedral–octahedral–tetrahedral units, the Y cations form the octahedral layer and Z is Si or Al in the tetrahedra. The ideal end-member compositions of a number of micas are shown in the table below, according to the occupancy of the W, Y and Z positions.

Name	W	Y	Z
Muscovite	K	Al_2	$AlSi_3$
Paragonite	Na	Al_2	$AlSi_3$
Margarite	Ca	Al_2	Al_2Si_2
Phlogopite	K	Mg_3	$AlSi_3$
Annite	K	Fe_3^{2-}	$AlSi_3$
Lepidolite	K	Li_2Al	Si_4

More complex substitutions lead to a wide variety of minerals of similar structural type.

Because the surface symmetry of the mica sandwich is hexagonal or approximately so, further structural variations are possible by changing the way in which the sandwiches are stacked. Rotating adjacent layers by increments of 60° relative to one another does not affect the relationship between the surfaces of two units, but produces a number of stacking sequences, or polytypes. Most muscovites repeat after two sandwiches (Fig. 3.28), but in others the repeat is after every layer or else after three unit layers. Furthermore disordered arrangements of sandwiches are also possible and are found in low-temperature sedimentary micas. This stacking disorder, coupled with the wide range of cation substitutions possible, presents many problems in the identification and classification of micas.

3.6 Framework silicates

In the framework silicates all tetrahedra share all their corners with others, forming a rigid three-dimensional structure. If no ion substitutes for Si, the entire framework has the composition SiO_2 and there are no bonds available for the inclusion of interstitial cations. The various forms of silica, SiO_2 are framework structures, in which the tetrahedra are connected in different ways. The feldspars, which form the most important group of minerals in the earth's crust, are also framework structures. In these 25–50% of the tetrahedral positions are occupied by Al^{3+} instead of Si^{4+}, requiring the presence of interstitial cations within the framework to maintain charge balance.

3.6.1 THE SILICA MINERALS

The minerals quartz, tridymite, cristobalite and coesite are polymorphs of SiO_2 whose fields of stability have been illustrated in Fig. 1.1. We will direct our attention to the relationships between quartz, tridymite and cristobalite, all of which have both high and low modifications. In the high–low transformations only slight distortions occur, with no breaking of Si—O bonds. The transformations between the three mineral species however involves a major rearrangement of the silica tetrahedra and the breaking of Si—O bonds. Notice that in Fig. 1.1 the low forms of tridymite and cristobalite do not appear on the diagram. This is a significant point which will be discussed in more detail in a later chapter.

The linkage of the silica tetrahedra in quartz, tridymite and cristobalite is fundamentally different, producing more open arrangements in the higher temperature forms. The high-temperature form, high cristobalite is stable only above 1470°C. Below this temperature a transformation to high tridymite should take place, but as this requires a major reorganization of the tetrahedra, the change is very sluggish. During fairly rapid cooling, tridymite will not form and the high temperature form will be retained metastably down to low temperatures. At about 230°C high cristobalite distorts to the low-cristobalite structure. This involves a relatively minor change and takes place very rapidly.

A very similar description can be given for the behaviour of tridymite, stable between 1470 and 867°C. Failure to transform to quartz below this temperature retains the tridymite structure which at around 150°C undergoes a distortion to low tridymite. It is important to recognize that both of these high–low transformations occur within the stability field of a different phase (see Fig. 1.1). The transformations of silica are illustrated schematically by the diagram in Fig. 3.29.

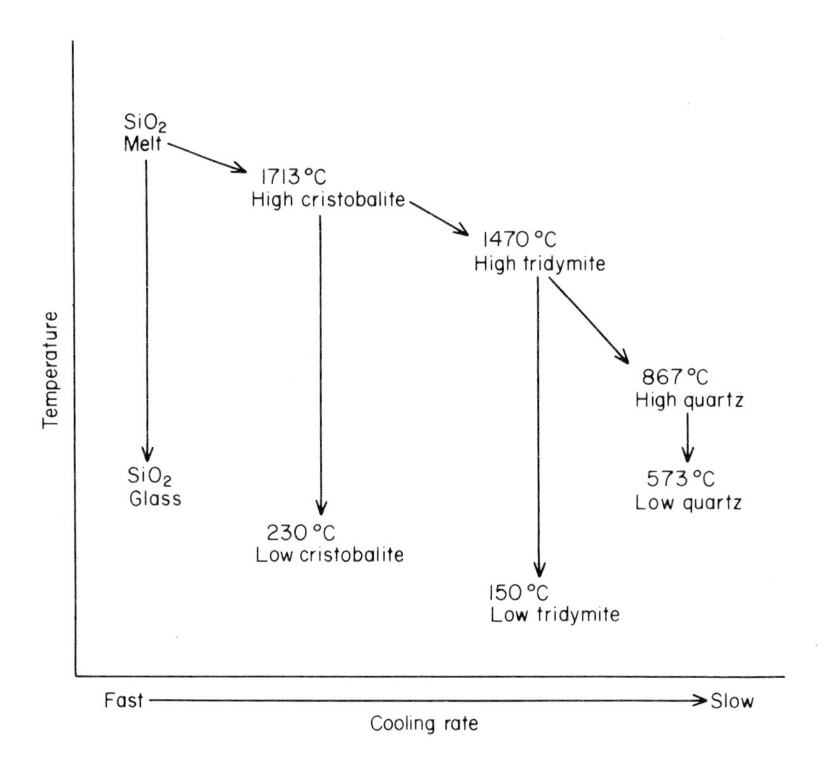

Fig. 3.29 Schematic diagram showing the transformations of silica as a function of cooling rate.

3.6.2 THE FELDSPARS

In the feldspars 25–50% of the tetrahedra in the framework contain Al^{3+} instead of Si^{4+}, and the interstices in the framework are occupied by Na^{+}, K^{+}, Ca^{2+} or Ba^{2+} ions. The basic construction of the framework is of rings of four tetrahedra with alternate pairs of vertices pointing in opposite directions. The rings are then joined in layers giving a structure such as that shown in Fig. 3.30(a). Fig. 3.30(b) shows the way in which these rings are joined normal to the plane of Fig. 3.30(a). The framework may be visualized in terms of these crankshaft-like chains extended along the *a* axis and linked to the adjacent chain by the tetrahedral apices. Tunnels between the linked chains are the sites for the cations in the structure. There is one interstitial cation site for each four tetrahedra, and the sites are always fully occupied. Therefore the ideal formula for a feldspar is MT_4O_8. When M is Na^+ or K^+ one of the T atoms is Al, and when M is Ca^{2+} or Ba^{2+} two of the atoms are Al. On this basis the end members of the feldspar minerals are obtained.

The most important feldspars are those containing K, Na or Ca, and the three end members at high temperatures are sanidine $KAlSi_3O_8$, albite $NaAlSi_3O_8$ and

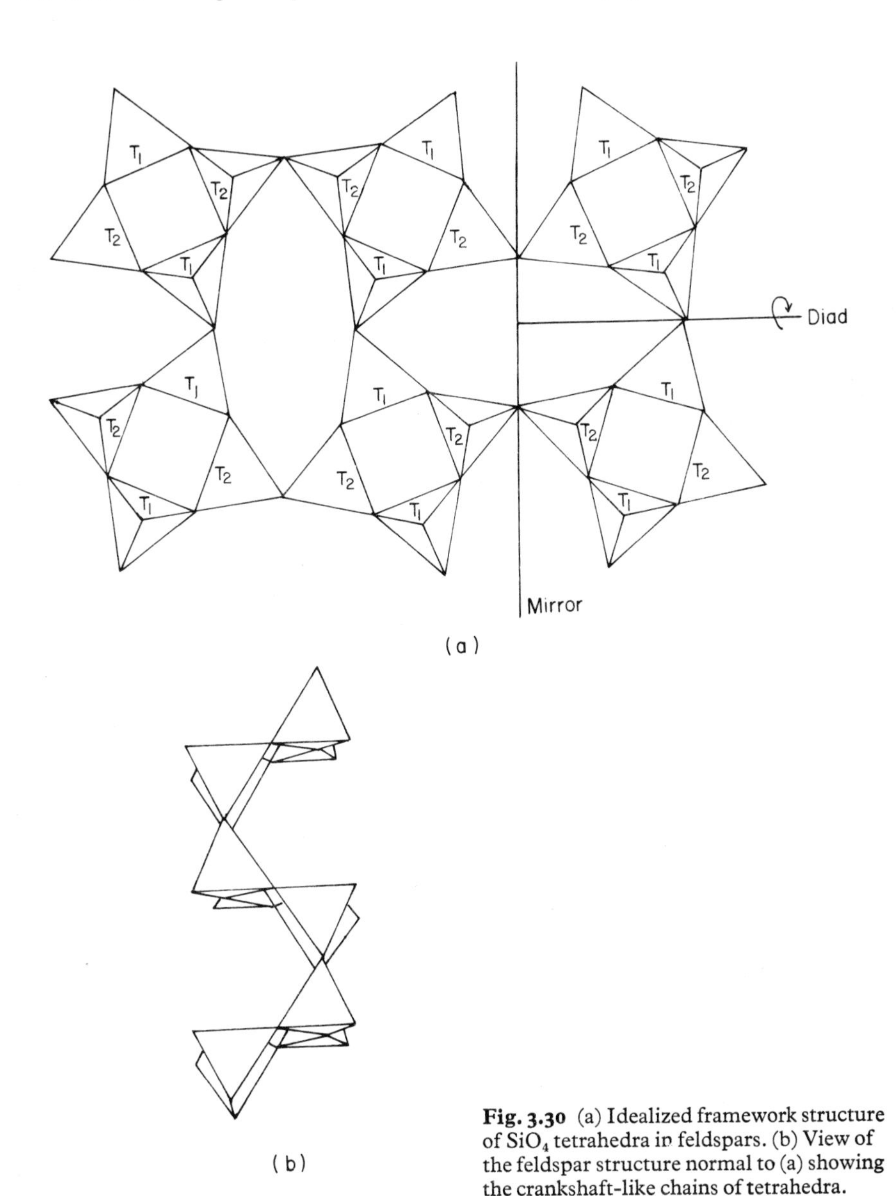

Fig. 3.30 (a) Idealized framework structure of SiO_4 tetrahedra in feldspars. (b) View of the feldspar structure normal to (a) showing the crankshaft-like chains of tetrahedra.

anorthite $CaAl_2Si_2O_8$. The extent of solid solution at high temperatures between these end members is shown in Fig. 3.31. The effect of high temperatures on the basic framework is to expand the crankshaft-like chains and increase the size of the cation site, allowing complete solid solution between the Na and K end members as well as the Na and Ca end members. Only partial substitution is possible between K and Ca, and hence there are two principal series of feldspars, the alkali feldspars and the plagioclases. Note that the smaller ions, such as Fe, Mg, Cr etc., which are commonly found in chain and layer silicates, do not occur in the feldspars to any

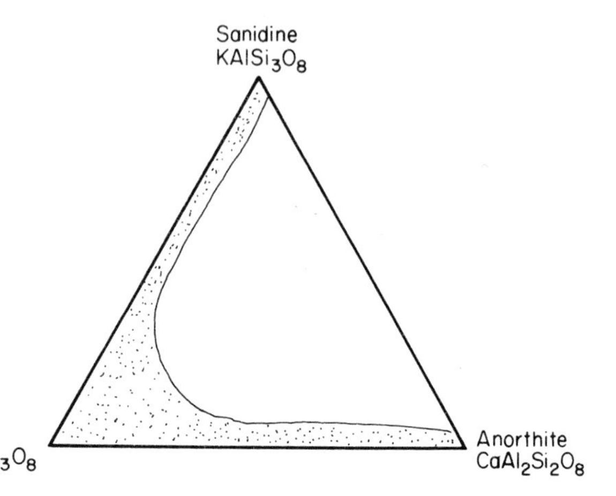

Fig. 3.31 Extent of high-temperature solid solution in the feldspars.

notable extent because the cation sites are too large and the framework cannot close in around the smaller ions. Of the cations found in the feldspars, K and Ba, due to their large size, are able to hold the framework apart as the structure cools; the feldspars containing the smaller Na and Ca ions tend to undergo structural collapse around the cation sites on cooling. The way in which the feldspar structure responds to heating and cooling is similar to the other silicates in the sense that the tetrahedra remain rigid while expansion and contraction is taken up by the changing size of the cation sites. Thus one of the aspects of the behaviour of feldspars on cooling is this tendency for the framework to distort to a lower symmetry by collapse about the relatively smaller Na and Ca cations.

The second aspect of the cooling behaviour of feldspars is concerned with the extent of solid solution between the end members as the size of the cation sites decreases. Simply on the basis of ionic radius we would not expect the solid solution between the K end member and the Na end member to persist to low temperatures. The fact that the framework structure collapses around the Na ion in the Na-rich feldspars further limits the extent to which Na feldspars can take up the larger K ion into solid solution as the temperature decreases. Thus in the alkali feldspars the general situation is one of complete solid solution at high temperatures leading to a breakdown into Na-rich and K-rich regions, i.e. exsolution at low temperatures.

Considering that the ionic radii of Na and Ca are very similar it might be expected that compositions in the plagioclase solid solution would respond fairly uniformly to a decrease in temperature, at least from the point of view of the Na, Ca-distribution. Under such conditions a disordered Na, Ca-distribution at high temperatures might be expected to give way to some kind of ordered distribution at low temperatures. However, the situation in the plagioclase feldspars is seriously complicated by the fact that the end members albite $NaAlSiO_3$ and anorthite $CaAl_2Si_2O_8$ have a different Al : Si ratio and that therefore this ratio will vary across the solid solution. In order to maintain charge balance locally in the structure there must be a linkage between Na—Si and Ca—Al. The Na and Ca ions are not free to move independently of the tetrahedral framework, and their behaviour will be intimately associated with the behaviour of the Si and Al atoms within the tetrahedra. In our previous examples of silicate structures we have not been concerned by behaviour within the tetrahedra and have considered the framework structure as a relatively inert skeleton only able to participate in the cooling behaviour by slight twists and distortions. In the feldspars the Si and Al within the tetrahedra play a vital role in the cooling behaviour and we must look again at the structure to examine the possible Si, Al-distributions.

The behaviour of Si and Al in the feldspars is dominated by two factors. The first, referred to as the aluminium avoidance principle, is that each tetrahedron containing Al will tend to be surrounded by tetrahedra containing Si so that no Al–O–Al linkages occur within the framework. In other words the Al and Si ions tend to order. The second factor is that any migration of Al and Si atoms within the structure to produce such order is *extremely* sluggish.

To appreciate, even in a general way, the complexity in the feldspars which arises from changing Al, Si-distributions we must consider the basic framework structure illustrated in Fig. 3.30(a). The topologic symmetry (i.e. the highest symmetry attainable by the framework when distortions and chemical contents are eliminated) leads to two sets of tetrahedra. Those marked T_1 in Fig. 3.30(a) are all equivalent to one another, as are those marked T_2. Under these undistorted conditions the framework has a unit cell which is monoclinic, and may be regarded as the type structure from which the lower temperature feldspar structures may be derived.

In order to show these two types of tetrahedra more clearly a schematic diagram can be drawn showing only the centres of the tetrahedra, i.e. the Si or Al atoms, in an orientation which shows a partial perspective view of the crankshaft-like chains of tetrahedra. Fig. 3.32 shows one such chain with the T_1 and T_2 types of tetrahedra

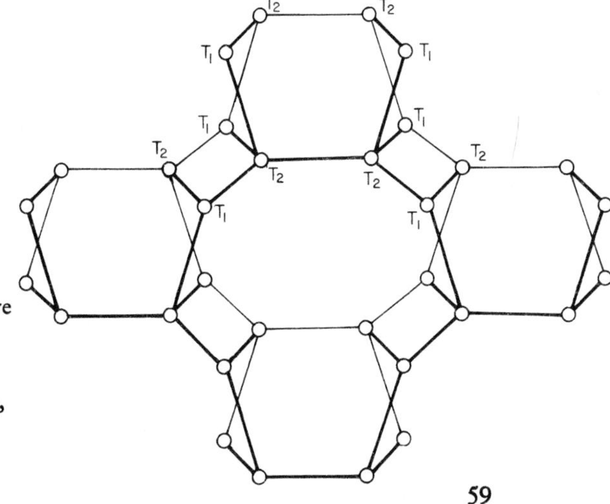

Fig. 3.32 Partial perspective view of the arrangement of tetrahedra in the feldspar structure. Only the centres of the tetrahedra are shown, and these are labelled T_1 and T_2, as in Fig. 3.30(a).

labelled in the monoclinic structure. High-temperature sanidine, $KAlSi_3O_8$, has such a monoclinic structure. If the Si and Al ions remain disordered the structure remains monoclinic and there is no tendency to collapse as the K ion is large enough to hold the framework apart on cooling. If however ordering of Si and Al occurs, as it will tend to do as the temperature falls, the structure cannot remain monoclinic and the symmetry must be reduced. This is because with a Al : Si ratio of 1 : 3 ordering cannot take place if there are only two types of tetrahedral sites, T_1 and T_2. Any preference for Al to occupy a particular site (say T_1) will result in it no longer being equivalent to its related site. The change of symmetry required by Si, Al-ordering reduces the symmetry to triclinic by distortion of the framework and results in the two types of sites being split into four, one occupied by Al the other three by Si (Fig. 3.33). The fully ordered form of $KAlSi_3O_8$ is called microcline. The migration of Al and Si during the ordering transformation in $KAlSi_3O_8$ is very sluggish and the local distortions produced in the structure, as it orders and therefore become triclinic, leads to severe problems with the result that even over geological times complete ordering may not be accomplished. Adularia and orthoclase are two forms of $KAlSi_3O_8$ which represent intermediate states in the ordering process. A further description of the behaviour of $KAlSi_3O_8$ will be given in Chapter 7.

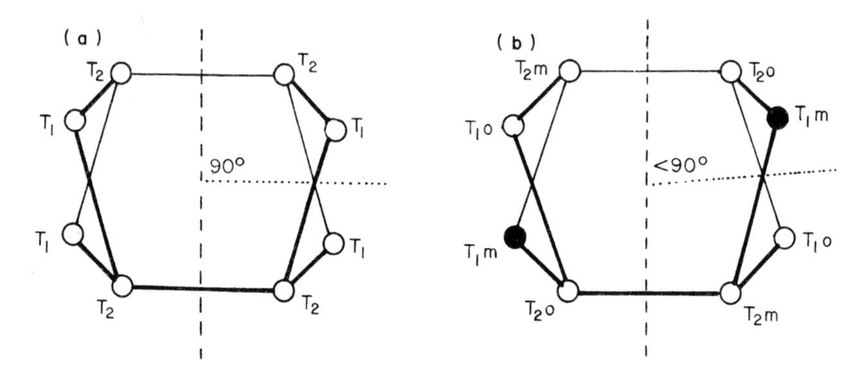

Fig. 3.33 (a) Undisorted monoclinic feldspar structure when both T_1 and T_2 tetrahedra have the same occupancies of Al and Si. (b) Ordered feldspar structure in which Al occupies the T_1m site resulting in a distortion to triclinic symmetry.

In the other end member of the alkali feldspars, $NaAlSi_3O_8$, we start off at high temperatures with a monoclinic structure similar to sanidine (Fig. 3.32). Before any Si, Al-ordering occurs however the framework collapses around the small Na ion and the result is a triclinic structure similar to that in Fig. 3.33(b) even though the framework is disordered. Within this triclinic structure the Si and Al atoms are able to order without the problems of producing local distortions, i.e. there is no further symmetry change associated with the ordering and hence the transformation, although still sluggish due to the slow migration of Si and Al, proceeds in a fairly uniform manner as the temperature decreases.

Therefore, considering the behaviour of the alkali feldspars in very general terms at this stage, we have a situation which on cooling involves three processes: firstly, the structural collapse around the $NaAlSi_3O_8$ composition; secondly, the segregation of Na- and K-rich regions as the extent of solid solution decreases; and finally the ordering of Al and Si which will present particular problems at compositions where the framework has not already collapsed to triclinic symmetry. Within the alkali feldspars the Al : Si ratio is constant at 1 : 3 and as both Na and K are monovalent there are no inherent difficulties in considering the behaviour of these cations separately from the behaviour of the Al and Si. In other words it is possible for Na and K to migrate through the framework structure without affecting local charge balance. This is not so for the plagioclase feldspars whose low-temperature behaviour is accordingly more complex. The general principles governing their behaviour can be understood by considering in turn Al, Si-ordering in the two end members anorthite $CaAl_2Si_2O_8$ and albite $NaAlSi_3O_8$.

In anorthite, unlike the alkali feldspars, the Al and Si atoms remain fully ordered right up to the melting point, because with an Al : Si ratio of 2 : 2 a simple ordering scheme is possible where the tetrahedra contain alternating Al and Si atoms (Fig. 3.34). This causes the unit cell to double in size. The ordered distribution of Al and Si results in a triclinic unit cell throughout the whole temperature range,

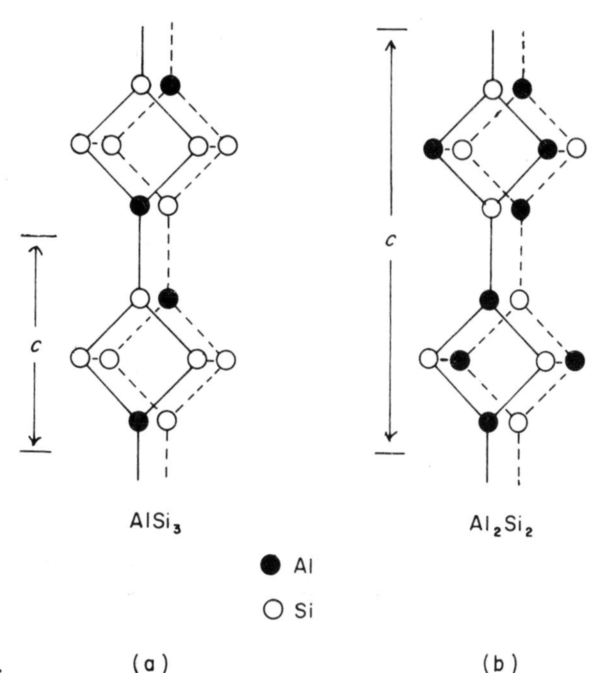

Fig. 3.34 (a) Perfect Al, Si-order in microcline and low albite with Al : Si = 1 : 3. (b) Perfect Al, Si-order in anorthite with Al : Si = 2 : 2. Note the doubling of the *c* axis in ordered anorthite.

although at lower temperatures there is a further collapse of the framework around the small Ca atom further reducing the symmetry of the triclinic structure. The main point to be made about the anorthite structure at this stage is the ordering scheme of Al and Si when the Al : Si ratio is 2 : 2. We have already seen that in albite with an Al : Si ratio of 1 : 3 there is a different ordering scheme.

In considering the possible behaviour of the plagioclase compositions between these two end members we have already pointed out that the migration of the Na and Ca atoms in the structure is not independent of the Al, Si-distribution. Therefore the behaviour of a disordered plagioclase of intermediate composition is dominated by the way in which the Al, Si atoms respond to cooling. Assuming that they will tend to order in some way we might expect one of two possibilities: either each intermediate composition will have a unique Al, Si-ordering scheme which obeys the aluminium avoidance principle, or else separate regions with either the albite or the anorthite ordering scheme will form. Neither of these two possibilities is actually observed for the following reasons. The ordering schemes in albite and anorthite are fundamentally different and it is not possible to pass from one to the other by substituting Al for Si. There are therefore no fully ordered structures possible between the end member compositions. The alternative suggestion of an exsolution to regions of albite and anorthite would involve considerable migration of Al and Si atoms through the structure and this process is extremely sluggish. Therefore even if a breakdown to albite + anorthite was the best solution to the ordering problem in the intermediate plagioclases, the rate of this breakdown would be too slow for it to be observable even at the slow cooling rates of geological processes.

The complexity of the microstructures found in the intermediate feldspars arises from the different ways that the different compositions find, or attempt to find, a solution to this problem. As the behaviour is dominated by kinetics the cooling history will be an important factor in determining the transformations which take place.

In the beginning of this discussion on plagioclase feldspars we pointed out that due to the similar size of Na and Ca ions we might expect that on cooling ordering rather than exsolution might occur. Taken alone, ordering of Na and Ca would tend to produce a homogeneous intermediate plagioclase. Ordering of the Al and Si, however, tends to produce exsolution to albite- and anorthite-like regions of order as there are no ordering schemes possible at intermediate compositions. Thus Na, Ca-ordering tends to have the opposite effect to Al, Si-ordering. The need to maintain local charge balance, however, dictates that these two processes are not independent. The final outcome of such a situation will be a compromise solution: the resulting structure will represent some interaction between Na, Ca-ordering on the one hand and Al, Si-ordering on the other.

References and additional reading

Battey, M.H. (1975) *Mineralogy for Students.* Longmans.

Deer, W.A., Howie, R.A. & Zussman, J. (1971) *An introduction to the rock forming minerals.* Longmans.

Ernst, W.G. (1969) *Earth Materials.* Prentice Hall. Chs 1–4.

Papike, J.J. & Cameron, M. (1976) *Crystal chemistry of minerals of geophysical interest.* Reviews of Geophysics and Space Physics. Vol. 14, No. 1.

4

Some Basic Thermodynamics

4.1 Stability and equilibrium

In the previous two chapters we described the state minerals find themselves in at high temperatures, and suggested that at lower temperatures a different state may become more stable. Another way of saying this is that at higher temperatures a certain atomic configuration may be thermodynamically favourable, i.e. have the highest thermodynamic probability, while at lower temperatures the thermodynamically most probable configuration may be different. We related this thermodynamical probability to the free energy, F, which is a minimum when the probability is a maximum. Expressed in this way, we can see that in any mineral transformation the free energy will tend to a minimum value.

The most stable state at any given temperature will be that with the lowest free energy, and the stability of any particular composition or structure of a mineral will be one of our primary concerns. From the outset we have pointed out that many natural minerals are not thermodynamically stable: changes which will decrease the mineral's free energy are possible, but the slow kinetics of such changes may effectively maintain the unstable structures.

There are two types of possible instability, one of which is genuine instability and the other normally described as metastability. We can illustrate this by a simple mechanical analogy of a stone rolling down a hillside (Fig. 4.1). At the top of the hill the stone is in an unstable situation, a small change in its position will reduce its free energy (in this case potential energy). At the bottom of the first hollow it has reached a free-energy minimum, but it still has a higher energy than it would have at the bottom of the hill. It is said to be in a metastable state. Notice that to achieve

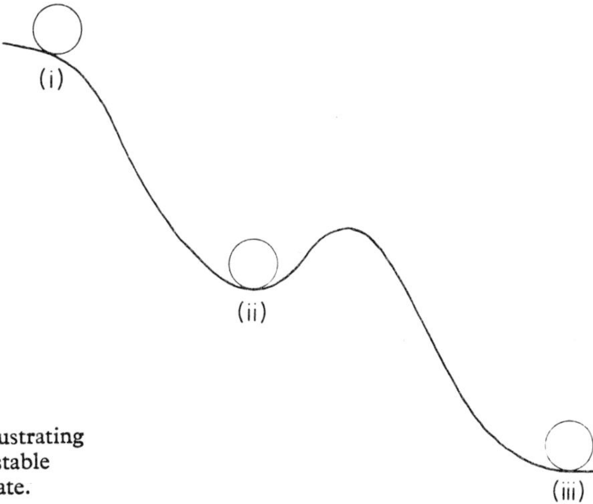

Fig. 4.1 Mechanical analogy illustrating the meaning of the terms (i) unstable (ii) metastable and (iii) stable state.

its most stable state, the stone must first pass through an intermediate, less stable and higher energy state; this acts as a barrier to the change unless some extra energy (the activation energy) can be provided. There is no such barrier to a change from the unstable state at the top of the hill.

The change in free energy of a mineral as it takes part in a transformation may be represented by Fig. 4.2 in which the 'reaction co-ordinate' is any variable which defines the progress of the transformation. The situation described could be, for example, the transformation from a high-temperature state to the low-temperature state at a temperature below the transformation temperature, T_c. The high-temperature state is therefore metastable with a higher free energy, F_1, than that of the low-temperature state, F_2. $\Delta F(=F_2-F_1)$ is negative and is termed the driving force for the transformation. The change from state 1 to state 2 is however opposed by an energy barrier (ΔF_a), and so unless the mineral can temporarily acquire the necessary extra energy to carry it over this barrier it will remain in its initial metastable state. The smallest free energy increment (ΔF_a) which will allow it to go over the barrier is the activation free energy of the transformation.

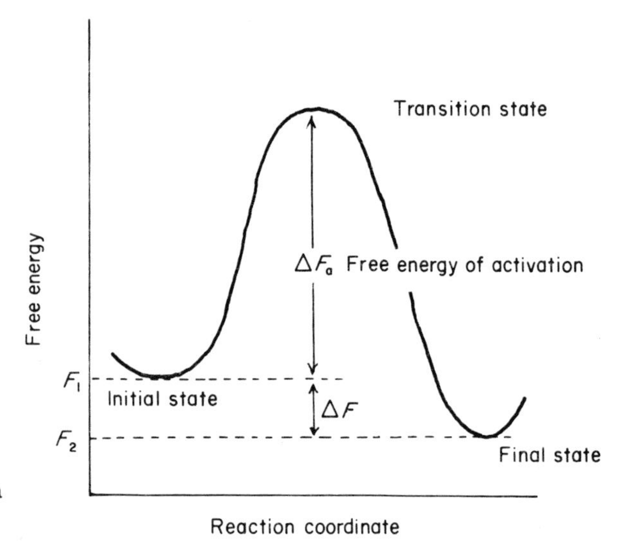

Fig. 4.2 The change in free energy of a mineral undergoing a transformation from the initial state to the final state.

The activation energy for such transformations is normally supplied by thermal fluctuations, so that the rate of transformations will be found to be strongly dependent on the temperature. At low temperatures, therefore, most transformations will take place at a negligible rate, and in most mineral transformations which require diffusion of atoms over distances greater than a few atomic distances, temperatures approaching half the melting-point temperature are required before the transformation rate becomes significant.

In considering a transformation, we have said that the free energy will tend to a minimum value. Another way of expressing this is to say that the mineral will tend towards equilibrium. At equilibrium the free energy is at a minimum, and so we have the term metastable equilibrium for the situation in the first hollow of Fig. 4.1, and stable equilibrium at the bottom of the hill.

In section 2.3 the definition of free energy in terms of internal energy and entropy was introduced:

$$F = E - TS$$

In the next section this expression will be examined in more detail, as this is the single most important expression that will be used in dealing with mineral transformations.

4.2 Free energy

The free energy, $F = E - TS$, is generally termed the Helmholtz free energy, and is used when there is no change in volume of the phase. The more general term is the Gibbs free energy, G, where

$$G = E + PV - TS.$$

This may be rewritten as $G = H - TS$, where $H(= E + PV)$ is the enthalpy. In solids at atmospheric pressure the PV term is negligible compared to the other thermodynamic quantities, E and TS. It is therefore reasonable to ignore this term in most cases. However, the Gibbs free energy function, G, will be used when free energy is referred to, and our equilibrium condition therefore will be that G must be a minimum.

At absolute zero the term TS vanishes and $G = H$. At absolute zero, therefore, the phase with the lowest internal energy will be the most stable. At temperatures above zero, both the internal energy and the entropy terms increase. The TS term, however, increases more rapidly than the internal energy term, so that at all temperatures above absolute zero, dG/dT is negative. The variation of free energy with temperature for a single phase is a curve such as that shown in Fig. 4.3.

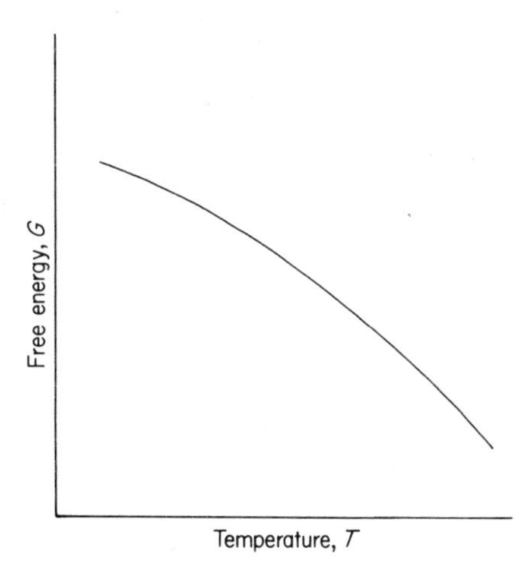

Fig. 4.3 The general form of the variation in free energy of a single phase with temperature.

The simplest type of phase transformation with which we are concerned is that between two polymorphic forms of a mineral. If a transformation from one state to another takes place, the G curves for each phase must intersect at some temperature, T_c, the transformation temperature. This is shown in Fig. 4.4. At temperatures above T_c, phase B has the lower free energy and is hence the more stable, while at

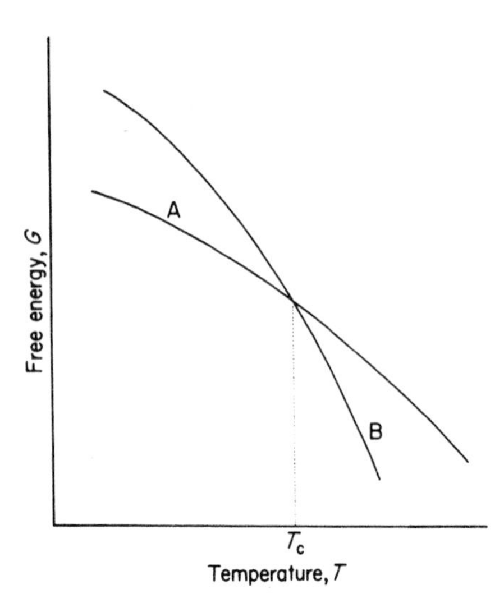

Fig. 4.4 G–T curves for two polymorphic forms A and B of a mineral. T_c is the transformation temperature for A ⇌ B.

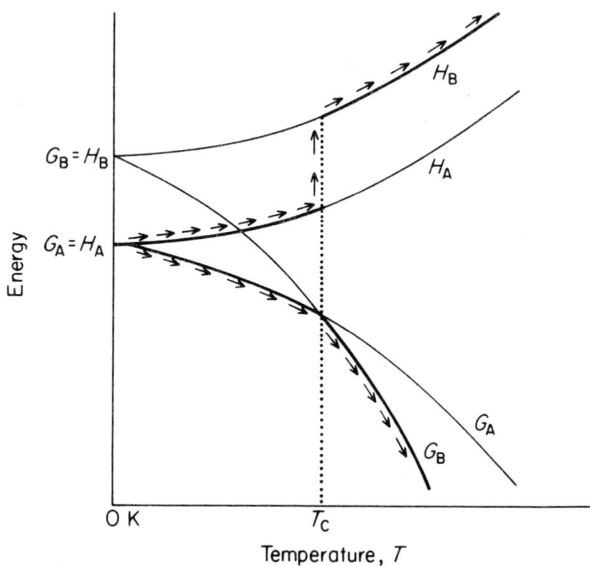

Fig. 4.5 Variation in internal energy (enthalpy) H and free energy G for two polymorphs A and B as a function of temperature. T_c is the polymorphic transformation temperature.

temperatures below T_c, phase A is the more stable. Ideally, at the transformation temperature T_c phase B should transform to phase A on cooling. Such a transformation is accompanied by an abrupt change in the internal energy, as can be seen from the following: at the transformation temperature the free energy of the two phases is equal. Therefore (neglecting the PV term),

$$E_1 - T_c S_1 = E_2 - T_c S_2$$

Therefore,

$$\Delta E - T_c \Delta S = 0$$

and

$$\Delta E = T_c \Delta S.$$

This change in internal energy at the transformation temperature is the latent heat of the transformation.

The slope of the G curve, $dG/dT = -S$. For one curve to cross another requires that with increasing temperature the high-temperature structure must have a higher entropy, and that in the situation shown in Fig. 4.4 this entropy change is discontinuous. The internal energy change at the transformation must also therefore be positive for increasing temperature. This is summarized in Fig 4.5.

4.3 Reversible and irreversible changes

For the two phases shown in Fig. 4.4 the thermodynamics require that on cooling phase B should transform to phase A at the temperature, T_c. In solids the mobility of the atoms is too slow for the kinetics to keep pace with the thermodynamic requirements, and this causes marked departures from ideal behaviour. Usually some degree of undercooling, ΔT, will be required before the transformation will take place. In some cases ΔT may be a fraction of a degree; in other cases it may be several hundred degrees, depending on the structural changes involved in the transformation. The ideal and the observed behaviour are contrasted in Fig. 4.6.

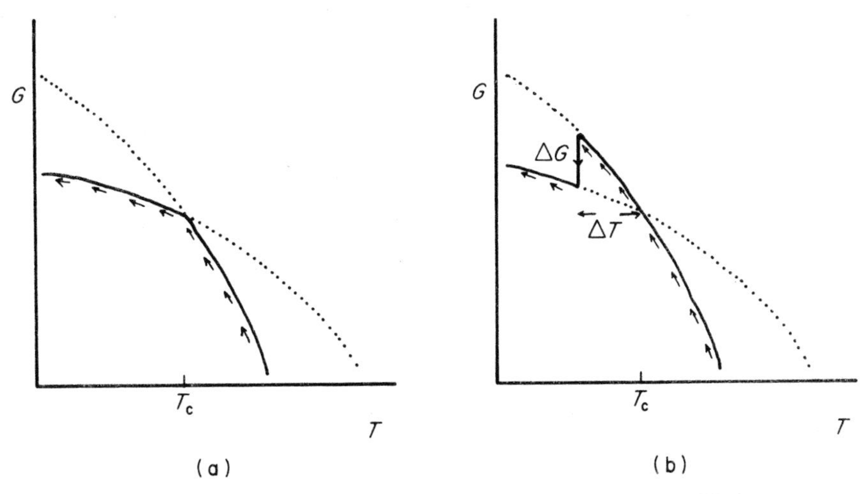

Fig. 4.6 (a) Reversible transformation from phase B to phase A under equilibrium conditions. (b) Transformation pathway from phase B to phase A under irreversible conditions involving undercooling ΔT.

When a transformation takes place at the transformation temperature, T_c, it is said to be reversible in that the same path will be followed on cooling and on heating [(Fig. 4.6(a)]. When substantial undercooling is required the transformation is irreversible, and will take place at different temperatures on cooling and on heating. For transformations in which the sluggishness causes a departure from thermodynamically ideal behaviour, the kinetics become dominant in determining the transformation temperature. The temperature at which the transformation actually takes place will depend on the cooling (or heating) rate. This is a point we will discuss in considerable detail in the next chapter, as it provides the key to understanding thermal histories of minerals by studying their transformations.

The effect of the undercooling temperature, ΔT, is to provide a driving force, ΔG, for the transformation. As ΔT increases, the driving force also increases, so that it might be expected that the transformation will take place more rapidly at lower temperatures. Atomic mobility, however, decreases very rapidly as the temperature decreases, so that the increase in ΔG is opposed by the increase in the sluggishness of the transformation. For transformations with increasing temperature, the degree of superheating which may be required will be considerably less, as increasing temperatures increase the atomic mobility and so favour the transformation.

This deviation from ideal behaviour which has been described for a simple polymorphic process will be found for all of the processes which are mineralogically interesting. Therefore the emphasis will be on the mechanism by which transformations take place, and in almost all of the cases that are described the actual behaviour of the mineral will be irreversible. Such transformations are said to take place under non-equilibrium conditions in that the free energy is not at a minimum throughout the process.

In the previous chapters we found that the most common characteristic of mineral structures at high temperatures is the existence of solid solutions. The behaviour of such solid solutions on cooling will be our primary concern, and in the next section we will describe a simple thermodynamic model to describe solid solutions and their cooling behaviour.

4.4 Solid solutions

When discussing the thermodynamics of solid solutions we are concerned with the free energy changes associated with mixing various proportions of the two end members of a solid solution pair. For example, in the olivines the two end members are forsterite Mg_2SiO_4 and fayalite Fe_2SiO_4. How does the free energy change on mixing these pure end members to give a composition $(Mg, Fe)_2SiO_4$? In any such problem we consider two aspects: firstly the change in enthalpy or internal energy* on mixing; and secondly the entropy change associated with mixing. The free energy change on mixing is then found from the formula

$$G_{mix} = H_{mix} - TS_{mix}$$

In a perfect solid solution between two end member minerals both minerals will have the same structure and very similar cell dimensions, so that an interchange of a pair of atoms imposes little strain on the structure. This pair could be, for example, Fe and Mg in the olivines or Ca and Mg in the pyroxenes, to name just two. We will designate the pair of interchangeable atoms A and B. In the solid solution A and B are distributed at random throughout the sites in the crystal. The simple thermodynamic model which follows will enable us to predict the behaviour of the solid solution on cooling.

4.4.1 THE FREE ENERGY OF SOLID SOLUTIONS

Firstly, we will consider the entropy change on mixing A and B. Assume that there are a total of N sites in the crystal, and a total of N atoms to be distributed. If the number of atoms of A is xN, then the number of B atoms must be $(1-x)N$, where x can be any number between 0 and 1. The number of ways in which xN atoms of

* For a solid at atmospheric pressure there is no significant difference between the Gibbs free energy, G, and the Helmholtz free energy, F, or between the enthalpy, H, and the internal energy, E, as the PV term is negligible. We will thus talk about the internal energy change on mixing but designate it H, to allow use of the Gibbs free energy, G, throughout.

A and $(1-x)N$ atoms of B can be arranged over the N sites is a measure of the configurational entropy of mixing (ΔS).

$$\text{Number of arrangements} = \frac{N!}{(xN)!\,[(1-x)N]!}$$

$$\text{Configurational entropy } (\Delta S) = k \log_e \omega = k \log_e \frac{N!}{(xN)!\,[(1-x)N]!}$$

We can simplify this expression using Stirling's theorem: $\log_e N! = N \log_e N - N$.

$$\text{Configurational entropy } (\Delta S) = -Nk\{x \log_e x + (1-x)\log_e(1-x)\}$$

If we take a mole of atomic sites, N is Avogadro's number and we can substitute $R = Nk = 1.98 \text{ cal K}^{-1}\text{ mol}^{-1}$ (the gas constant).

Notice that S is positive since x and $(1-x)$ are fractions. The value for S above is the change in configurational entropy on mixing As and Bs, as the values of S for the pure end members are both unity. The expression is then a measure of the extra entropy introduced by mixing, and is always positive. The term $-TS_{mix}$ in the equation for the free energy is therefore negative. The curve for the entropy of mixing as a function of composition is given in Fig. 4.7. The curve is symmetrical about the point $x = 0.5$, at which the entropy of mixing is a maximum.

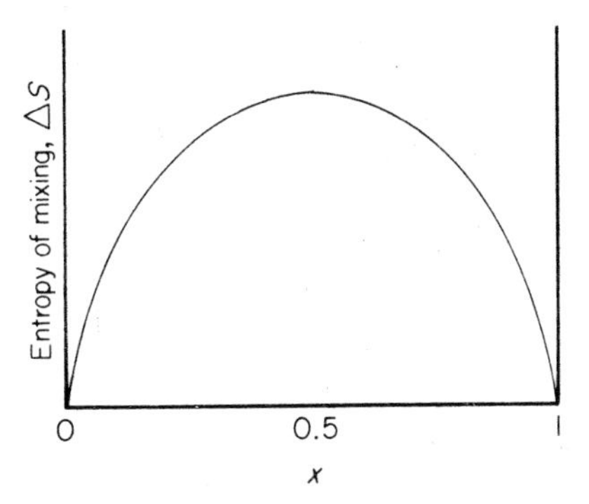

Fig. 4.7 General form of the curve for the entropy of mixing ΔS as a function of composition of a solid solution.

Next we will consider the change in internal energy or enthalpy of our mineral on mixing the two end member atoms A and B. We will use a model which assumes that the internal energy of an intermediate composition, containing both A and B atoms, stems entirely from the interaction between nearest neighbour pairs, and that this interaction between pairs is independent of what type of other nearest neighbours the given atom has. Strictly speaking this is a fairly gross approximation, but it will serve our purpose.

For this nearest neighbour model three interaction energies can be defined: V_{AA}—interaction energy of nearest neighbour A atoms, or A—A bond; V_{BB}—interaction energy of nearest neighbour B atoms; V_{AB}—interaction energy of nearest neighbour A—B atoms. To evaluate the internal energy of mixing, we need to know the number of each type of pair. This involves knowing the amount of A and B present and the total number of nearest neighbours around any particular atom. Again it will be assumed that of a total of N sites there are xN atoms of A and $(1-x)N$ atoms of B, and that the number of nearest neighbour atoms around each atom is Z. As a simple illustration consider the two dimensional arrangement of As and Bs below:

```
A—B—A—A
|  |  |  |
B—A—B—A
|  |  |  |
B—A—A—B
|  |  |  |
A—A—B—A
```

In this case $N = 16$, $x = 0.625$, so that there are ten atoms of A and six atoms of B. The nearest neighbour co-ordination is four.

In a random solid solution the number of atoms of type A next to any given atom will be Zx, and the number of atoms of type B next to any given atom will be $Z(1-x)$. Since there are a total of N atoms, and hence xN of A and $(1-x)N$ of B, the number of A—A, B—B and A—B pairs is as follows:

$$n_{AA} = \tfrac{1}{2}.\, xN.\, Zx = NZx^2/2$$
$$n_{BB} = \tfrac{1}{2}.\, (1-x)N.\, Z(1-x) = NZ(1-x)^2/2$$
$$n_{AB} = xN.\, Z(1-x) = NZx(1-x)$$

In the formula for n_{AA} the factor $\frac{1}{2}$ appears because we have counted, for each individual A atom, the number of nearest neighbours which are also A atoms, i.e. each A—A bond has been counted twice, so that we need to divide NZx^2 by 2. Similarly for the BB pairs.

If it is assumed that energy of mixing is the sum of all the energies of interaction of nearest neighbours,

$$\text{Total energy} = \frac{ZN}{2}\,[x^2 V_{AA} + (1-x)^2 V_{BB} + 2x(1-x)V_{AB}]$$

$$= \frac{ZN}{2}\,[xV_{AA} + (1-x)V_{BB} + x(1-x)(2V_{AB} - V_{AA} - V_{BB})]$$

The first two terms in this expression give the energies $\frac{1}{2}NZxV_{AA}$ and $\frac{1}{2}NZ(1-x)V_{BB}$ of the pure end members before mixing. The sign of the third term $\frac{1}{2}NZx(1-x)(2V_{AB} - V_{AA} - V_{BB})$ therefore determines whether the energy of the solid solution is higher or lower than that of the separate components. The term $(2V_{AB} - V_{AA} - V_{BB})$ relates to three possibilities for the behaviour of A and B atoms in a solid solution : that similar atoms will attract one another; that dissimilar atoms will attract one another; and that both types are attracted equally to one another.

Case 1 occurs when $2V_{AB} > V_{AA} + V_{BB}$; in other words, replacing A—A and B—B bonds by A—B bonds raises the internal energy of the solid solution, and H_{mix} is positive. Case 2 arises when $2V_{AB} < V_{AA} + V_{BB}$; this indicates a tendency for the formation of an ordered compound with the maximum number of A—B bonds, and H_{mix} is negative. If $2V_{AB} = V_{AA} + V_{BB}$, then $H_{mix} = 0$, and the solid solution is said to be ideal, with the internal energy independent of the distribution of atoms.

Combining the equations for the entropy change on mixing and the internal energy change on mixing gives the net free energy change using this model. As the free energy change is a function of the composition x it is possible to write:

$$G(x) = H_{\text{mix}} - TS_{\text{mix}}$$

$$G(x) = \frac{ZN}{2}\,[xV_{AA}+(1-x)V_{BB}+x(1-x)(2V_{AB}-V_{AA}-V_{BB})]-$$
$$-RT\,[x\log_e x+(1-x)\log_e(1-x)].$$

This rather cumbersome looking equation can be best understood by plotting the curves for H_{mix} and $-TS_{\text{mix}}$ and summing them graphically. Before doing this it it is worthwhile to note the difference in the slopes of the two curves at small concentrations:

$$\frac{dH_{\text{mix}}}{dx} \propto 1-2x$$

$$\frac{dS_{\text{mix}}}{dx} \propto \log_e(1-x)-\log_e x$$

For values of x near 0 and 1 the slope of S_{mix} is much larger than that of H_{mix} and so the free energy change is dominated by the entropy terms and is invariably reduced by a small concentration of the second component of a solid solution. This explains why it is difficult to obtain pure end member compounds without

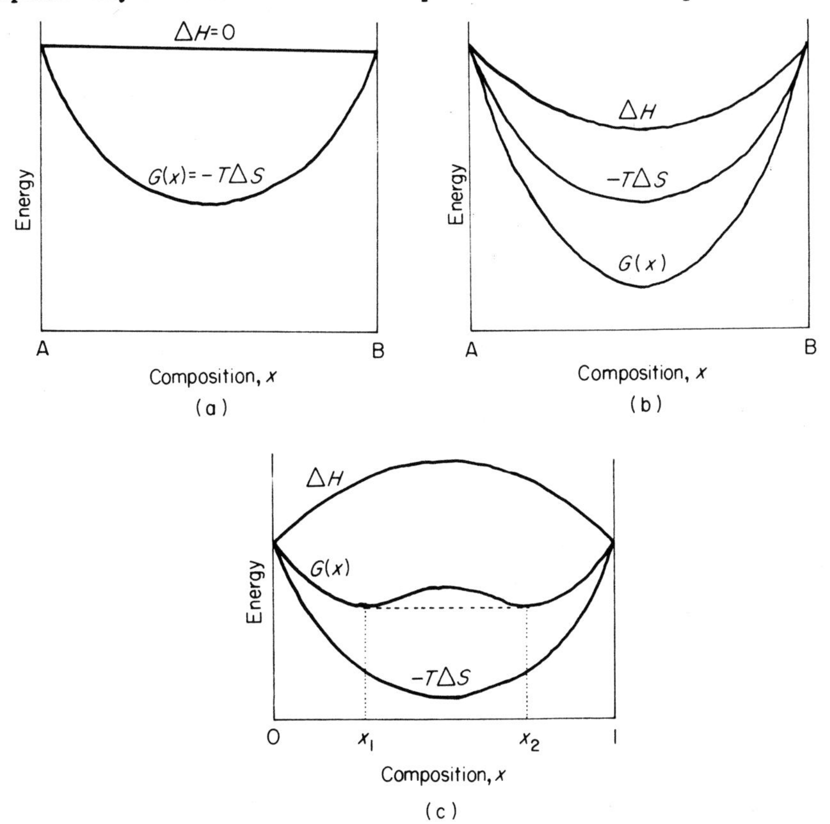

Fig. 4.8 Curves for ΔH, $-T\Delta S$ and $G(x)$ as a function of composition in three different cases: (a) when $2V_{AB} = V_{AA} + V_{BB}$, i.e. $\Delta H = 0$; (b) when $2V_{AB} < V_{AA} + V_{BB}$, i.e. $\Delta H < 0$; (c) when $2V_{AB} > V_{AA} + V_{BB}$, i.e. $\Delta H > 0$. Note: ΔH and H_{mix} are equivalent in this context.

contamination with other atoms: a small amount of impurity lowers the free energy making the mineral more stable.

Fig. 4.8 shows the two parts H_{mix} and $-TS_{mix}$ and the summation giving $G(x)$ for the three cases $H_{mix} = 0$, $H_{mix} < 0$ and $H_{mix} > 0$. In the first two cases the free energy of the solid solution is everywhere lower than that of the end members separately, and the solid solution will remain homogeneous. In case 2, the solid solution will not be ideal as there will be an attraction between A and B atoms. If this attraction is very strong an ordered compound may form such that A and B atoms alternate in position in the structure. For a case such as this the model being used here breaks down, as initially an ideal entropy of mixing was assumed. If As and Bs order in some way the entropy is very greatly reduced, and the smooth curve, such as in Fig. 4.7, no longer applies. We will return to the case where ordering occurs in a later section, and will now move on to the important case when $H_{mix} > 0$, and therefore unlike atoms tend to repel one another.

As shown in Fig. 4.8(c), the $G(x)$ plot is no longer always negative. For some range of compositions the free energy of the solid solution is higher than for others, and there will be a tendency for these compositions to unmix. Consider Fig. 4.8(c). For values of x between 0 and x_1 the solid solution remains homogeneous because the G curve is concave upwards; the same applies to compositions with x between x_2 and 1. For values of x lying between x_1 and x_2, however, the free energy will be reduced if the initially homogeneous solid solution unmixes or breaks down to regions of composition, x_1, and regions of composition, x_2. The tangent to the G curve at these two compositions defines the minimum free energy possible for any bulk composition between x_1 and x_2.

This is a very important general result and to clarify it further consider Fig. 4.9. For a composition, x_0, the free energy of a homogeneous solid solution is G_0. The common tangent to the two minima of the G curve defines the minimum free energy for compositions lying between these minima. Thus for the composition x_0 the minimum free energy would be G_1. To achieve a free energy value of G_1 the solid solution must break down to regions of composition x_1 and x_2.

The proportions of the two phases x_1 and x_2 formed in this unmixing process depends on the composition x_0. A simple lever rule may be used to calculate such proportions. From Fig. 4.9:

$$\frac{d_1}{d_2} = \frac{\text{Amount of phase with composition } x_2}{\text{Amount of phase with composition } x_1}.$$

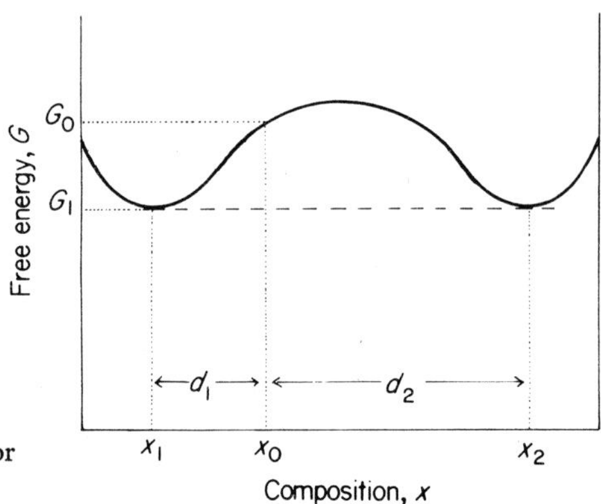

Fig. 4.9 The form of the $G-x$ curve of a solid solution at a particular temperature for the case when $H_{mix} > 0$.

From Fig. 4.9 it was possible to derive the most stable composition of the phases present at a single temperature. The effect of a change in temperature is to change the shape of the $-TS_{mix}$ curve and hence that of the $G(x)$ curve. At high temperatures the entropy term dominates and the $G(x)$ curve may everywhere be concave upwards, i.e. a homogeneous solid solution will be stable over all compositions. As the temperature falls the entropy term decreases and TS_{mix} is only greater than H_{mix} in dilute solutions (with x near 0 or 1). Thus there is a region at intermediate compositions for which unmixing will tend to occur. As the temperature falls further unmixing tends to occur over a greater compositional range and only very dilute solid solutions are stable as a single phase. In Fig. 4.10(a) $G(x)$ curves are plotted at three temperatures $T' > T'' > T'''$.

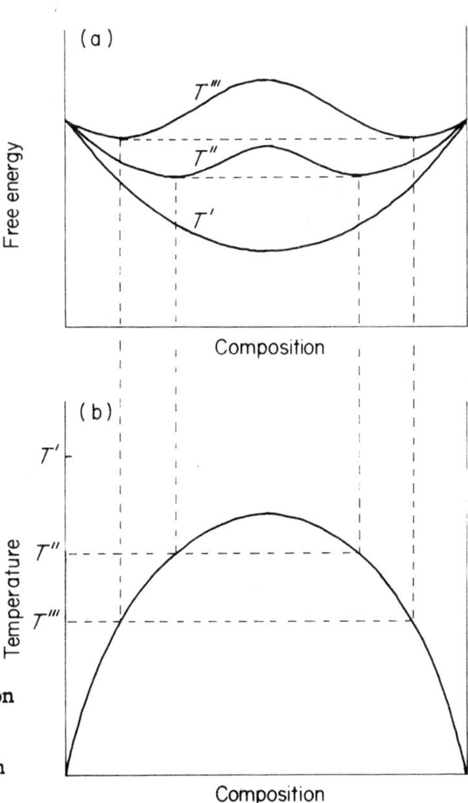

Fig. 4.10 (a) G–x curves for a solid solution with $H_{mix} > 0$ at high temperatures (T'), medium temperatures (T'') and low temperatures (T'''). (b) The phase diagram for the system described in (a).

This model illustrates a general principle that was discussed in the previous chapter. If a solid solution at high temperatures tends to break down to two phases at lower temperatures (due to say incompatibility in the sizes of substituting atoms) the extent of the solid solution will decrease with temperature. Physically speaking, this was attributed to a decrease in the vibrations of the structure with a consequent decrease in the sizes of the cation sites and a more limited tolerance for atomic substitution. This model expresses this in simple thermodynamic terms.

We will return to the $G(x)$ curves in a later chapter when we discuss the actual mechanism by which a homogeneous solid solution breaks down into two phases of different composition. Next, however, we will introduce another diagram which derives directly from the $G(x)$ curves, and is more directly useful in describing the temperature dependence of solid solutions.

4.5 The phase diagram

The relative positions of the free energy curves at various temperatures define the composition limits of one-phase and two-phase regions at these temperatures. As shown in Fig. 4.10(b) the points which limit the two-phase field at any temperature may be directly plotted from the G curve at that temperature, and the locus of such points is called the phase diagram. The limiting curve on this diagram which defines the two-phase region is called the solvus.

A phase diagram drawn in this way represents the ideal situation whereby phase changes are allowed an infinite time in which to occur so that the given material will always attain the ultimate condition or equilibrium state. It is therefore termed an equilibrium phase diagram. To illustrate an ideal sequence of processes in a system with a simple solvus we will consider the changes which take place in a single phase material of composition C_0 as it cools from a temperature above the solvus (Fig. 4.11). At a temperature T_1 the solid solution of composition C_0 is in equilibrium

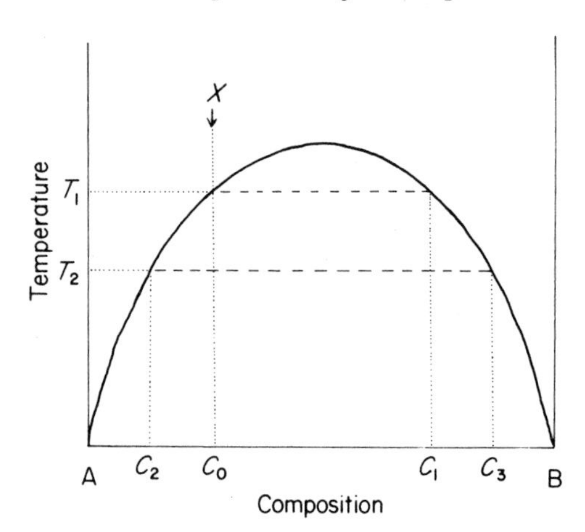

Fig. 4.11 Phase diagram with a simple solvus.

with an infinitesmal amount of solid solution of composition C_1 (as shown by the dotted line). As the free energies of these two solid solution compositions are equal at this temperature there is no driving force for any change to take place in C_0. At a slightly lower temperature T_2 a small amount of material of composition C_3 will exsolve from the initial composition leaving it depleted in the B component (C_2). As further cooling takes place the two phases continuously adjust their compositions by reacting with one another, so that at any given temperature the compositions of the two coexisting solid solutions are shown by the dashed horizontal lines (called tie-lines). At the lowest temperatures the final phase mixture will consist of regions of almost pure A and pure B, the proportion of each being determined by the lever rule.

This is the ideal situation. In practice it is very rarely achieved in solids. If one examines the sequence described above it is possible to see why. Firstly, at the temperature T_2 a small amount of the second phase of composition C_3 should form.

There are two problems here. As the composition of this second phase is considerably different to the original solid solution, its cell dimensions are also likely to be different even if the structures are basically the same. This is because of differences in the relevant ionic radii, which may be presumed to be the cause of the exsolution. The formation of a small region with different cell dimensions within a matrix of the host phase imposes a strain at the interface between the two regions. The presence of this strain will tend to suppress the formation of the phase. To counteract this a greater driving force for exsolution needs to be produced by increasing the degree of undercooling. This already means that we are departing from the equilibrium phase diagram, and the temperatures at which processes actually take place will be different than those predicted by the phase diagram. The second problem is that to form the second phase with composition C_3 requires the diffusion of considerable amounts of the B component to the site where the new phase has nucleated. Diffusion in solid is generally very slow, and in many minerals particularly so, so that temperatures have to be maintained for a very long time for the equilibrium compositions to be achieved.

Both the processes of nucleation and diffusion are functions of temperature and time, and so, unless the time scale of the cooling event is infinite, any proper description of the real behaviour of a system will also need to include time as one of the important parameters. For this reason the equilibrium phase diagram has a limited use in describing the real behaviour of minerals. Its use lies in the fact that it describes the state to which the phase or phases will ideally tend to move. This is important if we are to appreciate the significance of the real behaviour.

The temperature dependence of the processes implies that the degree to which equilibrium is attained depends on the position of the solvus curve on the temperature scale. If the solvus occurs at high temperatures the chances of achieving a situation at least approaching equilibrium are much greater than if the solvus occurs at low temperatures where diffusion may be practically arrested. A solvus at low temperatures arises when the H_{mix} term has a small positive value, which in turn may arise if the substituting cations have fairly similar ionic radii. In such a case solid solutions will persist metastably for indefinite periods at room temperatures. Often very high-resolution techniques, such as electron microscopy, will provide evidence for a fine scale breakdown in minerals which appear to be single-phase solid solutions.

Before considering real processes in any further detail, we will investigate some additional aspects of equilibrium phase diagrams.

4.5.1 SOLID SOLUTIONS WITH A SIMPLE SOLVUS

A simple solvus of the type shown in Fig. 4.11 occurs in a system where the two end members have the same structure and the solid solution is formed by a substitution of cations on the same types of sites. In this case we have made the assumption that the shape of the H_{mix} *VS* composition curve is symmetrical, i.e. that the enthalpy of mixing associated with atoms of A surrounded by B is the same as that for B surrounded by atoms of A. Another way of expressing this is that the interaction term $V_{AB} = V_{BA}$. This produces a symmetrical solvus.

An example of a binary system with a fairly symmetrical solvus is cassiterite (SnO_2)–rutile (TiO_2). Both minerals have a similar structure and the substitution

$Sn^{4+} \rightleftharpoons Ti^{4+}$ takes place on octahedrally co-ordinated cation sites. The similar ionic radii (0.71 Å for Sn^{4+} and 0.68 Å for Ti^{4+}) allow the formation of a complete solid solution at temperatures above about 1430°C (Fig. 4.12).

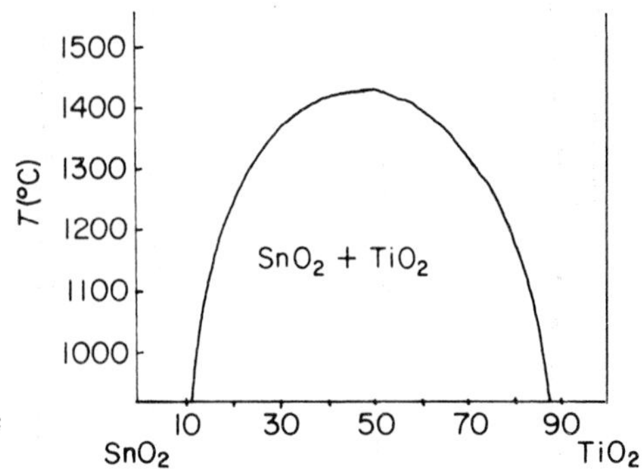

Fig. 4.12 Subsolidus phase diagram in the SnO_2–TiO_2 binary system.

Considering that the generation of a symmetrical solvus [Fig. 4.10(b)] depended on the validity of the simple nearest neighbour interaction model, it might be expected that in many systems some deviations will occur. The substituting cations will never be identical and differences in cation electronegativities, or in the structures of the end members, will result in an enthalpy of mixing which is not a symmetrical function of composition, i.e. $V_{AB} \neq V_{BA}$. The solvus will then be assymmetric.

A rather extreme example is provided by the system periclase (MgO)–zincite (ZnO), shown in Fig. 4.13. While Zn can enter the NaCl type structure of periclase fairly easily, only limited substitution is possible for Mg in the wurtzite type structure of zincite. This causes a very marked asymmetry in the MgO–ZnO solvus.

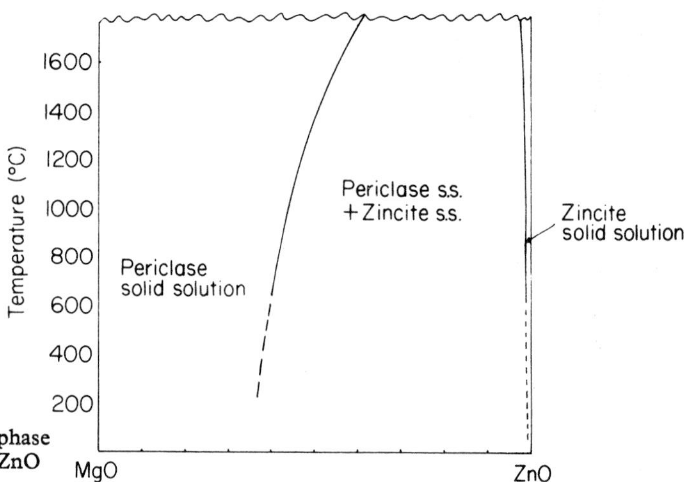

Fig. 4.13 Subsolidus phase diagram in the MgO–ZnO binary system.

In most binary mineral systems of interest one or both of the end members undergo polymorphic transformations on cooling. A further degree of complexity is then introduced into the phase diagram, and this will be discussed in the following sections.

4.5.2 THE CASE WHERE ONE END MEMBER UNDERGOES A POLYMORPHIC TRANSFORMATION

So far we have considered the simplest system where the two end members, A and B, have very similar structures and the degree of solid solution between them decreases uniformly with temperature. As an example of a more complex situation we will consider the case where one of the end members (e.g. A) undergoes a polymorphic transformation to a different structure at some temperature T_c. Above T_c we will denote the high-temperature polymorph by A_a, and the low-temperature polymorph formed below T_c by A_b. At high temperatures there is complete solid solution between A_a and the other end member B, just as in the previous example. Two consequences of a change in structure to A_b are that firstly a single G curve can no longer be drawn for the compositions between A_b and B, and secondly that the degree of solid solution between A_b and B is likely to be less than that between A_a and B, even at similar temperatures, because of the structural differences between these end members.

We can illustrate this situation by schematic G curves. There are two curves here, one for the A_a–B solid solution and the other for the A_b solid solution. We will assume that the A_a–B solid solution is complete at high temperatures, but is limited by a solvus at lower temperatures, whereas the extent of solid solution between A_b and B is much less. At temperatures above T_c [Fig. 4.14(a)] the G curve for the A_a–B solid solution lies below that of the A_b–B solid solution for all compositions, i.e. the A_a–B solid solution is stable over the whole composition range. Notice the different shapes for the two curves indicating the extent of solid solution.

As the temperature decreases the relative positions of the two curves change, such that at a temperature T_1, which is below T_c, the A_b–B curve lies below the A_a–B curve for some compositions [Fig. 4.14(b)]. The single-phase or two-phase mixtures which minimize the free energy at each composition may be found by constructing the line tangent to the two curves. In Fig. 4.14(b) this tangent defines two compositions x_a and x_b between which a two-phase mixture is stable. Compositions richer in the A component than x_a and richer in the B component than x_b are stable as single phases. Using similar construction to that in the previous section we may relate these G curves to the phase diagram at temperature T_1, as shown in Fig. 4.15.

As the temperature is dropped below T_1 to T_2 the G curve for A_b–B continues to drop relative to that for A_a–B. This has the effect of rotating the common tangent to the two curves, moving the tangent points closer to the centre of the diagram [Fig. 4.14(c)]. On the phase diagram this is shown as an increase in the width of the two-phase region, with the solubility of B increasing in A_b and decreasing in A_a. Note that B is still much more soluble in A_a than in A_b. This two-phase region in a binary system associated with a polymorphic transformation is called an inversion interval, where the term inversion refers to the transformation of A_a to A_b.

The effect of a solvus in the A_a–B solid solution is to change the curvature of its G curve below the maximum solvus temperature and so produce a two-phase region between compositions x_c and x_d in Fig. 4.14(d). The G curves drawn here at a temperature T_3 may be compared to the phase diagram at this temperature. At the temperature T_E, one common tangent touches the G curves at three positions

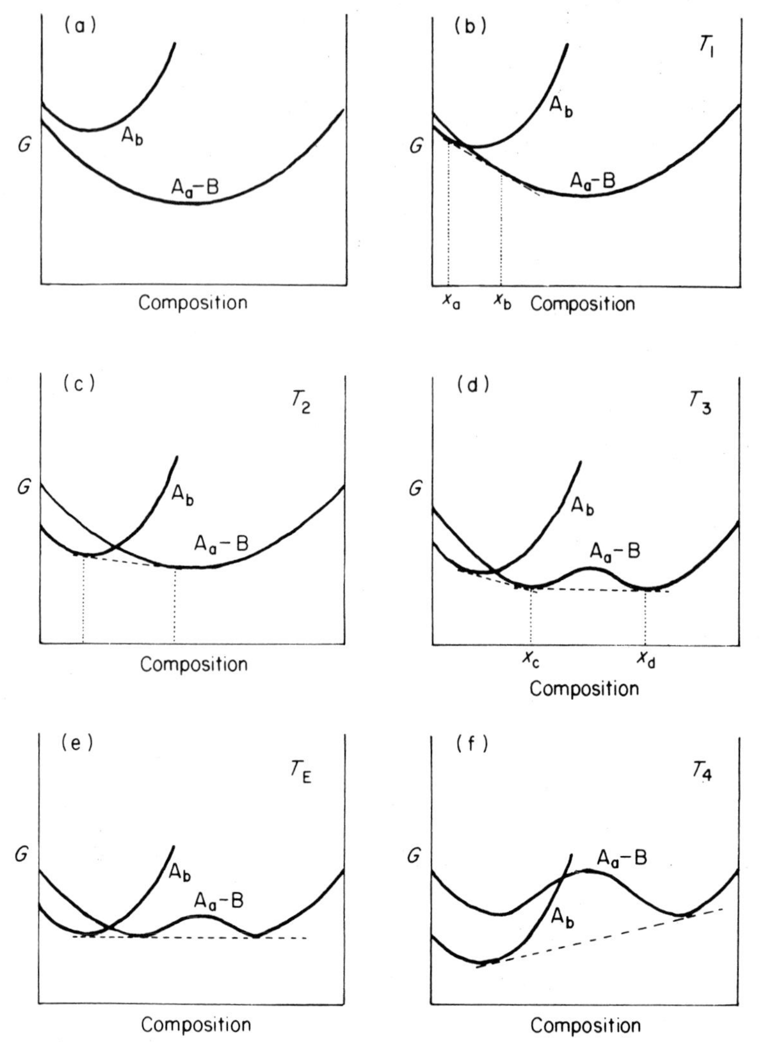

Fig. 4.14 Sequence of free energy–composition curves for the two phases A_b, and the A_a–B solid solution at a series of temperatures $T_1 \rightarrow T_4$ leading to the phase diagram in Fig. 4.15.

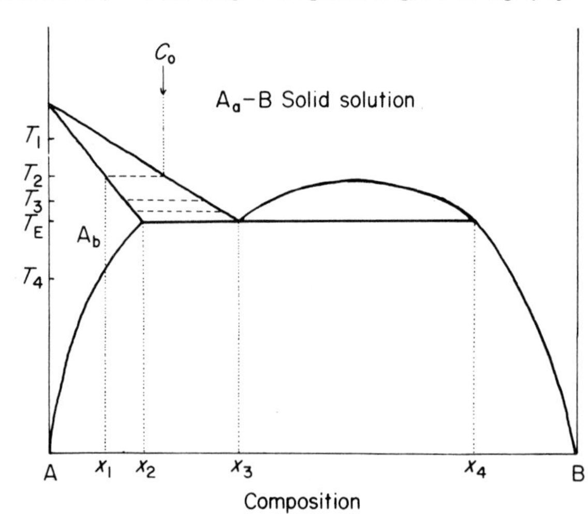

Fig. 4.15 Equilibrium phase diagram in which one of the end members undergoes a polymorphic transformation. The inversion interval intersecting the solvus leads to a eutectoid at T_E.

[Fig. 4.14(e)] indicating that three phases may coexist in equilibrium at this temperature. This is a unique temperature for, as can be seen from the phase diagram, at any other temperature a given composition will be stable either as a single-phase or a two-phase mixture. The combined effect of the inversion interval with the solvus produces this situation and the characteristic phase diagram of Fig. 4.15. T_E is known as the eutectoid temperature.

At a temperature T_4 below T_E the position of the G curves is as shown in Fig. 4.14(f), with the free energy minima moving further apart as the temperature drops even further. On the phase diagram this results in a reduction in the extent of solid solution, just as in a simple solvus.

Note that the effect of the polymorphic transformation on the A component was to introduce the inversion interval into a phase diagram, and to produce a markedly assymetric solvus, due to the abrupt decrease in the solubility of B in A when the structural change took place.

The more complex the phase diagram, the more complex will be the microstructures produced as a single-phase solid solution cools. To illustrate this consider the cooling behaviour of the single-phase A_a–B solid solution of composition C_0 in Fig. 4.15. Ideally, at T_2 a small amount of the A_b–B phase of composition x_1 should form. This involves both a structural and a compositional change. As cooling continues down to T_E the compositions of the two coexisting solid solutions A_b–B and A_a–B are given by the horizontal tie lines (dotted) until at T_E they are x_2 and x_3, the proportions of the two phases given by the lever rule.

At T_E the A_a–B solid solution of composition x_3 begins to break down into A_b–B solid solution with composition near x_2, and A_a–B solid solution with composition x_4. This involves a structural change in one of the phases as well as considerable diffusion to produce the chemical changes. Further cooling results in relatively minor readjustments in the compositions of the phases.

The microstructures observed during such a sequence of transformations will depend on the nature of the structural change in A, and more importantly on the extent to which equilibrium is maintained during the cooling process. In a general way Fig. 4.16 illustrates a sequence of microstructures which could form on cooling

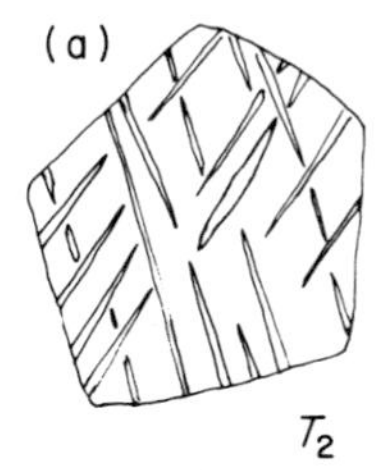

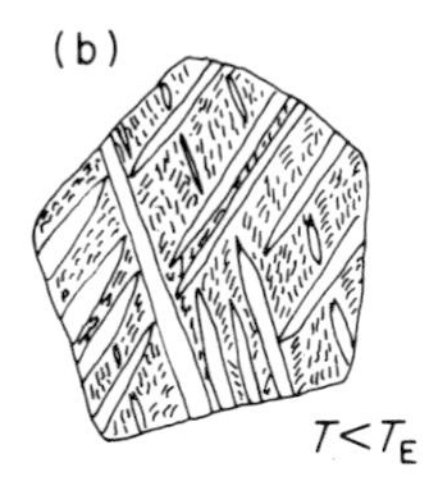

Fig. 4.16 General form of the equilibrium microstructure formed during cooling of a phase of composition C_0 in Fig. 4.15. Lamellae of A_b form within the solid solution which at temperature T_E decomposes to an intergrowth of A_b and B.

in such a system. The A_b phase has been drawn as oriented exsolution lamellae within the matrix A_a–B solid solution simply for the sake of illustration. The morphology of such exsolution features depends on the structures involved, and will be discussed in a later chapter. However, the figure does show the two-stage formation of the A_b–B phase.

As the equilibrium phase diagram becomes more complex, the likelihood of a deviation from equilibrium also increases, due to the greater number of transformations involved. To predict the nature of microstructures produced when the system does not behave according to the thermodynamic predictions, the nature and possible mechanisms of every transformation which may be encountered on cooling must be examined, and the effect of different cooling rates on the result must be evaluated. As shall be seen later, the ultimate microstructure produced may bear little resemblance to that which may be inferred from the equilibrium phase diagram.

4.5.3 THE CASE WHERE BOTH END MEMBERS UNDERGO POLYMORPHIC TRANSFORMATIONS

Where both end members undergo polymorphic transformations, inversion intervals occur at both ends of the phase diagram as shown in Fig. 4.17. To illustrate the effect of cooling on the free energy of the phases, three $G(x)$ curves are needed, one

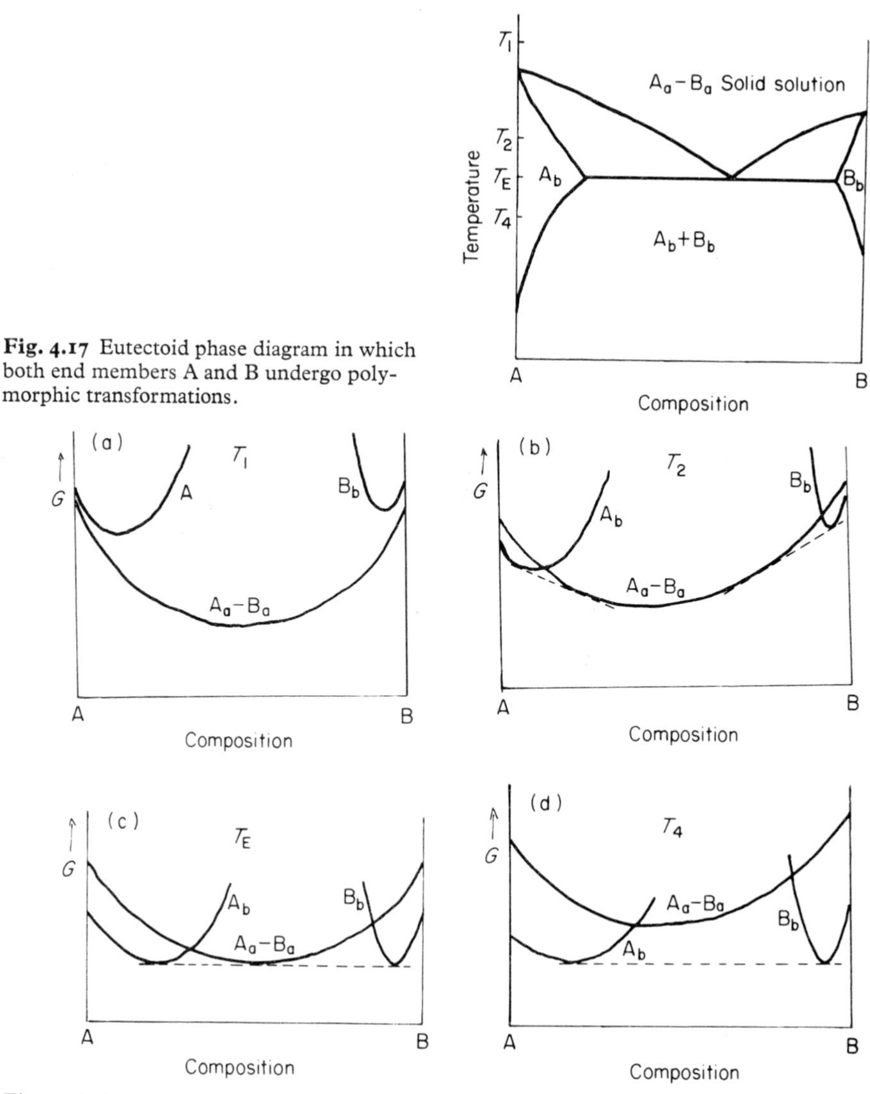

Fig. 4.17 Eutectoid phase diagram in which both end members A and B undergo polymorphic transformations.

Fig. 4.18 A sequence of free energy–composition curves at temperatures T_1, T_2, T_E and T_4 corresponding to the equilibrium phase diagram in Fig. 4.17.

for the high temperature A_a–B_a solid solution and two more for the lower temperature A_b and B_b phases. In Fig. 4.18 (a–d), the relative positions of these curves are shown at the temperatures T_1, T_2, T_E and T_4 on the phase diagram.

At T_1 the $G(x)$ curve for the high-temperature solid solution is lower than that for the other two phases at all compositions. At T_2 five different compositional regions are defined by the common tangent construction. The composition limits for the one- and two-phase fields are labelled on Fig. 4.18(b). At T_E all three phases are in equilibrium, and at this eutectoid temperature the compositions of the three phases are also unique. At a lower temperature T_4 the most stable assemblage over most of the composition field is a two-phase mixture of A_b and B_b with the compositions of each defined by the common tangent drawn in Fig. 4.18(d).

Example : the alkali–feldspar phase diagram

The alkali feldspars provide a mineralogical example of a phase diagram in which both end members undergo polymorphic transformations on cooling. At high temperatures the two end members, monoclinic albite $NaAlSi_3O_8$ and sanidine $KAlSi_3O_8$ are isostructural and a complete solid solution exists between them (Fig. 4.19). As we described in the previous chapter, the behaviour of alkali feldspars on cooling involves two factors. Firstly, the collapse of the Si, Al-framework around the small Na atom in the Na-rich feldspars leads to the formation of a triclinic structure. In pure Na feldspar this transformation occurs at a high temperature (≈ 1100°C). As more K^+ is substituted for Na^+ in the monoclinic structure this temperature falls. The second factor is the ordering of Al and Si in the tetrahedral sites. In the Na feldspars this ordering occurs within the triclinic structure. In K feldspars there is no structural collapse analogous to that in Na feldspars due to the large K^+ ion, and Si, Al-ordering takes place within the monoclinic structure. Ordering of Si and Al leads to a reduction in the symmetry of K feldspar to triclinic. This occurs at around 500°C.

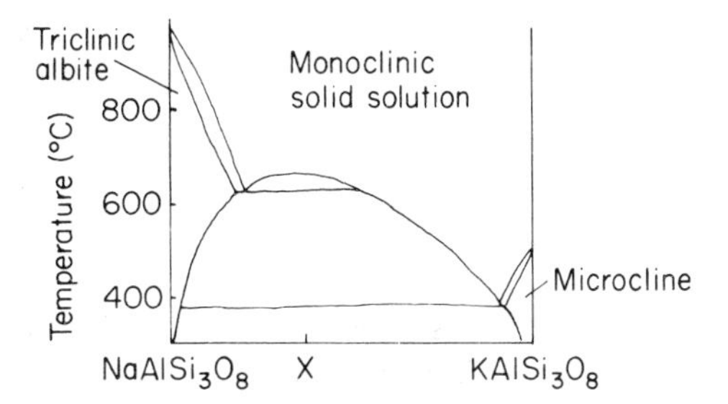

Fig. 4.19 Equilibrium subsolidus phase diagram for the alkali-feldspar system.

Both of these polymorphic transformations in the end members and their associated inversion intervals are shown in the phase diagram. Note that as soon as the Na-rich feldspar collapses to triclinic symmetry, the extent of solid solution with K feldspar is greatly reduced. In the triclinic form the alkali cation site in Na feldspar is reduced in size, effectively preventing substantial substitution of the larger K^+ ion.

The phase distribution or microstructure which results when the high-temperature solid solution is cooled depends firstly on the bulk composition, and secondly on the extent to which equilibrium is maintained during cooling) i.e. on the cooling rate. Under equilibrium conditions we can see from the phase diagram that for a composition X the first process involves the segregation of potassium and sodium atoms to form a two-phase intergrowth. At a lower temperature the Na-rich regions will transform to triclinic symmetry—the microstructure produced during this transformation will be described later. Finally, the K-rich regions will tend to transform to the triclinic as well, due to the tendency for the Al and Si atoms to order within the framework. The resulting microstructure will reflect the spatial and temporal relationships between these processes.

When the effect of cooling rate on the microstructures is considered the possibility of non-equilibrium processes may result in the formation of very complex textures which can only be unravelled by a knowledge of both the equilibrium phase diagram and the mechanisms of the processes involved. Such textures can, however, be potentially extremely useful as indicators of geological conditions. We shall return to this theme after we have discussed the relevant mechanisms and the types of microstructures which are formed during phase transformations.

4.6 The Gibbs phase rule

At this point it is appropriate to mention the phase rule which is often quoted in the standard treatment of phase diagrams. The phase rule relates the number of components* in a system, the number of phases* present and the number of variables which define the state of the system. In our examples we have dealt with a two-component system, i.e. with components A and B, and the only variable has been the temperature. When the pressure is constant the phase rule is

$$P+F = C+1$$

where P is the number of phases, F is the number of variables (or degrees of freedom) and C is the number of components.

From this rule it can be seen that there will be only one temperature at which three phases can coexist: $3+0 = 2+1$. The system then has no degrees of freedom, for if the temperature changes, F becomes 1 and the number of phases must be reduced to two. The invariant temperature is the eutectoid temperature shown in the phase diagram (Fig. 4.17).

Adherence to the phase rule is generally used as a criterion for equilibrium, and a departure from the phase rule indicates that one or more of the phases present are metastable. For example, in the phase diagram in Figure 4.17 only two phases can stably coexist below T_E. If due to sluggish kinetics some of the A_a–B_a solid solution still persists below this temperature the number of phases will be three, a situation not allowed by the phase rule. The phase rule can therefore be used to check the correctness of a phase diagram constructed from experimental data, as well as to tell at a glance whether an observed assemblage of phases is in equilibrium.

* The terms 'phases' and 'components' are defined in the following way. A heterogeneous system consists of two or more homogenous parts which have different physical or chemical properties. These homogenous regions are termed 'phases' and each phase is separated from other phases by some interface. The number of 'components' of a system are the smallest number of independently variable constituents which may be used to express the composition of each phase.

4.7 Free energy–composition curves and departures from behaviour equilibrium

G curves are very useful as a guide to understanding the behaviour of systems which are not at equilibrium. As a simple illustration of this point consider Fig. 4.20 which is essentially the same as Fig. 4.18(d). The stable equilibrium assemblage is a two-phase mixture of A_b with composition x_1, and B_b with composition x_2. A phase with the high-temperature solid-solution structure A_a–B_a and a composition C_0 (marked X in Fig. 4.20) is clearly not stable with respect to this pair. If however such a high temperature solid solution is retained down to T_4 due to very rapid cooling, for example, and then held at this temperature there are two possible modes of behaviour.

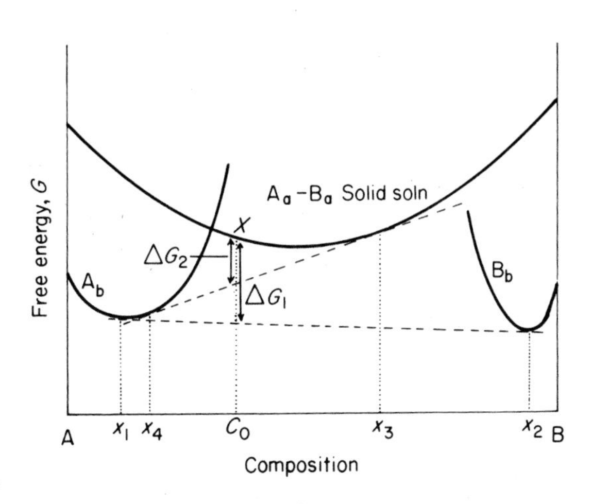

Fig. 4.20 Free energy–composition curves for the phases at temperature T_4 of Fig. 4.17. A solid solution of composition X may reduce its free energy by ΔG_1 by the breakdown into $A_b + B_b$, or alternatively by ΔG_2 by the exsolution of only one of the end member phases A_b.

The most stable behaviour involves the breakdown into the composition x_1 and x_2. This requires a considerable rearrangement of the atoms due to the large compositional differences involved as well as the structural changes required to produce the two new phases. The reduction in free energy for this process is shown as ΔG_1. An alternative mode of behaviour which involves a smaller reduction in free energy, but also less of a rearrangement of atoms (smaller compositional differences and a change in structure in one of the phases only) is the decomposition of C_0 to A_b with composition x_4 and the A_a–B_a solid solution with composition x_3. These compositions are defined by the common tangent to the two phases. The free energy reduction for this process is shown as ΔG_2.

The important point to note about this second process is that although the free energy reduction is less, the kinetics are likely to be more favourable than for the equilibrium process, as the extent of the changes is less. This is a theme which will be repeated time and again, and one that cannot be easily appreciated by considering the equilibrium phase diagram alone. In terms of the phase diagram, the second process represents an extension of the two-phase A_b–A_aB_a field below T_E, i.e. below the temperature at which the A_a–B_a solid solution is stable.

In practice there are many reasons for a system deviating from ideal behaviour, and we shall use G curves to illustrate the behaviour in such cases It is important therefore to appreciate the points made here and to investigate the possibility of non-equilibrium behaviour using diagrams such as in Fig 4.18.

As a second example of the application of free energy–composition curves to metastable systems, we will consider a similar situation to that in Fig. 4.20, but where the B_b phase forms as very fine particles as it exsolves from the high-temperature solid solution. The effect of the fine particles is to introduce an appreciable surface energy term into the free energy of B_b. This is shown schematically in Fig. 4.21(a) as the dotted line. This increase in the total free energy of B_b means that the position of the common tangent changes and the tangent point on the A_b curve moves to the right. The effect of this on the phase diagram is two-fold as can be seen from Fig. 4.21(b). Firstly the B_b phase is more soluble in the A_b phase when it occurs as fine particles (due to the phase boundary shift to the right), and secondly the eutectoid temperature T_E is reduced.

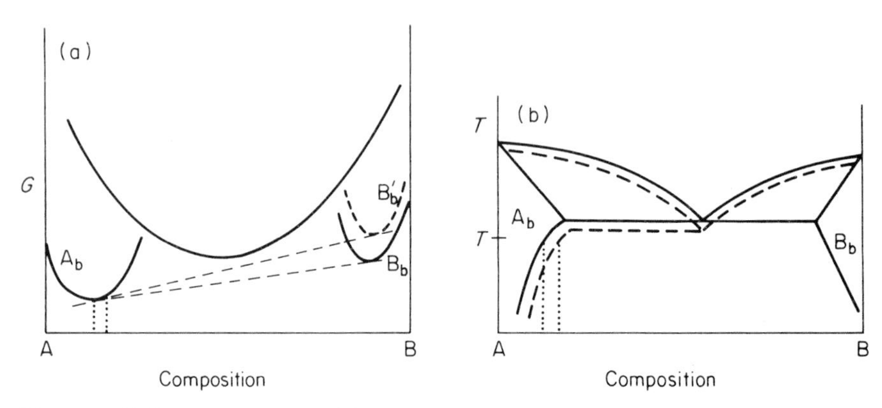

Fig. 4.21 (a) Free energy–composition curves indicating the change in free energy of phase B_b if it occurs as fine particles. (b) The effect of the increased free energy of phase B_b on the phase diagram indicating the change in solubility of B_b and the decreased eutectoid temperature.

Although these effects may appear to be insignificant they do illustrate some important points, namely that the morphology of the phases formed will have a bearing on thermodynamics, and secondly that less stable phases (such as the fine particles) are more soluble than the more stable phases. We shall return to these points when we discuss transformation processes in more detail.

4.8 Order–disorder transformations

In our derivation of the free energy–composition curves for solid solutions we considered three cases: when H_{mix} was positive; when H_{mix} was zero; and when H_{mix} was negative. The G curves that have been discussed so far arise from the first condition, H_{mix} positive, i.e. the two components A and B tend to separate on cooling. The second condition describes a perfect solid solution which remains disordered on cooling. The final condition, H_{mix} negative describes the tendency for A and B to attract one another and hence to form A–B nearest neighbours. As we mentioned in section 4.4.1, the case when H_{mix} has a large negative value leads to the formation of an ordered compound between the end members, and the model we have used to derive the G curves can no longer be used. Ordering of A and B atoms in a structure greatly reduces the entropy term, and the original assumption of an ideal entropy of mixing does not apply.

The kinds of ordered structures which may be produced from a solid solution depend on both the structure and composition of the high-temperature form. In general terms however, if we again consider our two component system, A–B, and assume that a fully ordered compound AB forms at low temperatures, we can draw a series of G curves for both the disordered solid solution and the ordered compound. These are shown schematically in Fig. 4.22, where curves are drawn for three temperatures. Fig. 4.22(a) is above the ordering temperature, Fig. 4.22(b) is at the ordering temperature, and Fig. 4.22(c) is below the ordering temperature. The resulting phase diagram (Fig. 4.23) shows a narrow one-phase field for the ordered compound separated from the disordered solid solution by two-phase fields whose limiting compositions are defined by common tangents to the G curves in Fig. 4.22.

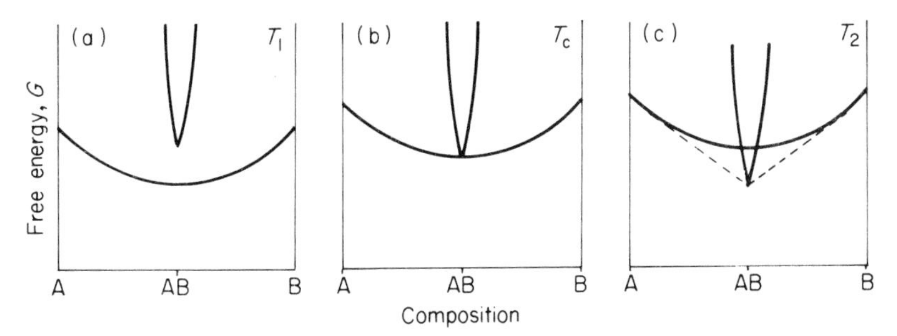

Fig. 4.22 Sequence of free energy–composition curves for the solid solution and the ordered compound AB at temperatures T_1, T_c, and T_2 corresponding to the phase diagram in Fig. 4.23.

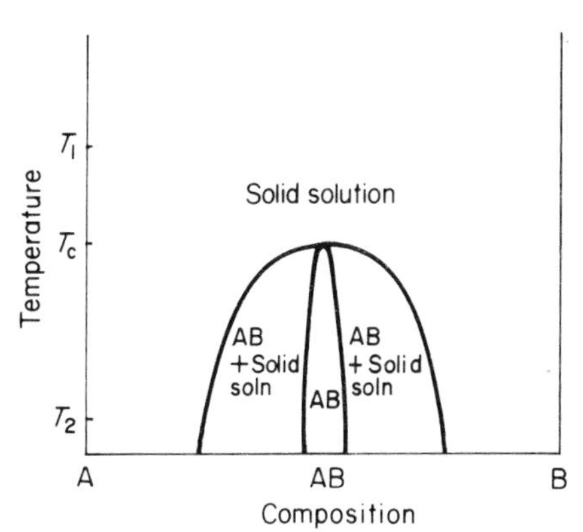

Fig. 4.23 Equilibrium phase diagram for the system described in Fig. 4.21.

4.9 Appendix—Activities of components in solid solutions

The final section of this chapter is added as an appendix because it describes a slightly different formulation of the thermodynamics of solid solutions to that described in section 4.4.1. This alternative approach, adopted in many petrology textbooks, describes the free energy of mixing of two components A and B in terms of their separate molal free-energy contributions.

Expressed in terms of the mole fractions of A and B (X_A and X_B), the configurational entropy associated with mixing A and B is

$$S = -R(X_A \log_e X_A + X_B \log_e X_B)$$

from equation 5.1, recalling that

$$X_B = 1 - X_A$$

For an ideal solid solution therefore, the free energy of mixing

$$\begin{aligned}\Delta G_{mix} &= H_{mix} - TS_{mix} \\ &= RT(X_A \log_e X_A + X_B \log_e X_B)\end{aligned}$$

as

$$H_{mix} = 0.$$

This may be written :

$$\Delta G_{mix} = X_A RT \log_e X_A + X_B RT \log_e X_B$$

The term $RT \log_e X_A$ is the partial molal free energy $\Delta \bar{G}_A$, i.e. the increase in free energy when one mole of A is dissolved in the solid solution. Similarly

$$\Delta \bar{G}_B = RT \log_e X_B.$$

Therefore, the free energy of mixing for an ideal solution may be written

$$\Delta G_{mix} = X_A . \Delta \bar{G}_A + X_B \Delta \bar{G}_B$$

This equation expresses the overall free-energy change associated with ideal mixing of A and B in terms of the mole fraction and the partial molal free energy contribution of each component.

When the enthalpy of mixing is not zero, i.e. if atoms of A and B do show some preference either for their own kind, or for their opposites, then a larger or a smaller free energy change ($\Delta \bar{G}_A$, $\Delta \bar{G}_B$) will be observed than that expected in an ideal solution. For a non-ideal solid solution the change in molal free energy is *not* given by the expression

$$\Delta \bar{G}_A = RT \log_e X_A$$

which holds only for an ideal solution. However, in order to retain the form of this simple relationship, we define a quantity 'a_A' known as the *activity* of A in the solution so that

$$\begin{aligned}\Delta \bar{G}_A &= RT \log_e a_A \\ \Delta \bar{G}_B &= RT \log_e a_B\end{aligned}$$

Physically speaking, if $\Delta H_{mix} > 0$, the stronger bond between similar atoms increases the effective concentrations of the two components in solution above their actual molal concentrations. The reverse is true when $\Delta H_{mix} < 0$. The *activity* expresses this effective concentration of a component in a solution.
Therefore, for a non-ideal solution

$$\begin{aligned}\Delta G_{mix} &= X_A . RT \log_e a_A + X_B . RT \log_e a_B \\ &= RT(X_A \log_e a_A + X_B \log_e a_B)\end{aligned}$$

For an ideal solution $X_A = a_A$; for a non-ideal solution we define the *activity* coefficient

$$\gamma_A = \frac{a_A}{X_A} \quad \text{and} \quad \gamma_B = \frac{a_B}{X_B}$$

Then $\Delta G_{mix} = RT(X_A \log_e X_A . \gamma_A + X_B \log_e X_B . \gamma_B)$

$$\Delta G_{mix} = RT(X_A \log_e X_A + X_B \log_e X_B + X_A \log_e \gamma_A + X_B \log_e \gamma_B)$$

or $$\Delta G_{mix} = RT(X_A \log_e X_A + X_B \log_e X_B) + RT(X_A \log_e \gamma_A + X_B \log_e \gamma_B)$$

$$= \Delta G_{mix}\,(\text{ideal}) + \Delta G_{mix}\,(\text{excess}).$$

The free energy of mixing of a non-ideal solution differs from that in an ideal solution by ΔG_{mix} (excess) which may be positive or negative, depending on whether $H > 0$ or < 0.

The activity of the components of a solution are indicators of the extent of the departure from ideality. For an ideal solution, $a_A = X_A$. This is termed Raoult's Law. A negative deviation from Raoult's Law with $a_A < X_A$ is associated with $H_{mix} < 0$, i.e. an attraction between A and B. A positive deviation with $a_A > X_A$ means that $H_{mix} > 0$ and indicates a tendency for phase separation. Activity–composition curves for ideal and non-ideal solutions are shown in Fig. 4.24. Note that at small concentrations of the impurity atom Raoult's Law is obeyed, in other words at small concentrations the entropy term is dominant.

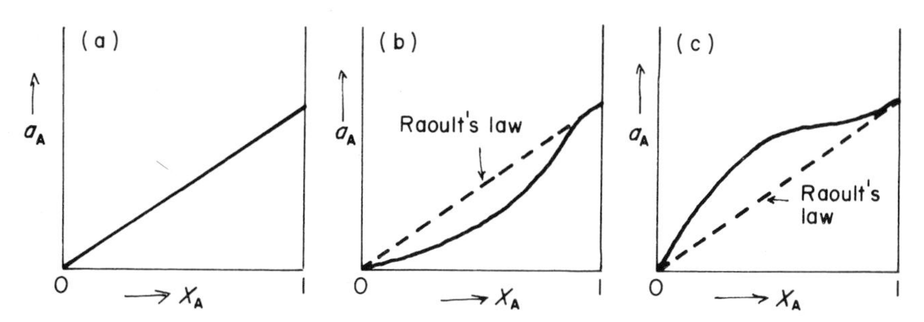

Fig. 4.24 Activity–composition curves for (a) ideal solution with $a_A = X_A$, i.e. $H_{mix} = 0$. (b) Negative deviation from Raoult's law with $a_A < X_A$, i.e. $H_{mix} < 0$. (c) Positive deviation from Raoult's law with $a_A > X_A$, i.e. $H_{mix} > 0$.

References and additional reading

Cottrell, A.H. (1965) *Theoretical Structural Metallurgy*. Edward Arnold. Ch. 10.
Ehlers, E.G. (1972) *The interpretation of geological phase diagrams*. Freeman. Chs 1–3.
Powell, R. *Equilibrium thermodynamics in petrology*. Harper and Row. Ch. 2.
Prince, A. (1966) *Alloy Phase Equilibria*. Elsevier. Chs 1 and 2.
Shewmon, P.G. (1969) *Transformations in metals*. McGraw-Hill. Ch. 4.
Smith, E.B. (1973) *Basic Chemical Thermodynamics*. Oxford University Press, Oxford.

5

Processes in Minerals

The thermodynamics of transformations in minerals is only one aspect which we must consider when describing the way minerals will behave as the external conditions change. The derivation of phase diagrams outlined in the previous chapter assumes that equilibrium is maintained throughout and takes no account of the actual process which takes place during the transformation. Clearly there are some changes which may require only minimal structural reorganization; others may involve a complete reconstruction of the atomic arrangement or the large scale diffusion of atoms through the lattice. Various 'degrees of difficulty' are implied here, and in the treatment which follows our ideas of the specific mechanism of the transformation and its kinetics will go hand in hand. Transformations which involve major reorganization of atoms will usually be more sluggish and hence departures from the equilibrium phase diagram will be more likely.

Thus whereas in Chapter 4 we were concerned with ideal behaviour in minerals, we now come to look at the factors which will affect their real behaviour. The kinetics of transformations will be treated more formally in Chapter 6 although it will be convenient to introduce some of the more general aspects of the relationship between kinetics and mechanism in the following section.

5.1 Classification of transformations

There are a number of ways in which the degree of structural or chemical change involved in a transformation can be described and this amounts to some type of classification scheme which can be used to characterize the mechanism by which one solid phase transforms to another. By the mechanism we mean the relationships between the parent and product phases: their spatial dimensions, distribution and crystallographic relationships. We shall now look more closely at the mechanism of transformations, initially defining broad categories which apply to both polymorphic transformations and those involving phase separation or exsolution.

5.1.1 DISCONTINUOUS AND CONTINUOUS TRANSFORMATIONS

At the very beginning of this book (section 1.4) we posed the question: how does a disordered AB alloy transform to produce an ordered compound AB, or else exsolve into distinct regions of A and of B? We implied that in the initial stages of the transformation there were two possibilities: either discrete particles of the final product could form within the disordered phase, or alternatively some gradual process could operate throughout the whole of the disordered parent phase until the final product was eventually reached.

The first of these mechanisms involves a change which is large in degree and small in extent, i.e. small regions of the product phase, which deviates quite strongly in structure and/or composition from the matrix, are formed at distinct (but unspecified) points within the matrix. Thus at some intermediate stage in the process the sample contains some regions which have transformed and others which have not. Clearly, some kind of nucleation event is implicit in this mechanism. A distinct interface separates each nucleus from the parent matrix. In a diffraction pattern, the diffracted rays from the new phase appear suddenly or discontinuously with time. We shall refer to such transformations, whether or not they involve compositional differences, as *discontinuous*.

A continuous transformation occurs simultaneously throughout the entire parent matrix by means of a mechanism which produces fluctuations in the composition or in the degree of order. Thus the change is small in degree but large in extent. The structural and chemical similarity between the parent and product phases implies that there is no distinct interface formed. The diffraction pattern of the new phase gradually develops from that of the parent and is very similar to it.

In the treatment which follows it can be seen that such fluctuations within the parent structure are strongly controlled by the symmetry and the elastic properties of the parent. Any fluctuation will in these circumstances be periodic, and in the initial stages of a continuous transformation we will speak of *modulations* within the parent. While these modulations are closely tied to the parent structure (i.e. no interfaces are formed) we should not, strictly speaking, refer to a new phase but to a modulation of the parent phase.

The distinction between a discontinuous and a continuous transformation is easiest to visualize in a transformation involving a compositional change. In Fig. 5.1, C_0 is the initial composition of the parent phase prior to an exsolution

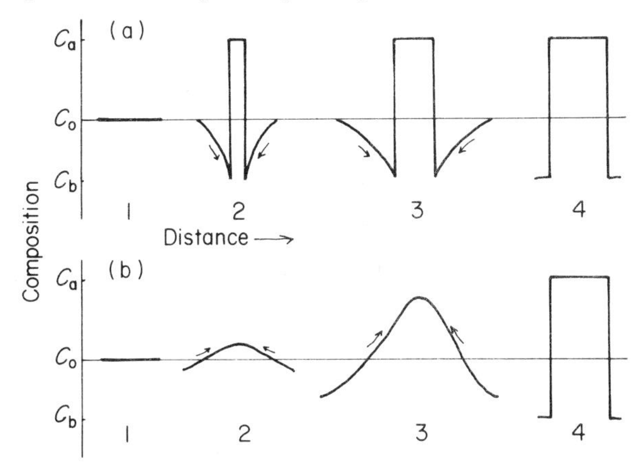

Fig. 5.1 (a) Discontinuous and (b) continuous exsolution processes. The initial and final states are the same in both cases, although the mechanisms are quite different.

transformation, and C_a and C_b are the equilibrium compositions of the final phases. The top diagram illustrates a discontinuous nucleation and growth process, while the lower diagram shows the development and subsequent increase in amplitude of a compositional modulation. Note that the final product is the same in both cases, and we need evidence of the early stages of the process before a mechanism can be defined unambiguously. The conditions under which the two processes may operate, however, may be quite different. Note also that in the discontinuous process growth takes place by diffusion from regions of high concentration to regions of low concentration (as shown by the arrows). This is termed 'down-hill' diffusion. In the continuous case, diffusion is 'up-hill'. We shall refer to this in a later section.

A similar diagram could be used for an ordering process, in which case the vertical axis is some measurement of the degree of order. Note that the diagram would then imply the existence of regions of 'positive' and 'negative' order on either side of the initial disordered state. The meaning of this will be discussed in section 5.5.

5.1.2 RECONSTRUCTIVE AND DISPLACIVE TRANSFORMATIONS

Next we introduce a classification scheme for structural transformations, developed by M. J. Buerger, which has some parallels to that above. It is based on the structural differences between the phases involved, specifically on the types of bonds (primary or secondary) broken during the transformation.

The two structures connected by the transformation may be so closely related that they are capable of being changed into one another without breaking primary bonds, as shown diagrammatically in Fig. 5.2. In this case, one structure is a distortion of the other, and the two forms are related by symmetry. This is called a

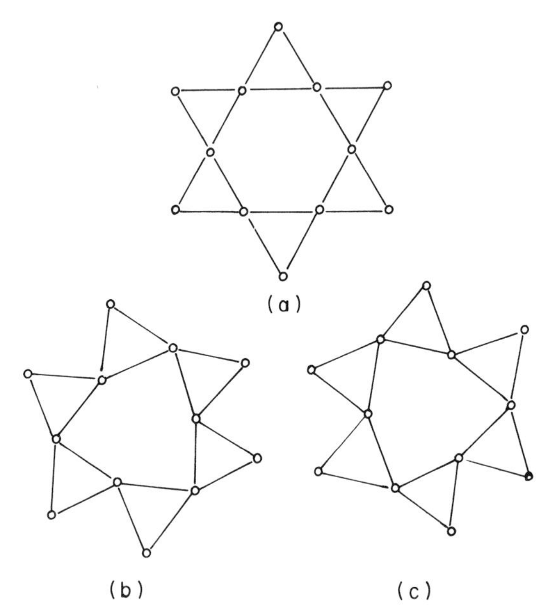

Fig. 5.2 (a) The undistorted high-temperature structure of high quartz, the circles representing the centres of the SiO_4 tetrahedra. (b) and (c) are two equivalent alternatives for the low-quartz structure which forms by a distortion from the high-quartz structure.

displacive transformation. Note that in Fig. 5.2 two possible alternatives exist for the distortion, resulting in two orientational variants related by a 180° rotation.

The second type of transformation takes place when the two structures differ from one another so much that no way exists of passing from one to the other without breaking primary bonds, as shown in Fig. 5.3. In this case the transformation requires the breaking of primary bonds, and the reconstruction of some of the structural units to the new form. This is called a *reconstructive transformation*. The two phases need not be related by any symmetry.

The breaking of primary co-ordination bonds in a reconstructive transformation requires an activation energy, and this imposes a barrier to the transformation.

Clearly, such transformations would be relatively sluggish. Displacive transformations, on the other hand, are very rapid and appear to involve no activation energy barrier.

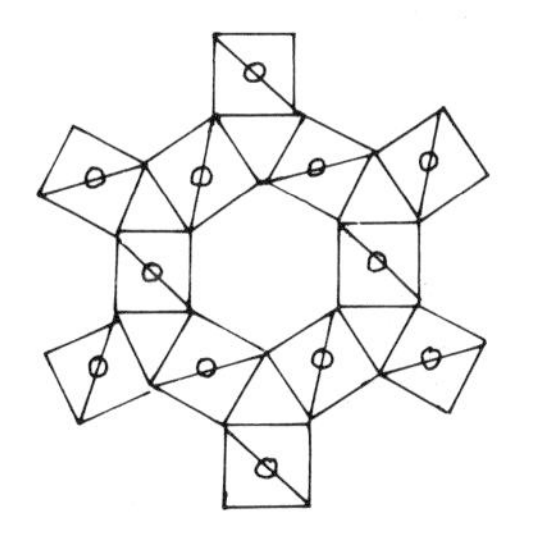
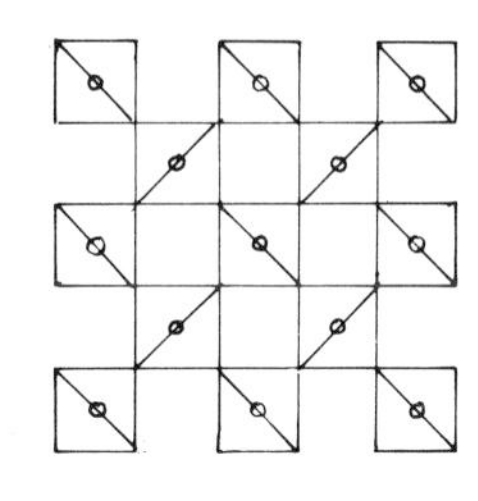

Fig. 5.3 A polymorphic transformation between these two structures, built up of similar structural units, requires a reconstructive process.

The connection between the two classification schemes we have mentioned is, at first sight clear. A reconstructive transformation will necessarily involve a nucleation event and hence may be equated with a discontinuous transformation mechanism. It is not as obvious to equate a displacive transformation with a continuous mechanism, although clearly they are both at the same end of the spectrum. It is important to realize that we are defining two extremes of behaviour, and that all transformations will not necessarily fit neatly into one of these pigeon-holes. Structural similarities between two phases may be of many intermediate levels and in practice the transformation mechanism in such cases may be far from clear-cut and must be investigated in every individual case. We shall however, find it useful to use terms such as displacive, reconstructive, continuous and discontinuous in a descriptive sense.

5.1.3 THE EXPERIMENTAL STUDY OF PHASE TRANSFORMATIONS—THE HIGH–LOW QUARTZ TRANSFORMATION

The mechanism of phase transformations may be studied by observing the course of the transformation in the electron microscope and determining the nature of the interface between the product and the parent phase. In some cases transformations may be rapid enough to be observed dynamically in an electron microscope equipped with a heated specimen stage. In this section we will describe some recent electron-microscope observations on the transformation from low to high quartz. This is the 'classical' example of a displacive transformation, and this elegant electron-microscope study introduces a number of features of the process and gives some insight into the relationship between a 'displacive' and a 'continuous' transformation.

Fig. 5.2(a) is a schematic illustration of the high-quartz structure, showing only the Si atoms at the centre of each SiO_4 tetrahedron. It has a six-fold rotational axis of symmetry normal to the page, and hence high quartz is hexagonal. Low quartz is a distortion of the high-quartz structure as shown by the two equally likely possibilities in Fig. 5.2(b) and (c). There is now only a three-fold rotation axis of symmetry, and low quartz is trigonal.

In transforming a single crystal of high quartz to the low form, it might be expected that the end product should be made up of equal proportions of the two orientational variants, and this is the case. The simple rotational relationship between the two variants defines them as *twins*, and their formation on cooling from the high form is referred to as *transformation twinning*. In quartz they are called Dauphiné twins.

In section 2.1.2 we described a similar situation in the description of distortional disorder. The question arose as to the true nature of the high-temperature form: could it be described as a space and time average of the two low-temperature alternatives? By observing the transformation mechanism we should be able to cast some light on this problem.

The experimental situation involves setting up a temperature gradient across a thin flake of quartz while observations are being carried out in a transmission electron microscope. The transformation temperature is 573°C, and in the flake shown in Fig. 5.4, the temperature on the left is just below 573°C and the phase is low quartz, while on the right the temperature is just above 573°C and the phase is high quartz. The contrast across the centre of the flake arises from the fine-scale Dauphiné twinning associated with the transformation.

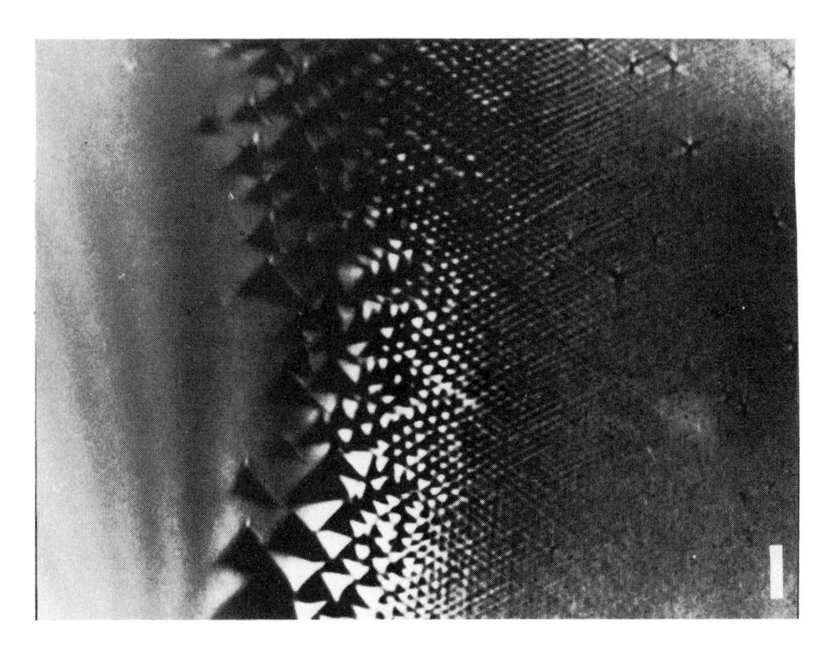

Fig. 5.4 Transmission electron micrograph of the boundary region between low and high quartz, obtained by setting up a temperature gradient across the crystal and observing the transformation dynamically. The boundary is defined by a fine-scale array of Dauphiné twins which develop in the low quartz (left) near the transition temperature and become progressively finer into the high-quartz region (right). The length of the scale bar is 0.2 μm (from Van Tenderloo *et al.*, 1976. Photo courtesy of G. Van Tendeloo).

The critical observation here is that a specimen of low quartz which originally had no twins in it, develops relatively coarse-scaled twins near the transition region on heating. These twins form simultaneously all over the specimen where the transformation temperature has been reached. As the temperature increases the twin domains flip from one variant to the other, and the scale of the twinning becomes finer and finer. As the high-temperature form is approached the frequency of the structure reversal increases until it eventually blurs imperceptibly into the apparently homogeneous high-quartz phase.

On cooling high quartz a twinned low-quartz phase would result, and the scale of the twinning would depend on the cooling rate.

The interpretation of these observations is that the high-temperature form is indeed a space and time average of the Dauphiné microtwinned low-temperature structure. We can envisage submicroscopic domains oscillating between the two variants with little or no spatial correlation between the oscillations. As the temperature is reduced the frequency of these oscillations decreases rapidly near the transformation temperature, and falls to zero, effectively 'freezing in' regions with one or other of the two possible distortions. If the cooling is slower the oscillations become spatially correlated over larger regions, resulting in a coarser twinned structure in the low-temperature form.

Regarded in this way the high–low quartz transformation is a change from a state of distortional disorder to an ordered state by a continuous process. Although there are many minerals which undergo displacive transformations we must be careful not to assume that a similar mechanism applies in every case. Furthermore, even in the case described here the experimental observations may be open to a different interpretation.

5.1.4 FIRST AND SECOND ORDER TRANSFORMATIONS

Finally we shall briefly mention a classification of structural transformations based solely on thermodynamics. The mathematical basis for this classification will not be described here.

First order transformations are those in which the first derivatives of the free energy function G are discontinuous at the transformation temperature. Thus there are discontinuities in entropy and volume as well as in the enthalpy, which appears as a latent heat of the transformation. In second order transformations, these first derivatives are continuous, but the second derivatives have discontinuities at the transformation temperature. The specific heat is one of the quantities to exhibit a discontinuity.

The main practical limitation of the direct use of this scheme is the unavailability of thermodynamic data. However we can again draw parallels between this scheme and the previous classifications. There is a good rough correlation between thermodynamically first order transformations and structural reconstructive transformations, as well as discontinuous transformations. Similarly, second order transformations equate fairly well with displacive transformations and continuous transformations. (We must recognize however the possibility of having 'mixed' order transformations which have features of both first and second order types.)

A reason for maintaining this classification, in the absence of the required thermodynamic data, is that a formal mathematical procedure exists for relating the thermodynamic order of a transformation to the degree of structural similarity between polymorphic forms (as exemplified by their symmetry differences). By comparing the symmetry elements of the low-temperature form to the high-temperature form, we can determine whether a second order transformation between the two forms is possible or not. Conversely we can predict the symmetry of structures derived from a high-temperature form by second order transformations.

5.1.5 TRANSFORMATION KINETICS—METASTABILITY AND ALTERNATIVE BEHAVIOUR

The practical importance of determining the nature or order of a transformation lies in the inherent association between the degree of change and the kinetics of the process. In the previous chapter, Fig. 4.6 illustrated the idea of a transformation hysteresis which occurs when a transformation takes place at a different temperature on heating and on cooling. This is due to the driving force ΔG which is required for the transformation to proceed, and the amount of hysteresis is related to the 'degree of structural change' involved. For a reconstructive process the hysteresis may be quite large and in the extreme case the transformation may not take place at all on cooling. For a displacive transformation the amount of hysteresis may be negligible.

Often in mineral systems we will find that both discontinuous and continuous transformations may be implicated in the cooling behaviour, and this is where the relationship between the transformation type and the kinetics becomes most important. We will illustrate this with three simple examples.

(i) Tridymite–quartz

We have already discussed briefly the transformations in SiO_2 (Fig. 1.1 and section 3.11.1). The particular aspect to be emphasized here is the behaviour of tridymite on cooling. As Fig. 1.1 shows tridymite is stable from 1470–867°C, below which quartz is the stable form. In Fig. 5.5 this situation is illustrated by schematic free

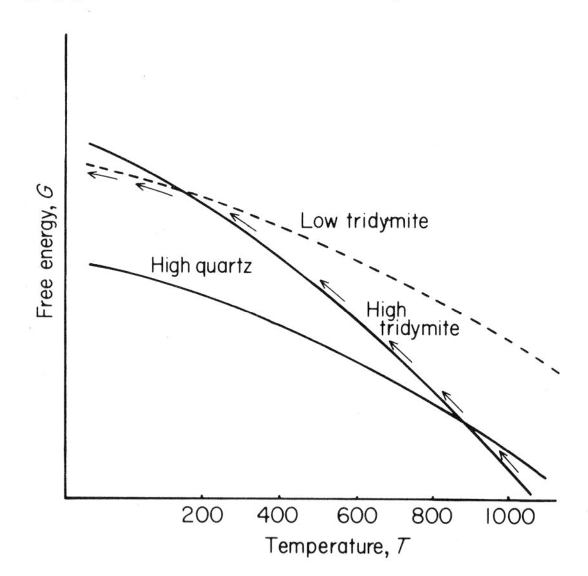

Fig. 5.5 Schematic G–T curves for high tridymite, high quartz and low tridymite, illustrating the metastability of low tridymite. The arrows show the course taken during rapid cooling of high tridymite.

energy–temperature curves for the two phases. The transformation from tridymite to quartz is a reconstructive process. Unless the cooling rate is extremely slow, quartz will not form due to the high activation energy of the process. The tridymite structure will persist down to about 150°C, at which a distortion of the structure takes place, to form low tridymite.

The high tridymite–low tridymite transformation is a simple displacive transformation and the structures are closely related. Thus there are no kinetic barriers to the process. Note that low tridymite is not a stable phase and represents a kinetically more favourable method of reducing the overall free energy, having failed in the thermodynamically more advantageous transformation to quartz.

This type of behaviour is termed *alternative metastable behaviour*. We will make considerable use of this term to describe transformations which occur under conditions when the thermodynamically ideal transformation is impeded for kinetic reasons. The metastable phase formed in this way does not usually have a stability field under any conditions. Fig. 5.5 illustrates the G–T curve for such a phase and the arrows show the course of the transformation during alternative metastable behaviour.

There are two important points to make about such behaviour. Firstly, the reduction in free energy involved in transforming to the metastable phase further decreases the driving force to the more stable phase. This makes it even less likely that the stable phase will form. In the case of tridymite this is fairly obvious because the equilibrium phase diagram is well understood. We shall meet more subtle examples in other systems where alternative metastable behaviour is not easily recognized as such and masks the ideal behaviour so effectively that the nature of the most stable phase(s) can only be indirectly determined by studying the nature of the transformations involved. Often the existence of continuous behaviour in a system will imply that under a different set of circumstances (e.g. cooling rate) a discontinuous process may have operated, and hence that the continuous process is alternative metastable behaviour.

The second point is that the metastable phase forms only on cooling. For example, on cooling high tridymite the first transformation may be to low tridymite, which after a long period of time may transform to quartz. If however quartz is heated, low tridymite will not form under any circumstances as such a transformation would involve an increase in free energy, which is clearly impossible. This can be seen by following the appropriate G–T curves in Fig. 5.5. Often this feature may be a criterion for recognizing alternative metastable behaviour in experimental work.

The transformation from the higher temperature form of SiO_2, cristobalite, to tridymite is another example of a situation where alternative metastable behaviour operates. Cristobalite is only stable above 1470°C, and below this temperature should transform to tridymite. Again this is a major reconstructive transformation with a large activation energy. An alternative to this transformation is the formation of low cristobalite at 270°C. The G–T curves for the phases will be similar in form to those in Fig. 5.5.

The examples we have chosen to illustrate the concept of alternative behaviour are among the most straightforward, and partly for this reason they will be of limited use as indicators of environments or cooling rates. Even here, however, we must approach any simple interpretation with caution. For example, the presence of low tridymite in an igneous rock may well indicate that it formed between 867 and 1470°C and then cooled fairly rapidly. However, as we shall see below, in many geological environments minerals may form *outside* their stability fields.

(ii) *Aragonite–calcite*

The persistence of a metastable polymorphic form commonly occurs in the system $CaCO_3$. Aragonite, which may form stably under high-pressure and low-temperature conditions in the earth, is thermodynamically unstable with respect to calcite under normal crustal and atmospheric conditions (Fig. 5.6). In these high pressure metamorphic rocks, however, aragonite is preserved after very long periods over which re-equilibration could have taken place. The transformation to calcite involves the rearrangement of the $(CO_3)^{2-}$ groups. Under dry conditions, when this rearrangement is very sluggish, aragonite does apparently make some attempt to lower its free energy by alternative means, although the nature of any minor structural readjustments is not known. There is evidence however that under aqueous experimental conditions, where the transformation to calcite is quite rapid, geologically older aragonite transforms to calcite at a slower rate than younger aragonite under the same conditions. Any structural adjustments which may have taken place in the older aragonite would reduce the driving force ΔG for the transformation to the more stable calcite phase and this may account for the slower kinetics.

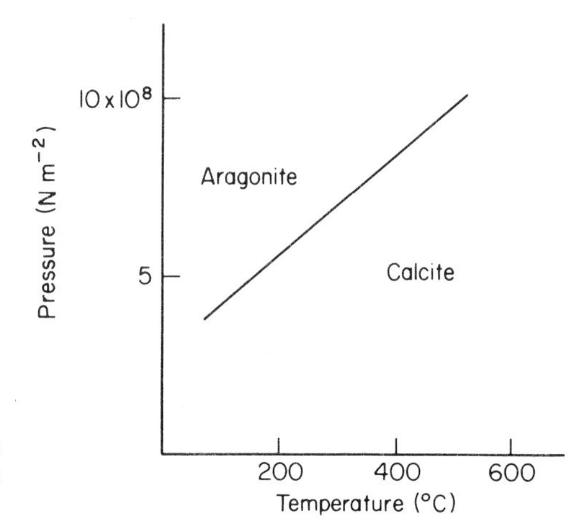

Fig. 5.6 Equilibrium *P–T* relationships between the two polymorphs of calcium carbonate, aragonite and calcite.

In the solid state, the aragonite–calcite transformation is sluggish because of the high activation energy involved in the reconstruction. During relaxation from the high-pressure, low-temperature metamorphic conditions under which aragonite forms, there is insufficient thermal energy available to surmount the activation energy barrier for the transformation to calcite. The transformation to calcite would take place at higher temperatures, or under conditions where the mechanism of the transformation was different. Experiments in the presence of aqueous fluids have demonstrated that the activation energy is lowered under these conditions, and the transformation proceeds very much more rapidly.

Aragonite is another phase which commonly forms outside its stability field so an interpretation of its presence in a rock must be made with due caution.

(*iii*) *Diamond–graphite*

The best-known case of the metastable persistence of a polymorph is that of diamond. The regions of stability of the two polymorphs of carbon are shown in Fig. 5.7. In the earth diamond is formed in deep-seated rocks at temperatures of around 1200°C and pressures around 5×10^9 N m^{-2}. Under atmospheric conditions diamond is a very 'stable' substance, and its transformation to graphite, the thermodynamically stable form under these conditions, is not capable of detection. The great structural differences between diamond and graphite result in a very large activation energy for the transformation, and it would only take place at high temperatures. At lower temperatures diamond is thus indefinitely preserved in a metastable state.

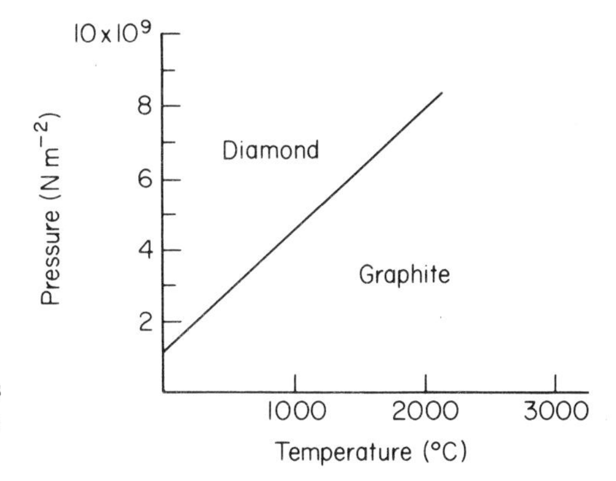

Fig. 5.7 Equilibrium *P–T* relationships between the two polymorphs of carbon, diamond and graphite.

The simple structure of diamond precludes the possibility of alternative behaviour taking place. In general the more complex the structure, the greater the likelihood of metastable low-temperature modifications and hence of alternative behaviour.

5.1.6 METASTABLE CRYSTALLIZATION

While discussing metastable *persistence* of phases it is convenient to describe briefly the common phenomenon of phases *forming* outside their stability field. We have already seen in our simple examples of alternative metastable behaviour that whenever the parent phase (or phases) in any potential transformation is in a highly metastable state, the nature of the product formed may depend on factors other than the attainment of the minimum free energy. We can extend this principle to the wider problems of crystallization of metastable phases in nature.

For example, when silica glass crystallizes, cristobalite may be the phase formed, even within the temperature range over which low quartz is stable. This is because the free energy difference between the parent phase, silica glass and any of the crystalline forms is high. Given that the driving force for a number of possible transformations is appreciable, the kinetics will determine which phases are formed. In general, the transformation which involves the minimum reduction in free energy is the kinetically most favourable, as the smaller ΔG implies a greater structural or chemical similarity to the parent. Thus in many cases, high-temperature polymorphs may be formed at low temperatures if the reactants are in a highly metastable state. In this context it is worthwhile noting that factors such as grain size may make an important contribution to the free energy of the parent phase. Fig. 5.8 illustrates how various polymorphs may crystallize outside their stability field from a highly unstable parent.

Crystallization from a highly metastable colloidal suspension is likely to form high-temperature phases in veins and other fluid pathways in rocks. Thus the disordered high-temperature form of potassium feldspar, sanidine, may be found in

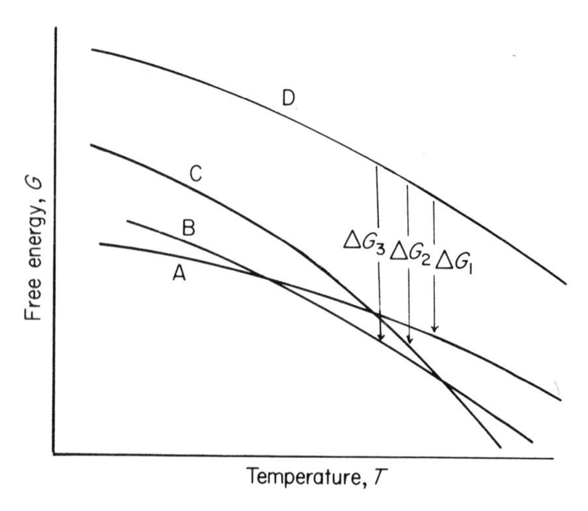

Fig. 5.8 Free energy–composition curves illustrating the similarity in the free energy reduction involved (ΔG_1, ΔG_2, ΔG_3) in the transformation from a highly unstable parent phase D to either of the phases A, B or C. Under these circumstances, the kinetics of the transformations may determine which phase is formed.

low-temperature veins. Similarly aragonite is widespread in veins associated with limestones, as well as in warm springs and in caves.

The general principle that in any transformation or reaction the kinetically most favourable sequence of phases will form, rather than those involving the greatest reduction in free energy, has been recognized by chemists for some time as the *Ostwald Step Rule*. This is illustrated schematically in Fig. 5.9. The activation energy for each step of the process is small relative to that for the formation of the stable form in a single step. Thus each step involves a relatively small deviation from the parent state. It must be remembered however that the end product of such a sequence may ultimately not lead to the stable form and in some cases may even inhibit its formation by narrowing the free energy difference involved.

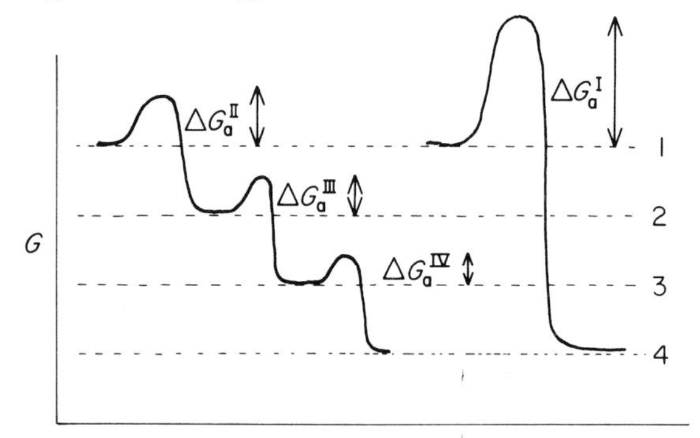

Fig. 5.9 Illustration of the Ostwald step rule. The direct transformation from state 1 to 4 involves a large activation energy (ΔG_1) and may be very sluggish. Transformation via a sequence of steps involves smaller activation energies and may be kinetically more favourable.

We will now return to the mainstream of this chapter which is a discussion of the mechanisms which may operate when one mineral phase transforms to another. This brief diversion into the kinetics of transformations has been necessary so that we can describe these mechanisms within a general kinetic framework, although we will return to a more detailed treatment of kinetics in Chapter 6.

5.2 Nucleation of new phases

A discontinuous transformation mechanism involves the formation of small particles or nuclei of the new phase within a matrix of the parent structure. The driving force for this transformation arises from the fact that below the transformation temperature, T_c, the new phase has a lower free energy than the parent. In the treatment which follows we derive some simple expressions which will illustrate the factors which affect the process of nucleation.

Consider the formation of small spherical particles of phase α within a matrix of the parent phase β. ΔG_V is the free energy difference, per unit volume of α, between α and β. At T_c the free energy of both phases is the same and hence $\Delta G_V = 0$. Below the transformation temperature ΔG_V is negative and there is a driving force for a transformation from β to α. This is opposed however by two positive free energy terms which accompany the creation of an interface between the new phase and the matrix. The first is a surface energy term, σ, per unit area of surface, and the second is a strain energy term, ε, per unit volume of the phase formed.

The overall change in free energy, $\Delta G(r)$, on forming a spherical nucleus of radius r is therefore:

$$\begin{aligned} \Delta G(r) &= \Delta G_{\text{volume}} + \Delta G_{\text{surface}} + \Delta G_{\text{strain}} \\ &= \tfrac{4}{3}\pi r^3 \Delta G_V + 4\pi r^2 \sigma + \tfrac{4}{3}\pi r^3 \varepsilon \\ &= \tfrac{4}{3}\pi r^3 (\Delta G_V + \varepsilon) + 4\pi r^2 \sigma. \end{aligned} \qquad (5.1)$$

For a nucleus smaller than some critical radius, r_c, an increase in its size will increase the free energy of the system because the positive surface and strain energy terms will predominate. The relationship between $\Delta G(r)$ and r at some temperature T

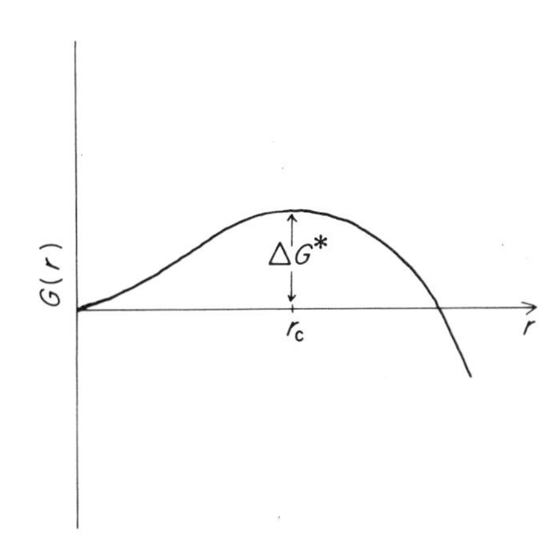

Fig. 5.10 Plot of the free energy of formation of a nucleus as a function of its radius r. r_c is the critical radius, and ΔG^* is the free energy of activation for nucleation.

below T_c is shown in Fig. 5.10. At this temperature particles below the critical radius in size will be unstable; such subcritical particles are referred to as embryos. The condition for continued growth of an embryo is that the radius should exceed r_c, at which point

$$\frac{\delta[\Delta G(r)]}{\delta r} = 0.$$

$$\frac{\delta[\Delta G(r)]}{\delta r} = 4\pi r_c^2 \, (\Delta G_V + \varepsilon) + 8\pi r_c \sigma = 0$$

$$\therefore \qquad r_c = \frac{-2\sigma}{(\Delta G_V + \varepsilon)} \qquad (5.2)$$

The activation energy barrier to nucleation, ΔG^*, is given by

$$\Delta G^* = \frac{16\pi}{3} \cdot \frac{\sigma^3}{(\Delta G_v + \varepsilon)^2}. \tag{5.3}$$

We are mainly interested in the temperature dependence of r_c and ΔG^*, and as σ and ε are virtually independent of temperature, the controlling factor will be the temperature dependence of ΔG_v. At the equilibrium transformation temperature T_c, ΔG_v is zero, and it becomes increasingly negative as the temperature is lowered below T_c, so that we can write

$$r_c \propto \frac{1}{T_c - T} \quad \text{and} \quad \Delta G^* \propto \frac{1}{(T_c - T)^2} \tag{5.4}$$

In physical terms, we can envisage that at temperatures at which atomic mobility is appreciable, local rearrangements of atoms occur continuously as a result of thermal agitation. Within the stability field of the high-temperature phase, in this case β, such regions can only have a transitory existence and rapidly disperse to be replaced by others elsewhere in the crystal. However once the temperature falls within the stability field of the α phase, such fluctuations become of prime importance as they are the potential source of embryos of α. Thus the formation of a viable critical nucleus depends on the chance event that an embryo acquires sufficient thermal energy to overcome the activation energy barrier ΔG^* and achieve the critical size.

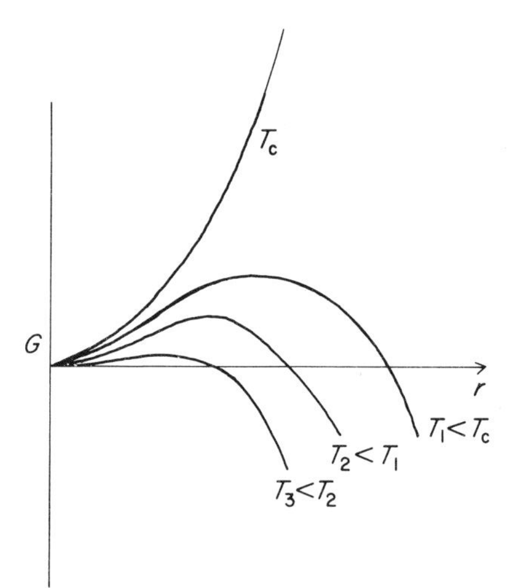

Fig. 5.11 Plots of the free energy of formation of a nucleus of radius r at a series of temperatures. T_c is the equilibrium temperature. At greater degrees of undercooling, both the critical radius and the activation energy decrease.

The equations above show that the larger the undercooling $\Delta T (= T_c - T)$, the smaller the fluctuation in free energy required to create a critical nucleus, and the smaller its size. At the transformation temperature, T_c, no nucleation is possible, as both ΔG^* and r_c will be infinite. Therefore for a discontinuous transformation some degree of undercooling will always be required. The variation in the activation energy of nucleation of spherical embryos as a function of their radius is shown in Fig. 5.11 for a series of temperatures.

Next we shall consider some of the factors which affect the magnitudes of the surface energy and the strain energy terms in equation 5.1. Both terms add a positive contribution to the free energy change and hence increase the activation energy barrier for nucleation. Any mechanism which decreases their value will therefore be kinetically more favourable. Broadly speaking, two separate factors are important: the nature of the nucleation site, and the structural relationship between the phases.

5.2.1 HOMOGENEOUS NUCLEATION

Homogeneous nucleation takes place within regions of perfect crystal structure and is independent of crystal defects such as vacancies, dislocations and grain boundaries. The theory presented above assumes that this is the case, and it provides the most general introduction. The extension to heterogeneous nucleation, which takes place at preferred sites, is made in the next section.

The structural relationship between the two phases will determine the nature of the interface between them. If both phases have a similar structure the new phase may be oriented within the matrix so that there is a good lattice matching across the interface. If the two structures are quite different it may be impossible to achieve any structural continuity across the interface. The degree of lattice matching across an interface is described by the term *coherency*. Fig. 5.12 illustrates the meaning of the terms coherent, semi-coherent and incoherent as applied in this context.

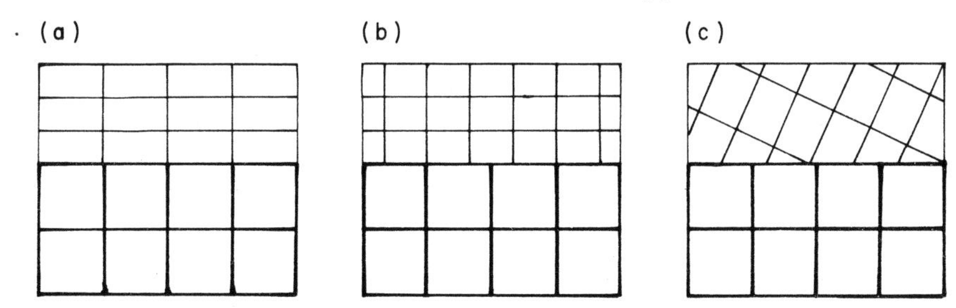

Fig. 5.12 Illustrations of the terms (a) coherent, (b) semi-coherent and (c) incoherent interfaces between two phases.

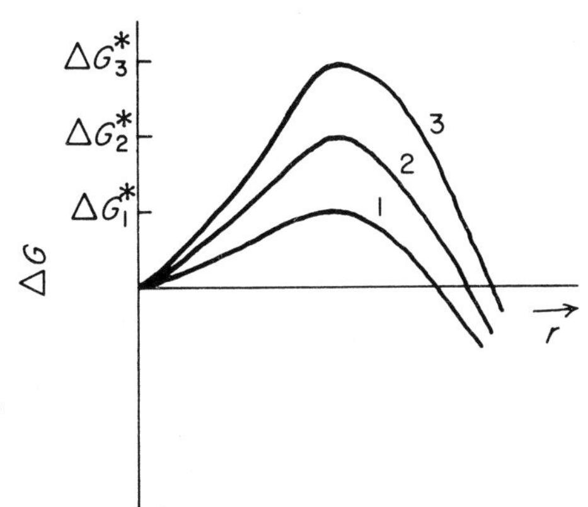

Fig. 5.13 The free energy of formation of a nucleus of radius r when the interface between the two phases is: 1, coherent; 2, semi-coherent; and 3, incoherent.

Interfaces with good lattice matching have lower surface energies than those across which lattice matching is poor. Therefore a coherent nucleus will have a lower activation energy for nucleation than a semi-coherent or an incoherent nucleus. This is shown schematically in Fig. 5.13. For any given pair of phases, the most favourable energetic situation exists when the orientational relationships between the two phases is such that the best possible lattice matching is achieved. As most minerals are anisotropic (the structure is different in different directions), some interfaces between the new phase and the matrix will have a lower surface energy than others. The new phase will tend to grow such that for any given volume formed the total surface energy is minimized by preferential growth of those surfaces with minimum energy. This results in a definite morphology and orientational relationship between the phases. There are many examples of exsolution in minerals where the exsolved phase takes the form of oriented lamellae within the matrix phase. This indicates that these lamellae formed originally with a high degree of coherency.

Finally we consider the origin of the strain term in equation 5.1 and the factors determining its magnitude. Again we are concerned with the nature of the interface between the phases, but in this case with the degree of mismatch across the interface. As an illustration, Fig. 5.14(a) shows a coherent interface with the lattices across the

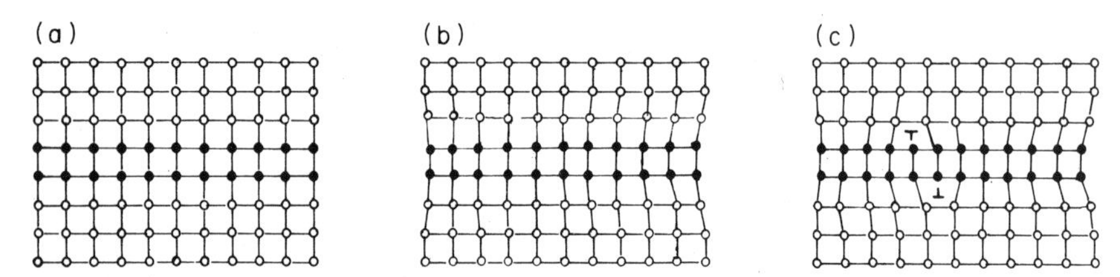

Fig. 5.14 Schematic drawings of a thin lamella (filled circles) within a matrix (open circles) for different degrees of mismatch across the interface: (a) is perfectly coherent, zero strain; (b) is coherent although elastically strained; and (c) is semi-coherent.

interface matching perfectly. In this case the strain energy term is zero. In Fig. 5.14(b) the interface is again coherent with the lattices continuous across the interface, but the slight difference in lattice spacing introduces elastic strain and hence a positive strain energy. As the degree of mismatch becomes too great, bonds are broken at the surface forming a semi-coherent interface [Fig. 5.14(c)]. This reduces the strain energy but increases the surface energy term. For a completely incoherent interface there is no elastic strain energy, although the surface energy term is a maximum.

Taken alone, a reduction in the strain energy term will reduce the activation energy for nucleation in much the same way as a reduction in the surface energy term. However, as can be seen from the discussion above, the two terms are not wholly independent of one another. A coherent interface is likely to have more strain and less surface energy associated with it than an incoherent interface where the situation is reversed. Therefore in any system a compromise must be reached between the two energy terms, and the situation which results in the smallest positive value for the sum of the terms will be the most favourable energetically.

The balance between the two terms will depend on the elastic constants and the size of the new-phase particles. Although a detailed treatment is beyond the scope of this book, we can appreciate the general situation by considering a typical case of a small plate exsolving (precipitating) from a parent solid solution. If it is assumed that the shape of the growing plate is such that the ratio of its radius (r) to its thickness (t) is constant, the surface energy term can be shown to be proportional to t^2.

The strain energy term is proportional to t^3. For a coherent precipitate the surface energy term is negligible, and for an incoherent precipitate the strain energy term is negligible, so that the energy inhibiting nucleation will be proportional to t^3 for a coherent nucleus and t^2 for an incoherent nucleus. As shown schematically in Fig. 5.15, the t^3 function is smaller for small values of t, and hence we would expect that in the early stages of formation of the precipitate a coherent interface would be the most likely. The thickness at which coherency is lost and an incoherent precipitate becomes more stable, depends on the difference in structure between the precipitate and the parent phase, as well as the elastic constants of the two phases.

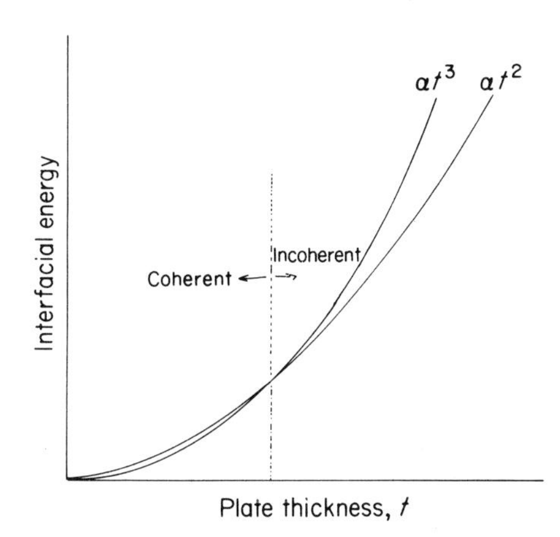

Fig. 5.15 Interfacial energy curves associated with coherent and incoherent lamellae of thickness t. Below a certain thickness a coherent lamella is energetically the more favourable.

The final point we shall make in this section is in regard to the shape and orientation of the nucleating phase. It was noted above that definite morphologies and orientations arise by preferential growth of surfaces with minimum energy. The strain energy term may also contribute to the final morphology and orientation due to the anisotropy of elastic constants in the matrix. For a completely coherent precipitate (for example, exsolution of a phase with the same structure as, but slightly different composition to the matrix) precipitation will take place on planes normal to elastically 'soft' directions in the matrix. For an incoherent precipitate, some strain energy arises from slight differences in volume between the phases, and in an anisotropic matrix the strain energy will depend on precipitate shape.

5.2.2 HETEROGENEOUS NUCLEATION

When the structural difference between the nucleating phase and the matrix is large, the activation energy for homogeneous nucleation will be very high, even when the magnitudes of the surface and strain energy terms are minimized as described above. For nucleation to take place under such circumstances, very large undercooling will be required to provide the driving force. At these lowered temperatures the atomic mobility will be decreased however, thus lessening the chances of random fluctuations producing a critical size nucleus.

Consequently, the transforming system will attempt to reduce this activation energy barrier by having nucleation occur on defects within the matrix structure. This is termed heterogeneous nucleation. These defects may be of many types: vacant sites within the structure, dislocations, grain boundaries, impurities and inclusions etc. All such defects have a certain energy associated with them. When nucleation occurs on a defect, the defect is partially or wholly obliterated and the energy of the defect becomes available to help overcome the activation energy barrier.

We shall not discuss the energy contributions of different types of defects here, but clearly the higher the energy of the defect, the more suitable it becomes as a potential nucleation site. Grain boundaries (especially between more than two grains, e.g. triple junctions) and deformation structures have high energies and are commonly sites for nucleation of new phases. Fig. 5.16 shows the nucleation of precipitates on twin planes.

Fig. 5.16 Transmission electron micrograph illustrating the nucleation of precipitates of hematite on twin boundaries in iron-bearing rutile. The length of the scale bar is 0.2 μm.

In every case heterogeneous nucleation will require a smaller activation energy and hence a smaller degree of undercooling. At temperatures close to the equilibrium transformation temperature it may be the only possible mechanism of nucleation. At lower temperatures both homogeneous and heterogeneous mechanisms may operate. The temperature gap between the onset of heterogeneous and homogeneous nucleation will depend on the magnitude of the surface and strain energy terms, i.e. on the structural differences between the two phases.

5.2.3 THE EQUILIBRIUM SOLVUS AND THE COHERENT SOLVUS

Having discussed briefly the problems of nucleating a new phase, we are now able to make our first simple modification to the equilibrium phase diagram. Fig. 4.10

illustrates the derivation of a solvus from the positions of the minima in the G curve as a function of temperature, for a solid solution with a positive enthalpy of mixing. A solvus derived in this way is referred to as an equilibrium (or chemical) solvus.

As we have seen, homogeneous nucleation necessarily involves a degree of undercooling. This results in a depression of the solvus curve to lower temperatures. The solvus relevant to homogeneous nucleation is termed the coherent solvus. Heterogeneous nucleation may take place at temperatures very near to the equilibrium solvus temperature given sufficient time, although some undercooling will still be necessary to provide a driving force.

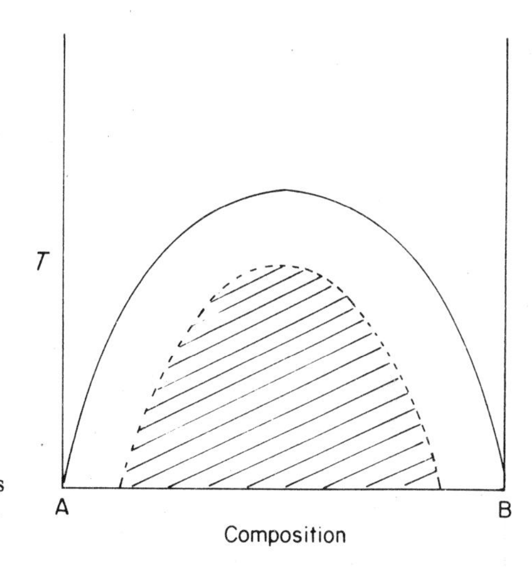

Fig. 5.17 Phase diagram showing the relative positions of the equilibrium solvus (full line) and the coherent solvus (dashed line).

Fig. 5.17 shows a schematic phase diagram in which the end members A and B have similar structures and form a complete solid solution at high temperatures. At lower temperatures the equilibrium solvus defines the limits of solubility and the extent of the two-phase region. The dashed line is the coherent solvus. Within the hatched region both homogeneous and heterogeneous nucleation are likely. Outside the hatched region, but below the equilibrium solvus line, only heterogeneous nucleation of the stable phase is likely to occur.

5.2.4 GROWTH OF PRECIPITATES

Once nucleation has been achieved, the free energy decrease associated with the new phase provides the driving force for its subsequent growth, which can be considered in terms of advancing the interface into the matrix. For an exsolution process this will involve the migration of atomic species to and from this interface, i.e. will depend on the relevant diffusion rates through the mineral. This aspect will be considered in the next chapter. At this stage, however, it is worth noting that the way in which a precipitate phase grows does depend on the nature of the interface.

An incoherent boundary does not in itself present any structural obstacle to growth, in that the diffusing atoms can be added to the interface in virtually any position. The migration of the interface will be mainly dependent on the relevant diffusion rates. If, however, there is some definite structural continuity across an interface, as is the case if it is coherent or semi-coherent, this represents a barrier to growth normal to it. Small coherent precipitates will therefore tend to grow along their length forming thin lamellar plates.

The problem of thickening these lamellae is often overcome by the formation of a stepped interface as shown schematically in Fig. 5.18. At each step there is a dislocation. The interface can then grow normal to itself by the lateral migration of these steps across the boundary plane. Such steps have been observed at semi-coherent precipitate/matrix interfaces in pyroxenes, amphiboles and feldspars. Fig. 5.19 shows a high-resolution electron micrograph of an augite lamella in an orthopyroxene matrix. The fringes are images of the lattice planes of the two phases, and steps can be seen along the interface boundary.

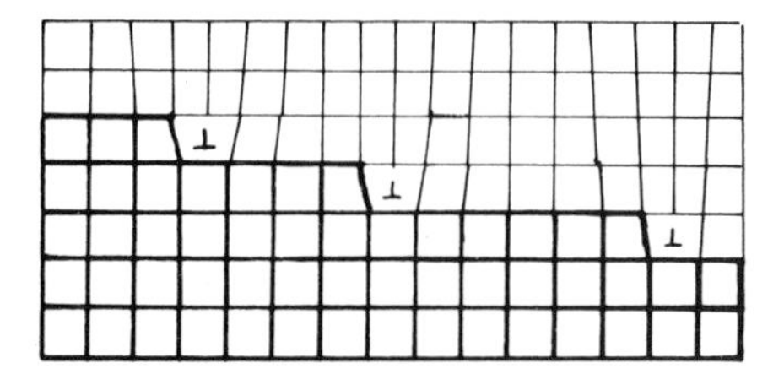

Fig. 5.18 A stepped, semi-coherent interface between two phases.

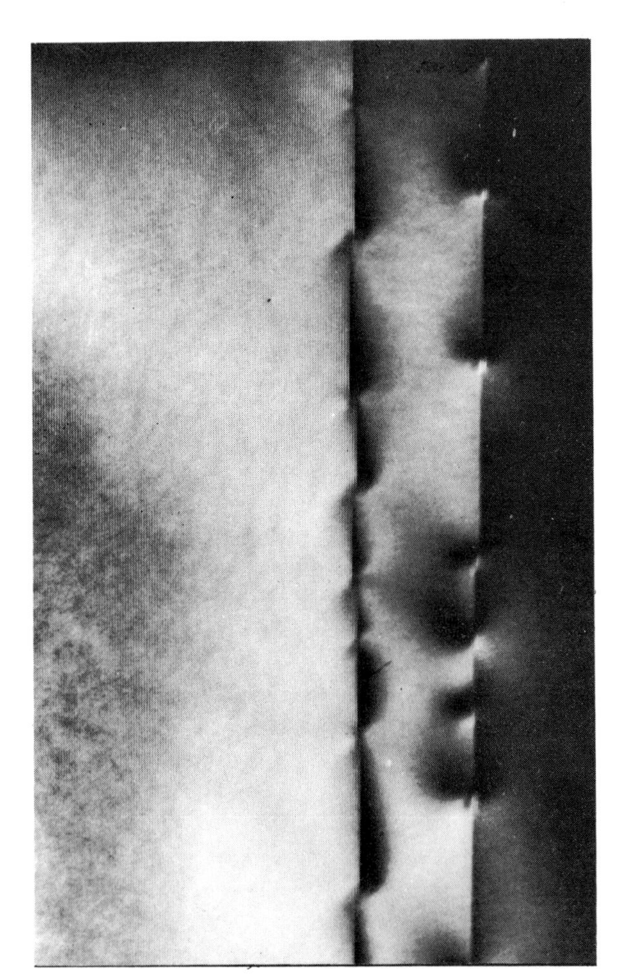

Fig. 5.19 High-resolution transmission electron micrograph of an augite lamella in an orthopyroxene matrix, showing the growth steps at the interfaces. The lattice fringes in the matrix are due to the 18-M repeat in the orthopyroxene structure. (From Champness and Lorimer, 1974. Photo courtesy of P. Champness.)

5.2.5 PRECIPITATE COARSENING

If homogeneous nucleation occurs uniformly throughout the matrix a large number of fine precipitates are likely to be formed. These may then thicken by a mechanism similar to that described above, until the concentration of the solute in the matrix reaches a near equilibrium level. It might be expected at this stage that the process has come to an end, with a microstructure consisting of a dense network of precipitates. This is not however the case, because if the number of precipitates is large, there is still a high interfacial area between the precipitates and matrix. The free energy of the system could be further reduced by producing a coarser microstructure, i.e. a fewer number of larger precipitates.

If temperatures are maintained at levels where atomic diffusion is appreciable, such coarsening will take place by the diffusion of solute from smaller particles to the bigger ones. Thus the higher energy particles dissolve at the expense of the larger, lower energy particles. Such a sequence can be seen in Fig. 5.20, in which coherent precipitates of an iron-rich phase in rutile undergo this coarsening process on annealing a thin specimen in an electron microscope.

5.3 Spinodal decomposition

At the beginning of this chapter we made the distinction between discontinuous and continuous transformation processes. The nucleation mechanisms described in the previous section were examples of discontinuous behaviour and involved the formation of a small discrete nucleus within a matrix of untransformed structure. Most of this theory can be applied to both polymorphic transformations and those involving phase separation.

In this section we will describe a fundamentally different type of process which is continuous and involves no discrete nucleation event. Initially we will apply it to transformations involving phase separation, but in a later section we will see that it may also be applied to polymorphic transformations of the order–disorder type.

The process by which a homogeneous solid solution decomposes continuously into regions of different composition is termed *spinodal decomposition*. It involves the evolution of sinusoidal composition waves simultaneously throughout the whole structure. Under appropriate conditions such compositional modulations may grow in both wavelength and amplitude until eventually the peaks and valleys of the wave become two discrete phases. The lower diagram in Fig. 5.1. illustrates this and compares it to the nucleation process. The modulations which develop in this way are on a very fine scale, typically between 50 and about 500 Å, and therefore are only observable by electron microscopy. Fig. 5.21 shows an electron micrograph of a spinodal modulation in an alkali feldspar.

5.3.1 COMPOSITIONAL LIMITS OF SPINODAL BEHAVIOUR

Spinodal decomposition can best be understood with reference to the shape of free energy–composition curves of supersaturated solid solutions. In section 4.4 we showed that a solid solution with a positive ΔH of mixing would, on cooling, develop a miscibility gap. The free energy curve is then of the form shown in Fig. 5.22, with two minima corresponding to the limits of miscibility. Within these minima the

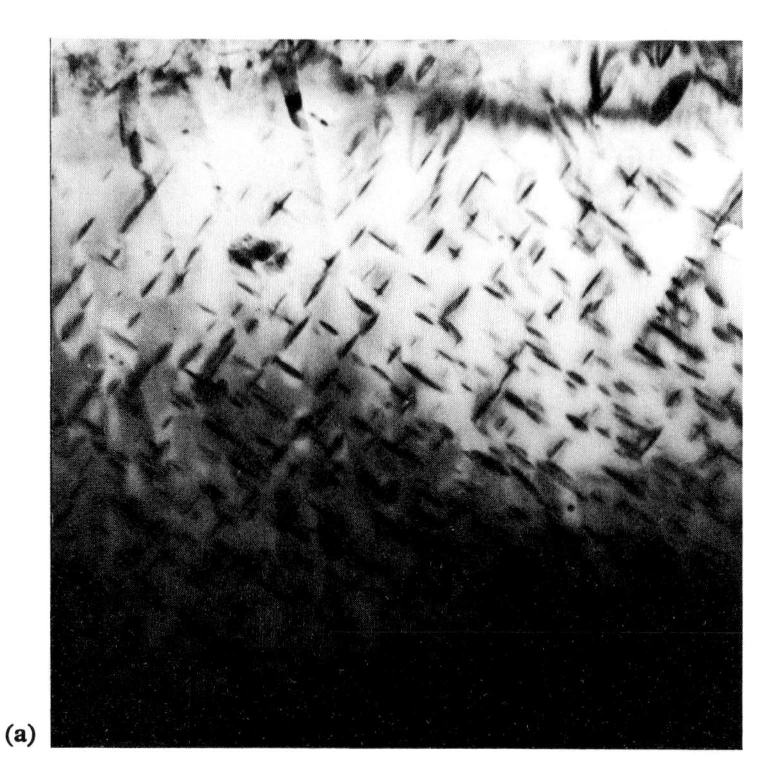

(a)

(b)

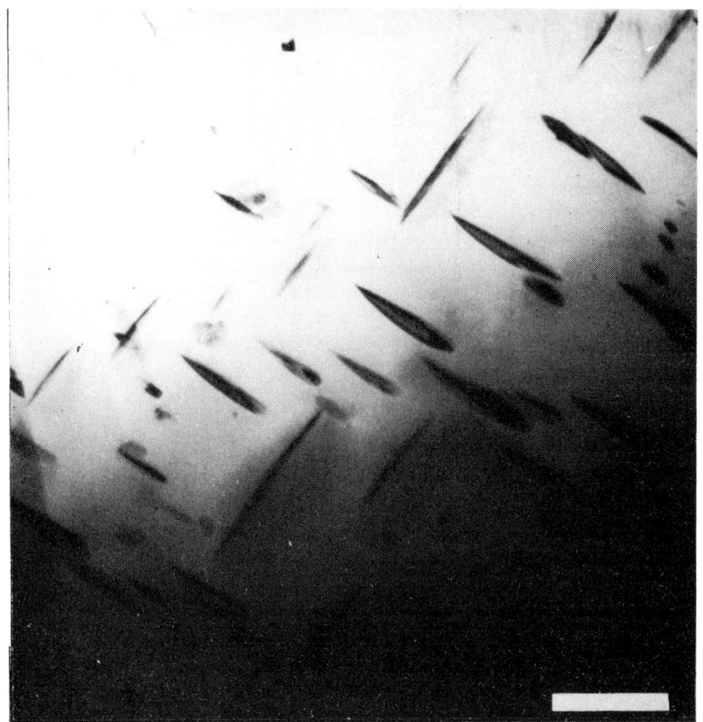

Fig. 5.20 A sequence of transmission electron micrographs showing the coarsening of coherent precipitates of an iron-rich phase in rutile. During experimental annealing the finer platelets dissolve at the expense of the larger platelets which increase their diameter. The length of the scale bar is 0.2 μm.

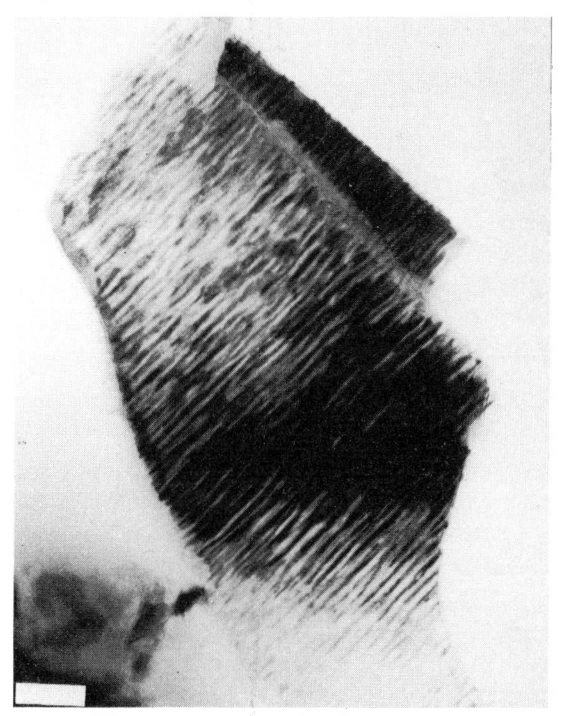

Fig. 5.21 Transmission electron micrograph of the modulated structure formed during the spinodal decomposition of an alkali feldspar. The length of the scale bar is 0.1 μm.

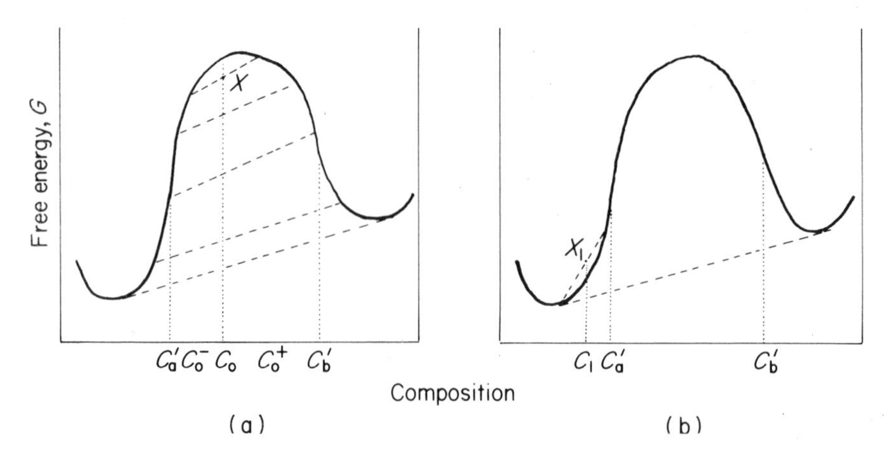

Fig. 5.22 Free energy–composition curves illustrating the free energy changes during the decomposition of (a) a phase with composition C_0 between the inflexion points of the curve, and (b) a phase with composition C_1 outside the inflexion point.

solid solution is not stable. The equilibrium-phase mixture in this case is determined from the common tangent to the minima, the points of tangency defining the compositions of the two phases.

Apart from these minima in the G curve, at which $\delta G/\delta c = 0$, there are two other important points on this curve. These are the points of inflexion at which $\delta^2 G/\delta c^2 = 0$, labelled C_a' and C_b' in Fig. 5.22. Within these points the G curve is concave downwards, while between the inflexion points and the minima the G curve is concave upwards. The fundamental difference between these regions can be appreciated by considering the possible behaviour of a homogeneous phase of composition C_0 compared to that of a phase of composition C_1.

If the phase at C_0 develops compositional fluctuations which result in the formation of some regions with composition C_0+ and hence others with composition C_0-, the new free energy of the system is given by the weighted average of the free energies of the two compositions, i.e. point X in Fig. 5.22(a). Clearly there is a decrease in the free energy of the system and hence any small separations in composition about C_0 should occur spontaneously. Once they have occurred they should not homogenize again for that would raise the free energy. Taking the same argument further, phase separation should continue on this scheme as indicated by the successive lines in the figure until the lowest free energy is achieved, which is the two-phase mixture given by the common tangent.

Contrast this with the possible behaviour of a composition outside the inflexion point. Any small compositional fluctuation about C_1 will raise the free energy to point X_1 in Fig. 5.22(b). The single-phase solid solution is therefore stable with respect to such small changes, and any fluctuations will spontaneously homogenize. In order to reduce the free energy of the solid solution a large compositional change in the right-hand direction of the diagram is necessary. The formation of fluctuations of sufficiently large compositional difference will involve the formation of an interface between this new region and the matrix. This corresponds to a discontinuous nucleation process and will involve an activation energy barrier. Ultimately the lowest free energy is again given by the common tangent, but the mechanism by which these two compositions C_0 and C_1 can achieve this is quite different.

In the same way that the locus of the minima in the G curve as a function of temperature defines the solvus (see Fig. 4.7), the locus of the inflexion points will define a curve within the solvus which limits the compositional region within which spinodal behaviour is thermodynamically possible. Such a curve is termed the chemical spinodal. Fig. 5.23 shows the equilibrium solvus and the chemical spinodal. Within the hatched region both spinodal decomposition and nucleation are possible, whereas outside the hatched region only a discontinuous nucleation mechanism can occur.

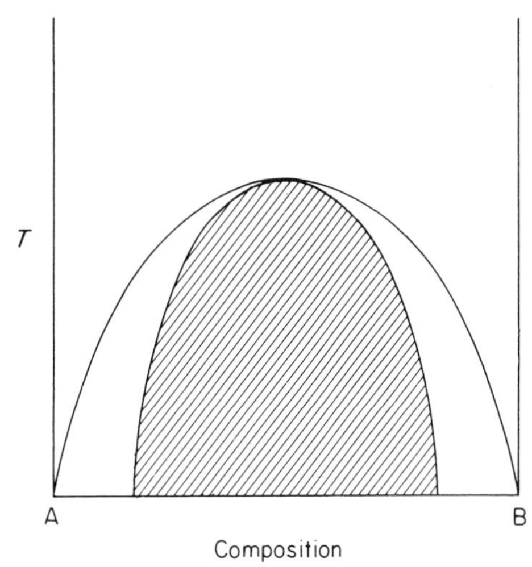

Fig. 5.23 Phase diagram showing the relative positions of the equilibrium solvus and the chemical spinodal curve.

5.3.2 SPINODAL DECOMPOSITION AS ALTERNATIVE METASTABLE BEHAVIOUR

Fig. 5.23 illustrates that the equilibrium behaviour on cooling the solid solution involves exsolution by a discontinuous nucleation process. In many mineral systems activation energies for nucleation are so great that, unless the cooling rate is extremely slow, nucleation will fail to occur under conditions of slight supercooling. When the temperature falls to intersect the spinodal curve, spinodal decomposition becomes possible. Being a continuous process it involves no activation energy barrier, and is therefore much more favourable kinetically than the nucleation process.

The decrease in free energy associated with small compositional fluctuations will be less than that for true nucleation. Therefore the spinodal process is an example of a transformation which takes place by virtue of its kinetics, rather than its energetic advantage. It thus falls within our definition of alternative metastable behaviour (section 5.1.5).

5.3.3 STRAIN ENERGY AND THE COHERENT SPINODAL

The development of a spinodal modulation involves the coexistence of coherent regions of slightly different chemical composition and hence lattice dimensions. Such a coexistence involves coherency strains in the structure and hence a positive strain-energy term which will oppose the spinodal decomposition. The effect of this is to depress the chemical spinodal curve to lower temperatures, i.e. undercooling is required to provide the extra driving force to overcome the strain-energy term. The spinodal term which takes into account the strain energy is termed the coherent spinodal, by analogy with the coherent solvus.

Fig. 5.24 shows a schematic diagram illustrating coherent and equilibrium solvus curves, and coherent and chemical spinodal curves. The relative positions of these curves will depend on the specific system. Interpretation of this diagram is the same as for the analogous Figs 5.17 and 5.23.

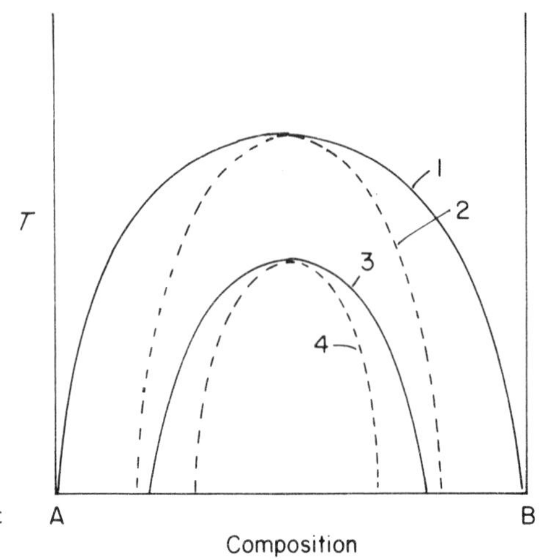

Fig. 5.24 Phase diagram illustrating the relative positions of (1) equilibrium solvus; (2) chemical spinodal; (3) coherent solvus; and (4) coherent spinodal.

The second effect of the strain-energy term is to control the orientation of the modulations which can grow. The primary control is imposed by the symmetry of the matrix phase as there is no change in symmetry associated with the development of a modulation. However, those fluctuations which occur in the elastically 'soft' directions in the matrix will be more favoured energetically and will develop at the expense of fluctuations in other orientations. In cubic crystals three sets of perpendicular modulations may develop, but in the lower symmetry silicate structures one or two modulations may occur on well defined-planes.

5.3.4 THE WAVELENGTH AND AMPLITUDE OF SPINODAL MODULATIONS

To appreciate the detailed description of a spinodal modulation requires a theoretical treatment which may be found in more advanced texts mentioned in the bibliography. We will outline however some of the main factors which define the character of a modulation.

The initial thermodynamic assumption which led to a G curve with a shape such as Fig. 5.22 was that the enthalpy of mixing was positive, i.e. like atoms tend to cluster. If such clusters, however, are very close together, or have large compositional differences, the region between the clusters will have a large compositional gradient. This introduces a gradient-energy term which becomes increasingly positive as the wavelength of the modulation decreases or its amplitude increases. Therefore such modulations will only be possible with undercooling below the spinodal curve sufficient to provide an increased ΔG term to balance the gradient energy. The gradient-energy term also puts a lower limit on the wavelengths which are possible.

The strain-energy term is independent of the wavelength but increases as the square of the amplitude of the fluctuation. Thus at small degrees of undercooling only low-amplitude modulations will be able to exist.

The net result is that at temperatures very close to the spinodal curve (i.e. involving only small negative values of the free energy) only very long wavelength and small-amplitude fluctuations can form. The distances over which atomic migration will need to occur are therefore large, and, if the diffusion rates are slow, such modulations will take a long time to develop despite the absence of a nucleation event. We shall return to this point in the chapter on kinetics after we have dealt with diffusion.

At greater degrees of undercooling shorter wavelengths and larger amplitudes become possible. As the amplitude and, hence, the coherent strain energy increases, it becomes energetically advantageous to decrease this strain by a loss of coherency. The situation is analogous to that for the growth of coherent precipitates described in section 5.2.1. The point at which the loss of coherency occurs will depend on the rate of change of lattice dimensions with composition in the system, and on the nature of the bonding. This latter point emphasizes a difference in spinodal behaviour in metals and in silicate minerals. The loss of coherency involves the breaking of bonds at the interface (see e.g. Fig. 5.12) and the nucleation of dislocations. In metals the energy required to do this is relatively small and hence loss of coherency occurs fairly early in the transformation sequence. In silicates the energy to form a dislocation is much larger and it is not unusual to find coherency maintained in coarse microstructures.

5.3.5 CONDITIONS NECESSARY FOR SPINODAL DECOMPOSITION

Our derivation of the spinodal curve (section 5.3.1) was based on the fact that a continuous G curve existed across the whole binary solid solution, in other words that the structures of the separating phases were essentially the same and differed only in composition. This limits the mineral systems in which spinodal decomposition may occur to those with a simple solvus or very similar end member structures. Many minerals do conform to this requirement as their solid solutions are often based on a variation in cation occupancy of the interstitial sites within a framework common to both end members. Thus the development of spinodal fluctuations involves only the movement of cations within a relatively inert framework which remains continuous through the crystal.

Many minerals therefore provide ideal conditions for spinodal decomposition. A further factor which may favour the development of spinodal fluctuations as an alternative to nucleation, is that many natural minerals which have not undergone deformation contain relatively few structural defects. Their long (up to 10^6 years) thermal histories provide the opportunity for defects to be annealed out of the crystal. The result is that relatively few potential sites are available for heterogeneous nucleation. While homogeneous nucleation remains a possibility in such circumstances, it is still an energetically difficult process compared with the spinodal mechanism. Therefore in minerals with broad, isostructural solid solutions, spinodal decomposition is a very convenient method of free energy reduction in circumstances where nucleation is difficult.

In the next section we will briefly consider the alternative mechanisms of phase separation open to a mineral system where the end members have quite different structures and only limited solid solution is possible at high temperatures.

5.4 Transitional phases

Fig. 5.25 shows part of a schematic phase diagram in which the phases A and B have different crystal structures and hence very little mutual solid solubility. A phase with composition C_0 in the figure would, according to the phase diagram, nucleate particles of phase B as the temperature decreased. As B is structurally quite different to A a large activation energy barrier will exist for this process. Heterogeneous nucleation may be possible given a sufficiently slow cooling rate, but structural differences would preclude the homogeneous formation of nuclei of B. We

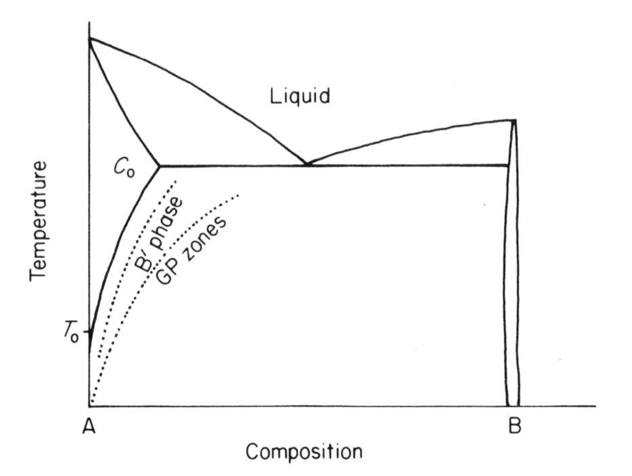

Fig. 5.25 Phase diagram in which phases A and B have different structures. Under metastable conditions, transitional phases may be formed, as indicated by the dotted curves.

may take it as a general rule that homogeneous nucleation of incoherent phases is highly improbable. Spinodal decomposition will not be possible in this system for the reasons outlined above, and so we are seemingly left with the situation that if heterogeneous nucleation did not have the opportunity to occur, the end result would be the persistence of a stranded solid solution phase of composition C_0 existing metastably at low temperatures.

There is one mode of alternative behaviour possible under such circumstances. This is the formation of metastable transitional phases which have a structure more similar to the matrix than is the equilibrium phase. Transitional structures may be coherent or semi-coherent with the matrix thereby lowering the activation energy for their nucleation. Coherent phases formed in this way often have the shape of small strained platelets which have a composition towards that of the equilibrium phase, but a structure like that of the matrix. They are clusters of solute atoms (B) on the parent crystal lattice. This type of coherent precipitate was first detected in the Al–Cu system by X-ray techniques in independent studies by Guinier and Preston in the 1930s. Consequently they are now termed Guinier–Preston (GP) zones, and the use of the term has been extended to analogous phases in other systems. Being coherent, such phases usually nucleate homogeneously.

Semi-coherent phases may also form in this way, with nucleation occurring heterogeneously on defects or on the strained boundaries of pre-existing GP zones.

A sequence of such transitional structures may form as a function of both time and the degree of undercooling, such that each stage of the process involves a further structural and/or chemical departure from the parent matrix, and hence a further reduction in free energy. In this respect such a sequence is an illustration of the Ostwald step rule (section 5.1.6).

Fig. 5.26 shows a free energy–composition diagram for the sequence of phases in our hypothetical system. A single curve is drawn for the phase A and the GP zones to emphasize the structural continuity between them. The composition C_0 is marked, and the tangent to the two minima in the G^A curve gives the composition of the A phase in metastable equilibrium with the GP zones. The formation of the GP zones involves the smallest free energy reduction, and takes place first under metastable conditions. A second transitional phase B′, which could be a semi-coherent phase, is shown with a separate G curve, reflecting the fact that a structural change is involved in its formation. The greatest reduction in free energy is associated with the formation of the equilibrium phase B. The relative positions of the curves for GP zones, semi-coherent, and equilibrium phases may differ from one system to another. In practice their compositions are very difficult to determine due to their small size.

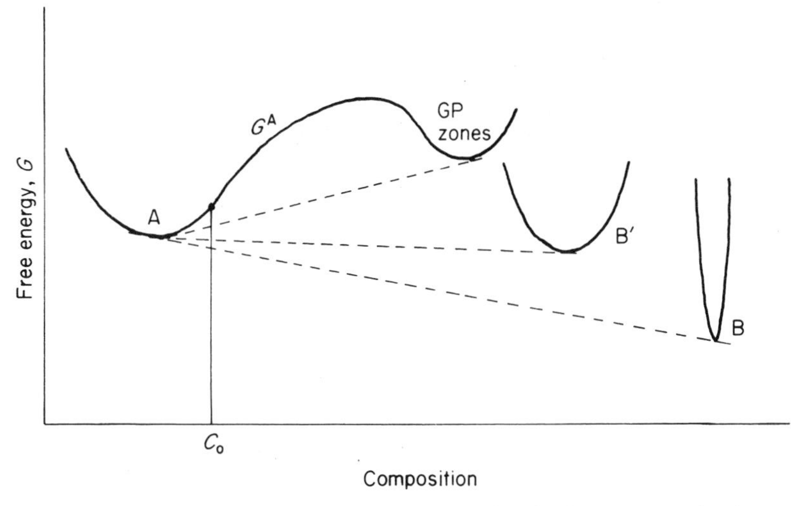

Fig. 5.26 Free energy–composition curves illustrating the free energy reduction associated with the formation of GP zones, the transitional phase B_1, and the equilibrium phase B, from a solid solution of composition C_0.

The relative free energies of the three phases determine the temperature at which each can form. Using a construction similar to that in Figs 4.14 and 4.18 it can be shown that the phase with the lowest free energy will always form at the highest temperature, and that the difference between the free energies will be reflected in the temperature intervals between the formation of the phases. Fig. 5.25 shows the dotted lines corresponding to the metastable formation of GP zones and the B′ transitional phase.

Transitional phases have been observed in many metal systems, and have technological importance in that their presence affects the mechanical properties of the alloys (precipitation hardening). Mineral systems are now being investigated in

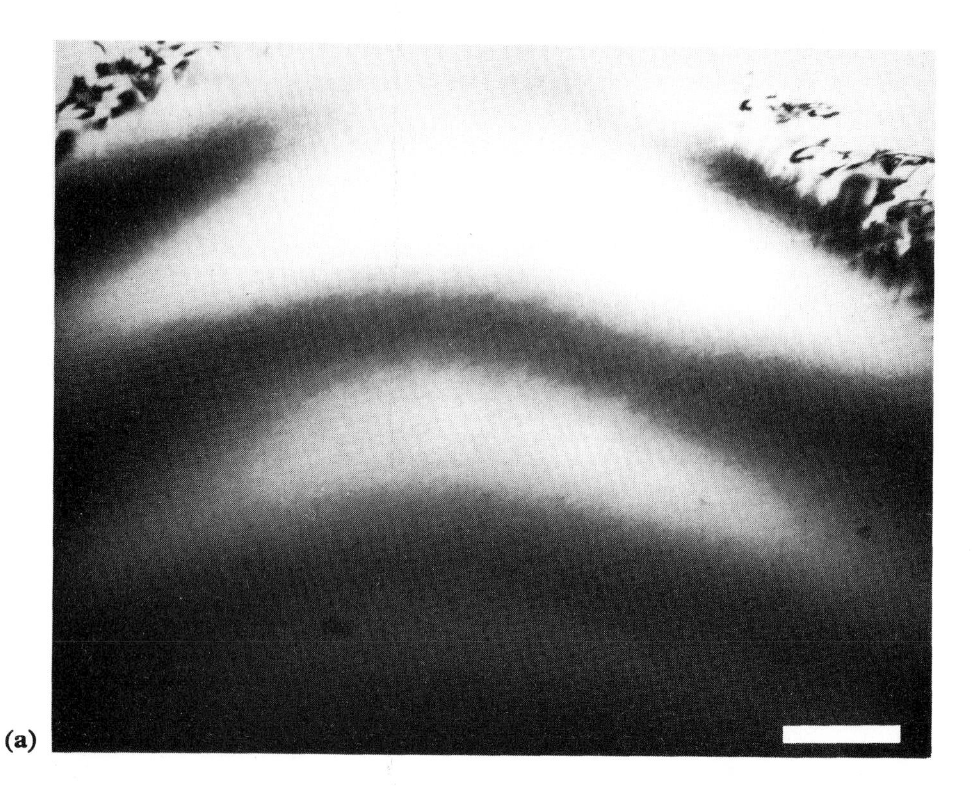
(a)

(b)

more detail and similar behaviour is found to occur. Fig. 5.27 shows an electron micrograph of the same area of a crystal of natural iron-bearing rutile (a) as a homogeneous solid solution, and (b) containing two sets of iron-rich GP zones which form in the early stages of the precipitation of hematite. This behaviour will be examined in more detail in a later chapter.

5.5 Order–disorder transformations

We have already discussed a number of features of the disordered state in Chapter 2, including various types of disorder and the relationship between disorder and configurational entropy. In high-temperature solid solutions we are mainly concerned with substitutional disorder (section 2.1.3) and in particular with the effect of falling temperature on the stability of the disordered state. The simple thermodynamic model developed in Chapter 4 broadly defined two possible modes of behaviour on cooling, depending on whether like atoms tend to occur as neighbours. If this is so (i.e. ΔH is positive) then the result is phase separation which has been discussed in the previous sections. In this section we will consider some of the consequences when unlike atoms tend to occur as neighbours, i.e. when ΔH is negative and ordering takes place. Some reference has been made to this already in section 2.1.3 and in the subsequent treatment of mineral structures. Here we will deal with the microstructures arising from order–disorder transformations and refer to some aspects of the process which are likely to affect the kinetics.

5.5.1 SOME CRYSTALLOGRAPHIC ASPECTS OF ORDERING

Most mineral structures can be considered in terms of a substructure consisting of a framework of anions or silicate tetrahedra within which the cations occupy certain interstitial sites. In the most common sulphides, for example, the sulphur atoms form a close-packed network while the cations are distributed among the interstitial tetrahedral or octahedral sites. In silicates the arrangement of SiO_4 tetrahedra forms the 'skeleton' of the structure, while cations again occupy interstitial sites. When describing substitutional disorder in such minerals we can define two broad categories:

1 Those in which only the interstitial cations take part in the ordering process.

2 Those silicates in which Si,Al-ordering also takes place within the silicate 'skeleton'. This category is much more complex because of the coupled charge balance relationships involved, such as $Si + Na \rightleftharpoons Al + Ca$ in the feldspars (section 3.6.2).

In the first category a distinction can be made between the case when the disordering species occupy topologically equivalent sites within the structure, and the case when disordering takes place over two or more non-equivalent sites. This difference can be illustrated diagrammatically. In Fig. 5.28, the square grid represents the substructure which in this case plays no part in the disordering transformation. Within the squares are the interstitial sites which are all topologically equivalent. In our example these interstitial sites are occupied by two types of atoms shown as black and white dots. Fig. 5.28(a) is the disordered state in which there is an

Fig. 5.27 (a) Transmission electron micrograph of a homogeneous grain of iron-bearing rutile. (b) After experimental annealing, two sets of fine platelets are precipitated. These have the characteristics of GP zones. The length of the scale bar is 0.2 μm.

equal probability that any interstitial site contains either a black or a white atom. It is therefore shown as a gray (stippled) dot. In this disordered state the interstitial sites retain their crystallographic equivalence. The unit cell of the structure is shown by the dotted lines.

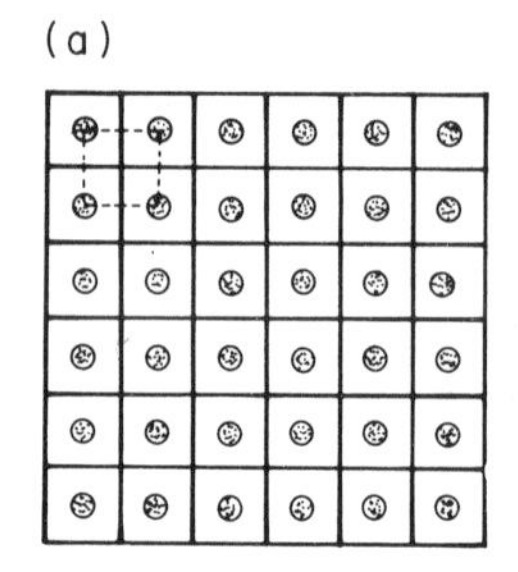

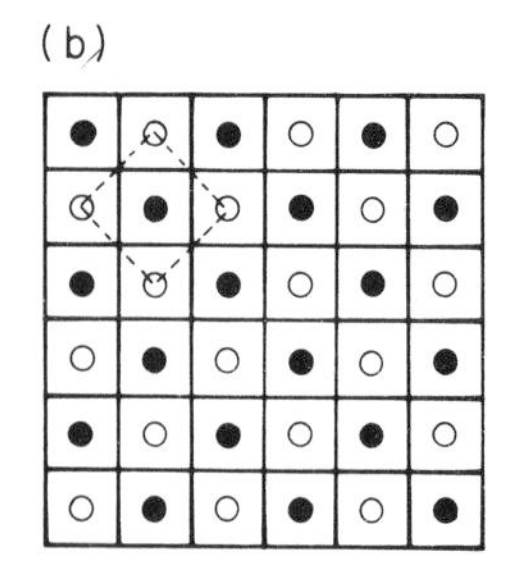

Fig. 5.28 The ordering process within topologically equivalent sites in a structure, leading to the formation of a superstructure in the ordered form. The dotted lines outline the unit cell in each case.

In Fig. 5.28(b) the interstitial atoms are now fully ordered. The interstitial sites are therefore no longer crystallographically equivalent and the size of the unit cell repeat is increased. This ordered structure is referred to as a *superstructure*. Whenever ordering between topologically equivalent sites reduces the translational symmetry of the structure by destroying the crystallographic equivalence of the sites, a supercell, which is a simple derivative of the disordered subcell, will form.

In the sulphides, where cation ordering takes place within the topologically equivalent interstitial sites, superstructures are formed on cooling. The existence of complex superstructures is one of the main features of low-temperature sulphide mineralogy.

The second case, where ordering takes place between two topologically non-equivalent sites, is illustrated in Fig. 5.29. There are now two types of interstitial sites: at the centres of the square and the octagonal interstices. In the disordered structure [Fig. 5.29(a)] the two sites are chemically equivalent but topologically quite distinct. Ordering therefore does not produce any change in the translational symmetry and the unit cell repeat remains the same [Fig. 5.29(b)]. Under circumstances such as these it is possible for an ordering transformation to take place without any change in the symmetry of the mineral.

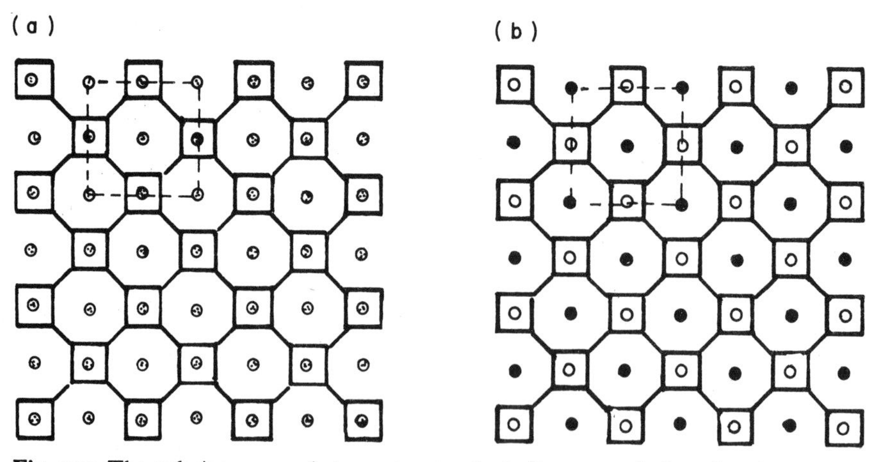

Fig. 5.29 The ordering process between two topologically non-equivalent sites does not affect the size of the unit cell of the structure.

In the orthopyroxenes (section 3.3) where there is the possibility of ordering Mg on the M1 sites and Fe on the M2 sites, partitioning of the two cations can take place continuously as a function of temperature without any structural changes involving symmetry. As in the schematic diagram of Fig. 5.29, the symmetry is determined from the structural topology and is independent of the cation distribution.

In the second category, in which ordering takes place within the framework structure, it can be seen that the loss of crystallographic equivalence of atomic sites during ordering need not always lead to a decrease in translational symmetry and the formation of a supercell. When other elements of symmetry are lost the unit cell may stay substantially the same. Fig. 5.30 is a schematic illustration of a structure in which the atoms within the framework are undergoing an ordering transformation. For simplicity no interstitial cations are considered. In Fig. 5.30(a) the framework is disordered, and as a result there are two sets of mirror planes present, as shown by the dotted lines. Atoms related by these mirror planes are crystallographically equivalent. In the ordered state [Fig. 5.30(b)] these mirror planes are lost, but the size of the unit cell as shown by the dashed lines, remains the same. In many cases such as this, the ordering may also lead to a slight distortion of the structure.

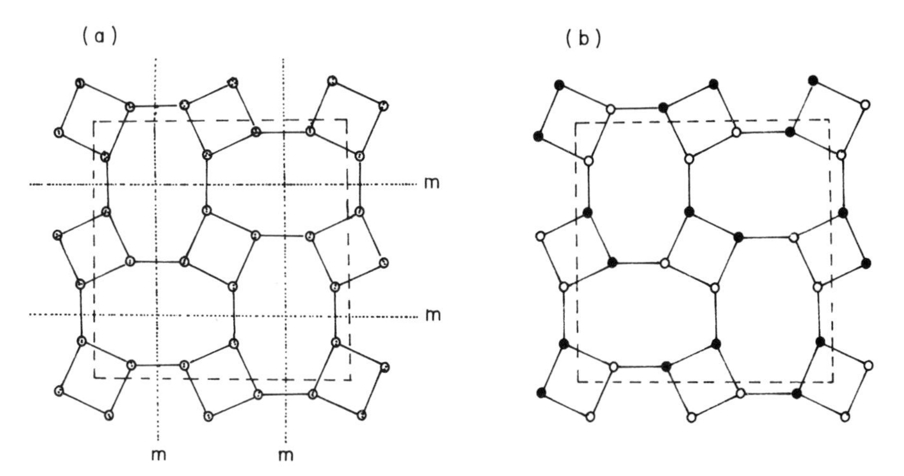

Fig. 5.30 Ordering of atoms within the framework structure leading to no change in the unit cell repeat, but a loss of symmetry.

A situation similar in principle to that described here is Si, Al-ordering in K feldspar, $KAlSi_3O_8$ (Fig. 3.33). Ordering results in the loss of a mirror plane and a diad rotation axis of symmetry, and hence a change in the crystal system from monoclinic to triclinic. The unit cell, apart from the small distortion associated with the ordering, remains substantially the same in both sanidine and microcline.

5.5.2 ORDERED DOMAINS AND BOUNDARIES

(a) *Orientation variants*

Whenever ordering results in a change of symmetry there may be a number of possible orientation variants of the new structure formed. For example, in pyrrhotite Fe_7S_8 (section 2.6.2) at high temperatures the vacancies are disordered over the octahedral interstices of a hexagonal close-packed sulphur structure. On cooling, the vacancies order within alternate cation layers in the manner shown in Fig. 2.15. The result is a change in the unit cell symmetry from hexagonal to pseudo-orthorhombic (pseudo, because there is in fact a slight distortion associated with the ordering which reduces the true symmetry to monoclinic). Fig. 5.31 shows that there are three orientational variants possible in forming an orthorhombic supercell from the hexagonal subcell (six if the monoclinic distortion was taken into account).

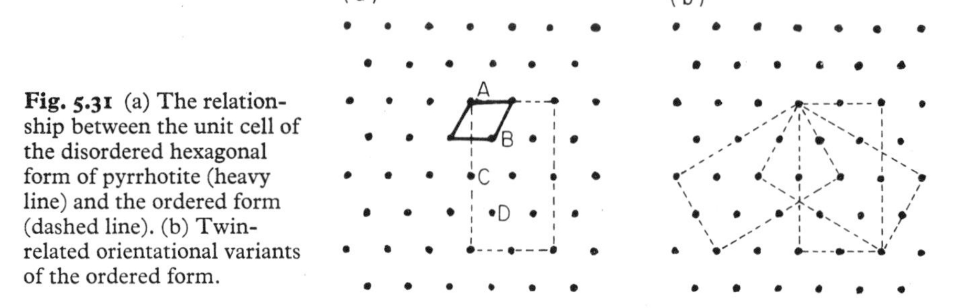

Fig. 5.31 (a) The relationship between the unit cell of the disordered hexagonal form of pyrrhotite (heavy line) and the ordered form (dashed line). (b) Twin-related orientational variants of the ordered form.

If the ordering process is one of random nucleation throughout the crystal, then the probability of a particular variant forming in any part of the crystal is one in

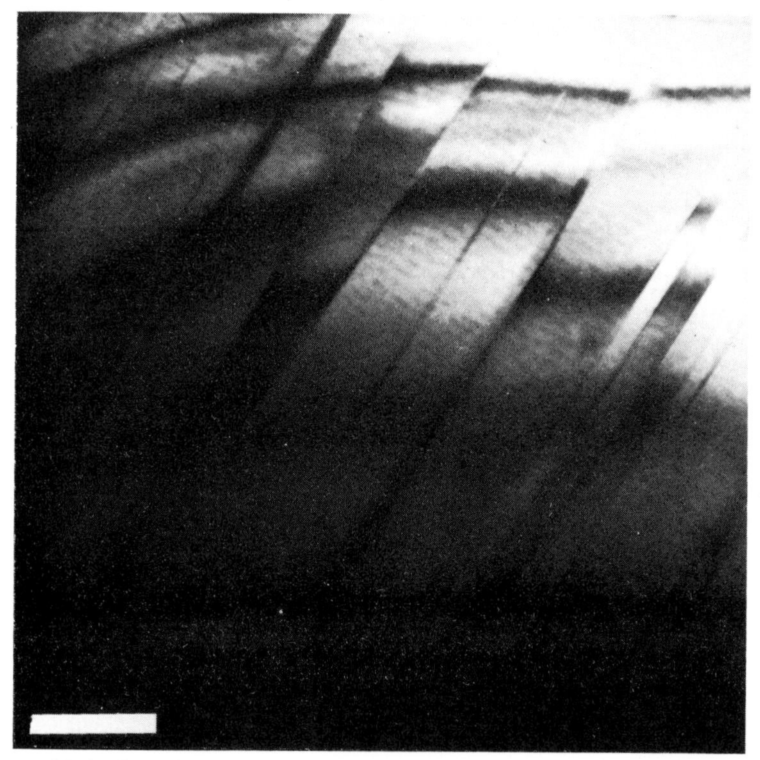

Fig. 5.32 Twin planes formed during the cation ordering process in pyrrhotite. Note the offset of the thickness fringes across alternate twinned regions due to the different crystallographic orientations [as illustrated in Fig. 5.31(b)]. The length of the scale bar is 0.2 μm.

three. The end result at low temperatures is therefore a domain structure made up of regions in which any one of the three possibilities may have nucleated and grown. As the unit cells within each domain are related to one another by a simple rotation they are described as *twin related*, and the boundaries between the domains are *twin planes*. Fig. 5.32 shows such twin planes formed during the ordering of pyrrhotite.

Twin planes may be coherent or semi-coherent depending on the degree of distortion from the parent phase. If the distortion is appreciable the strain energy associated with the twin plane will be high and a loss of coherency may be advantageous energetically.

Whenever orientation variants of a low-temperature form exist, the formation of such transformation twins is an inevitable consequence of a random nucleation process. We have already seen an example of this in the high–low quartz transformation (section 5.1.3) and will meet more examples later. The formation of a twinned structure is not limited to ordering transformations and may be formed in any transformation where orientation variants are possible.

The scale of the twinning depends on the rate of the nucleation process which in turn depends on the cooling rate. The relative scale of the twinned microstructure may therefore provide information on the thermal history of the mineral.

(*b*) *Translation variants*

Often there will be a number of different ways in which an ordered structure can be built within a given orientation variant. This refers to the number of possible origins from which the unit cell of the ordered structure can be drawn within the given orientation variant. In the case of pyrrhotite the situation is rather complex as there are four different translational variants possible within each orientation variant [corresponding to an origin at any of the positions labelled A, B, C, D in Fig. 5.31(a)]. This leads to the formation of a very complex domain structure with domains related by both orientation and translation.

To illustrate this point, therefore, a simpler example will be chosen in which the ordering process does not involve the formation of orientation variants. Fig. 5.33(a) illustrates the formation of independently nucleated regions of an ordered structure similar to that in Fig. 5.28, except that for simplicity the framework structure has been omitted. There are two translational variants possible in this case, as can be seen from Fig. 5.33(b) where the two initial regions have grown together. The two domains formed are out of phase with one another and related by a vector which is a lattice vector of the disordered structure. The domains are termed *antiphase* domains, and the boundaries which separate then are *antiphase boundaries*.

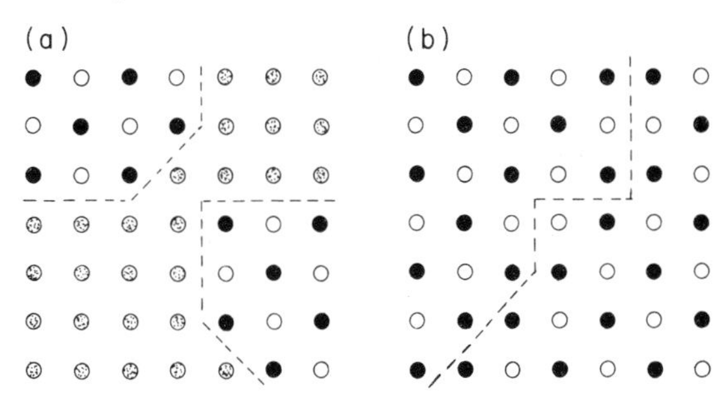

Fig. 5.33 The formation of antiphase boundaries during the growth of independently nucleated ordered domains.

Note that the structure within the antiphase boundary is disordered and hence at low temperatures the presence of such boundaries gives rise to an energy term which is proportional to the total area of the boundary. Thus a free-energy drive exists for destroying the boundaries by coarsening processes.

As there is no orientational difference across antiphase boundaries they cannot be imaged by light microscopy and electron microscopy must be used. Fig. 5.34 shows an electron micrograph of antiphase domains and their boundaries formed by vacancy ordering in the copper iron sulphide bornite Cu_5FeS_4. Again the scale of the microstructure depends on the thermal history, although a coarse microstructure may develop either by slow, widely-spaced nucleation, or by coarsening an initially fine-scaled texture by annealing below the transformation temperature. The two micrographs in Fig. 5.34 show the effects of a different cooling rate on the scale of the domains in the same bornite crystal which has undergone several experimental heating and cooling cycles.

(a) 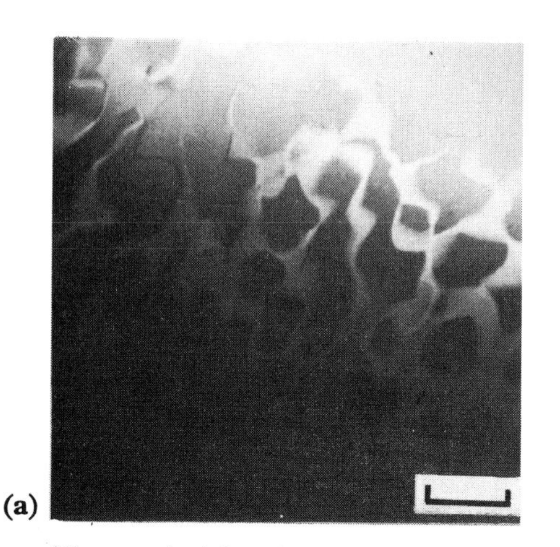

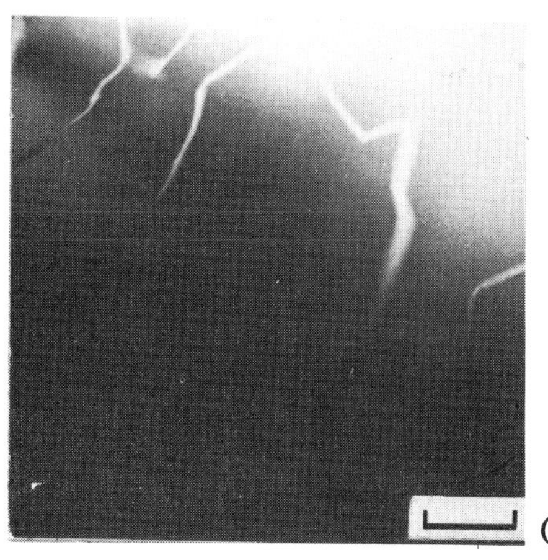 (b)

Fig. 5.34 Antiphase boundaries formed during cation ordering in bornite, Cu_5FeS_4, by (a) rapid cooling and (b) slower cooling of the grain from the disordered state. Note that the contrast on either side of the boundary is the same. The length of the scale bar is 0.1 μm.

Antiphase domains may be formed in any polymorphic transformation in which translational variants occur. Notice that in cases of ordering between topologically non-equivalent sites when the unit cell remains the same, as shown in Fig. 5.29, there is no possibility of forming antiphase domains. Therefore electron microscopy can provide no information on ordering processes in minerals such as orthopyroxenes and olivines.

5.5.3 CONSERVATIVE AND NON-CONSERVATIVE ANTIPHASE BOUNDARIES

When the displacement vector relating two antiphase domains is parallel to the antiphase boundary, no compositional change is introduced by the boundary. It is said to be *conservative* [Fig. 5.35(a)]. If the displacement vector does not lie in the boundary plane, the net effect is the same as a removal or insertion of a slab of material. If this slab has a composition which is not the same as that of the crystal, the antiphase boundary is non-conservative, and its presence causes a change in chemical composition. Fig. 5.35(b) illustrates such a case where the extra slab is a row of black atoms.

Across a non-conservative boundary there are many more like nearest neighbours, than across a conservative boundary, and so from nearest neighbour energy interaction terms alone we would expect that a non-conservative boundary would have the higher energy.

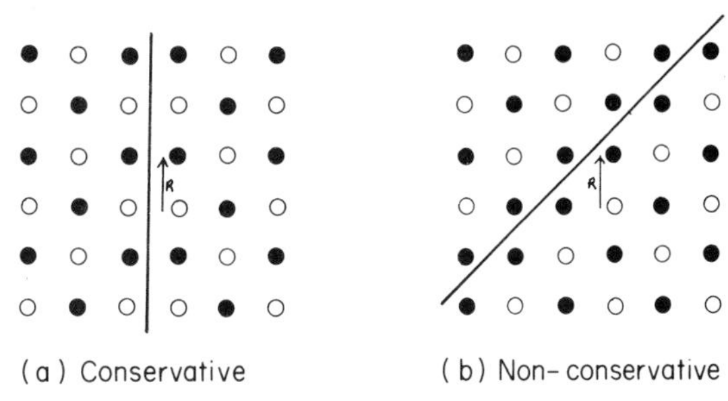

Fig. 5.35 Conservative and non-conservative antiphase boundaries.

5.5.4 NON-STOICHIOMETRY AND ORDERING

In all of the examples of ordering that have been considered so far we have assumed a bulk composition with a simple proportion of the two cations undergoing the ordering process. In solid solutions, however, there is a broad range of composition which will naturally include compositions which do not correspond to the simple stoichiometric proportions generally associated with low-temperature ordered compounds. For example, in the pyrrhotites $Fe_{1-x}S$, at high temperatures the value of x may lie anywhere between 0 and 0.125.

On cooling, the behaviour will be governed both by the ΔH of mixing in the solid solution and the relative kinetics of the possible processes. In general terms, the non-stoichiometric solid solution may either tend to form a single ordered structure or a mixture of ordered structures each with its own ordered pattern. This can be illustrated by considering the possible behaviour of a disordered phase of overall composition A_7B_5 within the binary solid solution A–B.

In Fig. 5.36 this is represented by 42 black atoms (As) and 30 white atoms (Bs) arranged on a square grid. In the disordered state (not shown) the probability of any site being occupied by a black atom is the same for all sites at 7 in 12. Although this is a grossly oversimplified model we can list five possible modes of behaviour.

1 As and Bs can separate entirely to form regions of the pure end members.

2 Ordering could occur with the excess A atoms accommodated on non-conservative antiphase boundaries. A boundary as shown in Fig. 5.35(b) changes the stoichiometry by the amount required here. If this occurred such boundaries would have to be regularly distributed throughout the ordered structure.

3 Ordering could occur with the excess A atoms exsolved to form regions of pure A coexisting with the ordered structure. This is shown in Fig. 5.36(a).

4 An ordering scheme could be available in which a single-phase ordered structure of the same bulk composition could form. An example is shown in Fig. 5.36(b), in which the extra A atoms are distributed in an ordered way throughout the structure. A characteristic of such a transformation is that in cases where ordering leads to non-equivalence of atomic sites, the unit cell will become progressively larger as the ratio of the elements deviates further from a simple fraction.

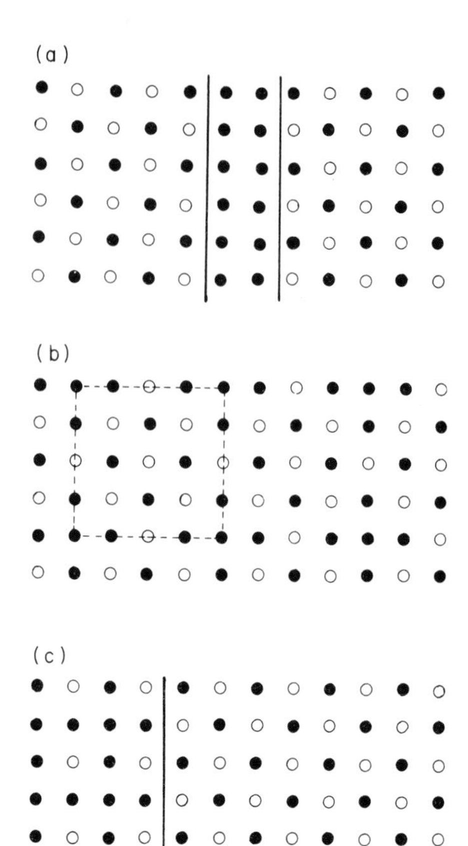

Fig. 5.36 Different ordered arrangements of black atoms (As) and white atoms (Bs) within an overall composition A_7B_5.

5 A compromise solution may be the formation of two ordered compositions which are able to achieve a 'better' ordering scheme than the original single composition. In our simple cubic arrangement we can illustrate this by a breakdown such as:

$$6A_7B_5 \longrightarrow 6A_3B + 24AB$$

as shown in Fig. 5.36(c). We have already discussed briefly the possibility of a similar situation in the intermediate pyrrhotites (section 2.6.2) where a composition such as $Fe_{10}S_{11}$ may break down into ordered regions of Fe_9S_{10} and $Fe_{11}S_{12}$, which may be more favourable energetically than attempting to order the vacancies within a single phase of composition $Fe_{10}S_{11}$.

These five possibilities are based on simple geometrical considerations, and the extent to which they apply in any particular case is strongly dictated by the structure itself and the flexibility available for different ordering schemes. In some structures it is geometrically possible to form an ordered compound at virtually every composition, whereas in others only a few compositions may be able to form ordered arrangements.

The question of the stability of the various arrangements is more difficult to answer. As has been pointed out, simple stoichiometries lead to fairly simple ordered superstructures. To form ordered compounds from the more complex compositions requires the formation of large superstructures, and it seems intuitively reasonable to assume that there is some relation between the size of a superstructure and its free energy. It could be argued that as a simple stoichiometric ordered compound has a lower free energy than a disordered compound at low temperatures, and as the successively larger superstructures required for a deviation from stoichiometry reduce the 'degree of order', then the difference in free energy between the disordered compound and the superstructure decreases as the size of the superstructure increases. We will assume on this basis that possibility 4 in Fig. 5.36(b) results in a higher energy structure than 5 in Fig. 5.36(c). The theme of the stability of ordered structures will be taken up again in section 5.5.7.

5.5.5 DEGREES OF ORDER

We have used the term 'degree of order' in a fairly loose way both in relation to configurational entropy and in reference to the general notion that whenever two or more types of atoms are distributed in a structure there will be some arrangements which are 'more ordered' than others. We will now outline some ways of describing the degree of order in a structure.

It is easy to recognize a fully ordered structure such as that in Fig. 5.37(a), because all of the atoms are in the 'right' place. In the fully ordered form there are two types of sites, α sites containing all N of the black atoms and β sites containing all N of the white atoms. If we begin to interchange atoms such that there are some

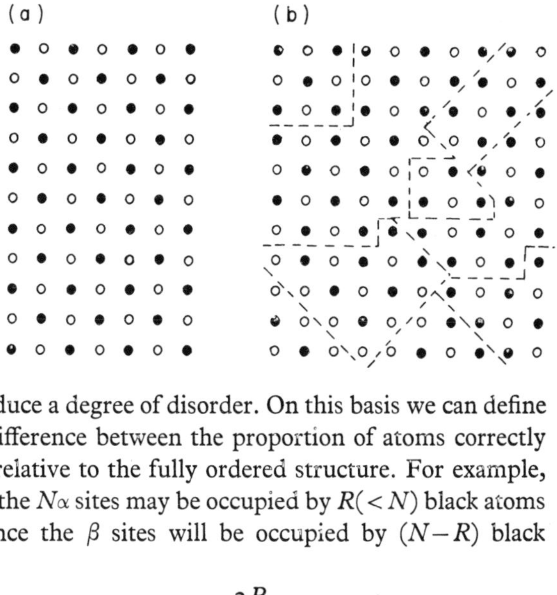

(a) (b)

Fig. 5.37 (a) A fully ordered structure. (b) Ordered domains separated by antiphase boundaries resulting in a high degree of short-range order but very little long-range order.

black atoms on β sites we introduce a degree of disorder. On this basis we can define an order coefficient S as the difference between the proportion of atoms correctly placed and incorrectly placed relative to the fully ordered structure. For example, in a partially ordered structure the $N\alpha$ sites may be occupied by $R(<N)$ black atoms and $(N-R)$ white atoms. Hence the β sites will be occupied by $(N-R)$ black atoms and R white atoms.

The proportion of atoms correctly placed is then $\dfrac{2R}{2N}$

The proportion of atoms incorrectly placed is $\dfrac{2(N-R)}{2N}$

Therefore

$$S = \frac{2R}{2N} - \frac{2(N-R)}{2N}$$

$$= \frac{2R - N}{N} \tag{5.5}$$

For complete order, $R = N$ so that $S = 1$. For complete disorder $R = N/2$ (i.e. there is an equal probability of finding either a black or a white atom on any site), and so $S = 0$. S is called the *long-range order coefficient* because the occupancy over all sites in the crystal is being considered.

If we look at the distribution in Fig. 3.37(b), however, we see that it is completely ordered in small domains separated by antiphase boundaries. Taken over the whole crystal our definition of long-range order would give a low-order coefficient because across each boundary our site labelling of 'right' and 'wrong' positions change places. In this case we have a high degree of *short-range order*, but very little long-range order. A coefficient of short-range order (σ) has been defined which is based on the proportion of atoms which have the correct nearest neighbours, so that the only disorder appears at the domain boundaries. Complications arise however as it is not necessary for the order within the domains to be perfect.

In some minerals there are more than two species to be ordered. For example, the solid solutions in the copper–iron sulphides have copper atoms, iron atoms and vacancies disordered over the tetrahedral sites in the close-packed sulphur structure. We can envisage a situation where, on cooling, the vacancies become ordered while the copper and iron atoms remain disordered over the occupied sites. This is a state of *incomplete order*. Complete order would be attained when vacancies, copper atoms and iron atoms were all ordered.

5.5.6 CONTINUOUS AND DISCONTINUOUS ORDERING PROCESSES

The discussion at the beginning of this chapter on the differences between continuous and discontinuous processes applies to ordering transformations in the same way as it does to those involving phase separation. An important point made in the discussion was that at temperatures above the transformation temperature random chemical fluctuations existed throughout the crystal. At lower temperatures there were circumstances under which such fluctuations could develop continuously (spinodal decomposition) while in different circumstances a certain critical sized fluctuation had to develop before the transformation could proceed (nucleation). The same concept applies to ordering transformations. Above the disordering temperature long-range order is absent, but short-range order persists in terms of random order fluctuations.

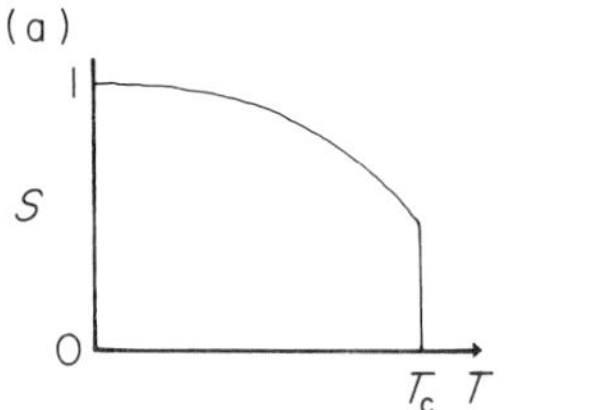

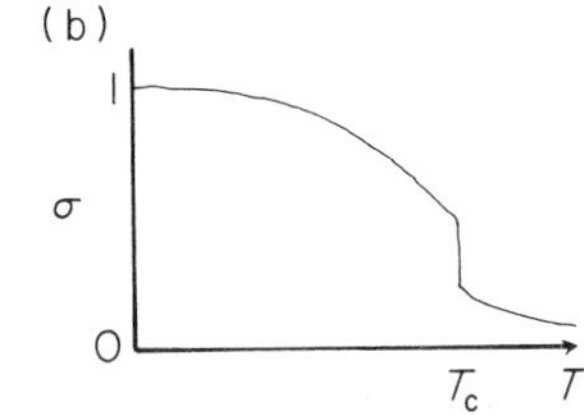

Fig. 5.38 The change in (a) long-range order and (b) short-range order as a function of temperature for a discontinuous transformation.

When the ordered form has a structure significantly different from the disordered parent a transformation to the ordered form must take place by a discontinuous process. Under equilibrium conditions local regions of highly ordered material, i.e. nuclei, must develop. This is shown in Fig. 5.38 in terms of the change in long- and short-range order as a function of temperature. Such a nucleation process would be subject to all of the constraints we have already mentioned for such processes, and therefore under relatively rapid cooling conditions the nucleation event may be impeded.

Under non-equilibrium conditions (i.e. at a greater degree of undercooling) the disordered form may become unstable with respect to some periodic order fluctuations which develop in much the same way as compositional fluctuations in spinodal decomposition. The term *spinodal ordering* is applied to such a case. Fig. 5.39 illustrates this schematically by a plot of the free-energy change as a function of the degree of order for a number of temperatures. Above the ordering temperature T_c any degree of order increases the free energy of the system. At temperatures below T_c but above T_s (the temperature below which continuous ordering can develop) a free energy of activation exists, corresponding to a certain critical degree of order (i.e. a nucleation event). Below T_s the degree of order can increase continuously from zero and result in a decrease in the free energy.

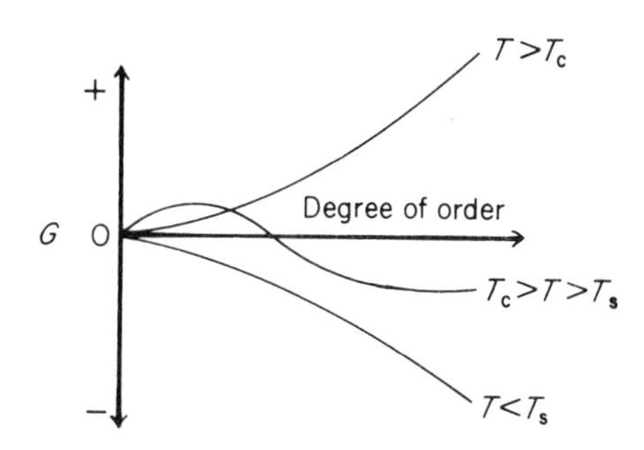

Fig. 5.39 The change in free energy as a function of the degree of order for a transformation which proceeds by a discontinuous mechanism at temperatures T where $T_c > T > T_s$, but may proceed by a continuous mechanism at $T < T_s$.

Within a periodic ordering fluctuation the peaks and valleys of the ordering wave correspond to alternative translational or orientational variants of the ordered form. In the early stages of the process no distinct interfaces exist between the alternatives and hence no antiphase or twin boundaries can be defined. If such an ordering modulation continued to develop by increasing its amplitude (i.e. the degree of order within the maxima and minima) and its wavelength, it is possible that ultimately the microstructure would be indistinguishable from one in which copious nucleation had occurred resulting in a fine twin or antiphase structure. It is likely however that an ordering modulation may become 'frozen in' as the temperature continues to decrease.

Fig. 5.40 contrasts the continuous and discontinuous ordering mechanisms. The signs + and − on the degree of order axis represent either of two variants of the ordered form. When only two variants exist, the microstructure will initially appear as a single modulation and may develop into a fine lamellar structure. Where more than two variants exist a 'tweed' texture initially forms and may coarsen into a cross-hatched domain structure. An example of a coarse cross-hatched structure is that of microcline in which the twinning arises from Si, Al-ordering (Fig. 5.41). We shall return to the problem of ordering in K feldspar, particularly in relation to continuous ordering processes, in a later chapter.

The above discussion relates to cases in which discontinuous ordering processes should occur at equilibrium, while continuous processes may operate as alternative metastable behaviour under non-equilibrium conditions. In cases where the structure of the ordered form is sufficiently similar to that of the disordered form, continuous processes may be possible from the outset. The variation in long- and short-range order as a function of temperature is shown in Fig. 5.42, which should be compared with Fig. 5.38. Continuous ordering processes may operate at any temperature below T_c as shown by the schematic free-energy curves in Fig. 5.43 (cf. Fig. 5.39).

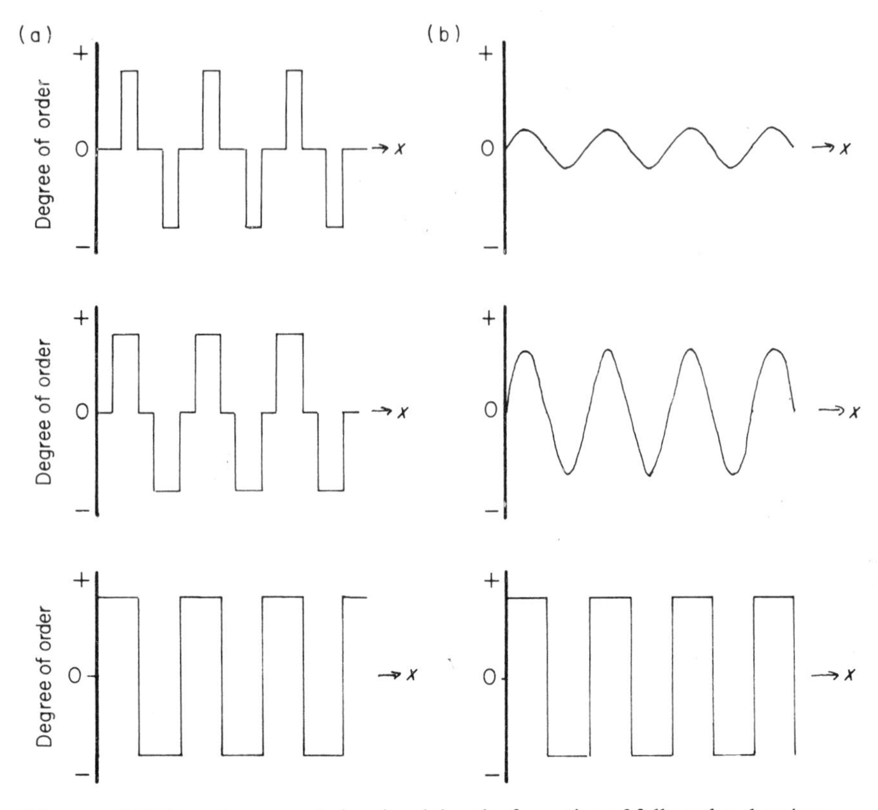

Fig. 5.40 (a) Discontinuous ordering, involving the formation of fully ordered regions within a disordered matrix. (b) Continuous ordering involving the growth of ordering modulations.

Fig. 5.41 Optical micrograph of a microcline crystal showing the typical cross-hatched twinning which is due to the two orientational variants which are possible in the formation of the Si, Al-ordered form. The width of the photograph is about 0.3 mm. (Photo courtesy R.H. Colston).

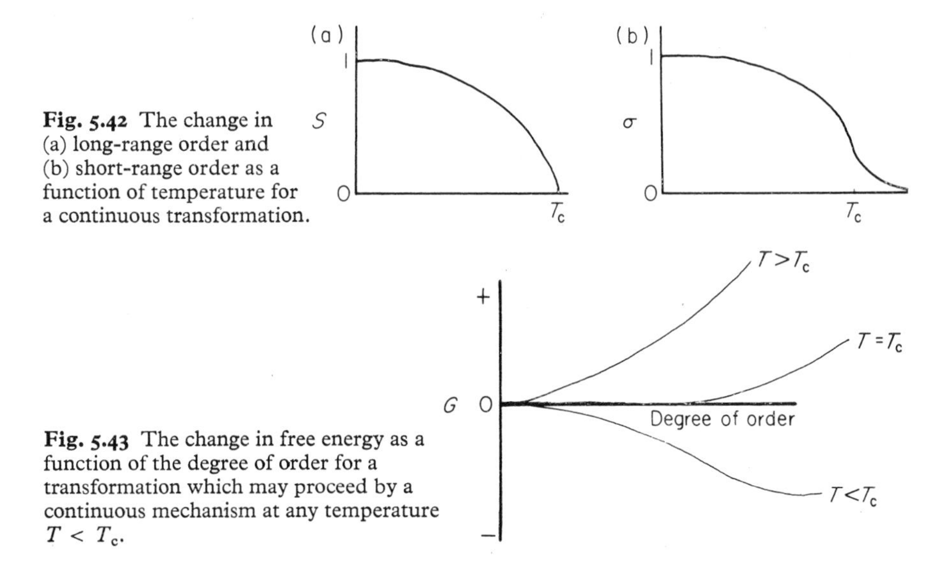

Fig. 5.42 The change in (a) long-range order and (b) short-range order as a function of temperature for a continuous transformation.

Fig. 5.43 The change in free energy as a function of the degree of order for a transformation which may proceed by a continuous mechanism at any temperature $T < T_c$.

5.5.7 ORDERING AS ALTERNATIVE METASTABLE BEHAVIOUR

Throughout our discussions of ordering and exsolution we have assumed that our simple thermodynamic model (section 4.4), involving the sign of the ΔH of mixing, defined which of the two processes would operate. The model has been quite successful in dealing with equilibrium situations where the first step in whichever process operates is to decide whether or not atom A will tend to prefer atom B as a nearest neighbour. Under grossly non-equilibrium conditions we find that such a simple nearest neighbour model becomes totally inadequate in predicting the likely behaviour of solid solutions on cooling, principally because it deals only with nearest neighbours and thus ignores the overall distribution of the elements in the system. Any distribution is associated with a certain free-energy curve, and under non-equilibrium conditions, where we are not dealing only with stable states, all such configurations of the system and the relative positions of their free energy minima must be considered.*

Consequently the distinctions between ordering and exsolution behaviour are no longer simple and well defined as we enter the non-equilibrium region. The further we move into non-equilibrium (i.e. the greater the undercooling), the higher is the free energy of the disordered or high-temperature form relative to that of the stable products. As the temperature falls, it becomes increasingly likely in a complex structure, that the free-energy curves of other metastable states will drop below that of the disordered form. These alternative states will almost invariably involve some ordered configuration of the elements. If the transformation to one of these ordered forms is kinetically possible, it will take place and so reduce the free energy of the system.

We will illustrate the general idea with a schematic example in which a solid solution phase should, under equilibrium conditions, break down to the end member phases A and B. We have assumed structural changes in both end members so

* A proper thermodynamic treatment of such a case involves the concepts of free-energy bands in k-space (reciprocal space) and is beyond the scope of this book.

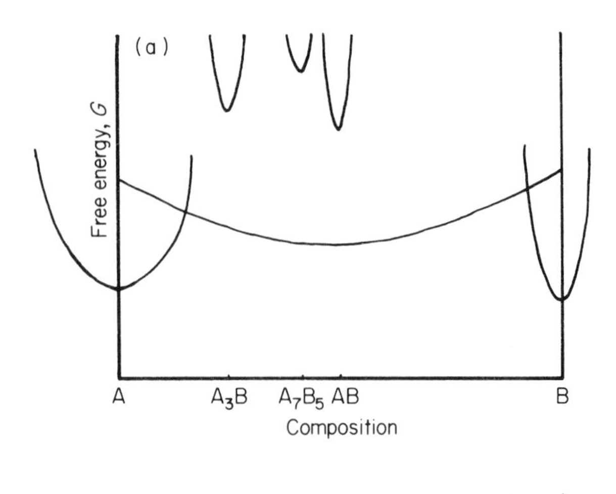

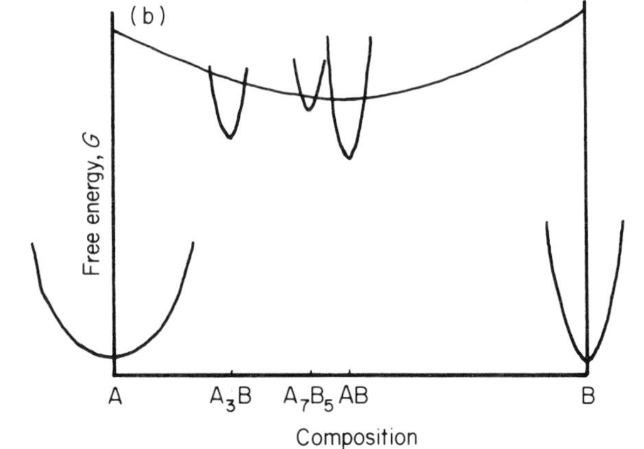

Fig. 5.44 Free energy at (a) moderately high and (b) low temperatures illustrating possible relationships between the free energies of ordered intermediate phases and end members.

that there is no longer a continuous free-energy curve between them. Fig. 5.44(a) illustrates the situation at near equilibrium temperatures. The tangent to the two minima at A and B lies below that of the solid solution, but we will assume that the activation energy of the transformation is high, and that on continued cooling the disordered state will be retained. Note that there are other free energy curves associated with ordered forms around compositions AB, A_3B and A_7B_5, but that at this stage they lie above the solid solution curve. This example has been deliberately oversimplified by choosing three compositions—in fact a large number of possible ordering curves may exist.

As the temperature falls, a stage is reached where the ordering curves drop below the solid solution curve [Fig. 5.44(b)]. We have now entered the regime where we have the possibly surprising result that an ordered (or partially ordered) form may occur as a metastable alternative to exsolution. When we consider that the ordering process will generally involve little structural change and relatively small diffusion distances, it can be concluded that kinetically it is more likely to take place than exsolution under these circumstances. From the relative positions of free-energy curves we can envisage a complex sequence of transformations taking place in a solid solution of composition A_7B_5. Initially a single ordered form may develop at the same composition, subsequently to break down into A_3B–AB. Ultimately other more stable states may be formed, perhaps eventually leading to the formation of the stable equilibrium assemblage. It must be remembered that each step in such a process further reduces the free-energy drive to the most stable state.

While we will see examples of such grossly metastable behaviour in a number of minerals, we must emphasize that diagrams such as Fig. 5.44 must not be taken too literally, nor must ordering as an alternative to exsolution be assumed to be some general rule. The general rule which can be accepted is that a disordered or high-temperature state will, on cooling, attempt to do something to reduce its free energy. Whatever this may be will depend on the structure itself. Very simple structures may have no alternatives available to the ideal transformation, but in general the more complex the structure the more likely it is that other metastable configurations will exist and that at least one of these will be a kinetically feasible alternative. The same general argument applies to any solid-state reaction taking place under metastable conditions.

5.6 Diffusion

All of the processes we have described in this chapter depend in some way on the movement of atoms from one site in a mineral structure to another. These movements take place as a result of thermal oscillations which may occasionally have amplitude sufficient to cause an atom to jump from one site to a new adjoining site. In this way random fluctuations or clusters may exist throughout the structure especially at high temperatures. If in a disordered solid solution there is no free-energy difference between various random configurations, such clusters will not persist. As soon as there is some other state more stable than the disordered one certain configurations of clusters will become more stable than others, i.e. there will be some potential gradients set up and hence a greater tendency for atoms to jump in certain ways than in others. The net result is a macroscopic flow of atoms down the potential gradient. Such a statistical drift is ultimately responsible for the phase changes we observe.

We have already noted in Chapter 1 that there are two ways of describing diffusion processes. The first is concerned with the nature of the atomic jumps (atomistic approach) whereas the second deals with the measurable parameters such as mass transport (phenomenological approach). In order to appreciate some of the factors which affect diffusion rates, our first step is to develop a simple model which relates the two descriptions.

5.6.1 ATOMIC FLUX, FICK'S LAWS AND D

Consider in Fig. 5.45 a sequence of atomic planes in a structure, and assume that a given atom may jump with a frequency Γ from one plane to another, with jumps in

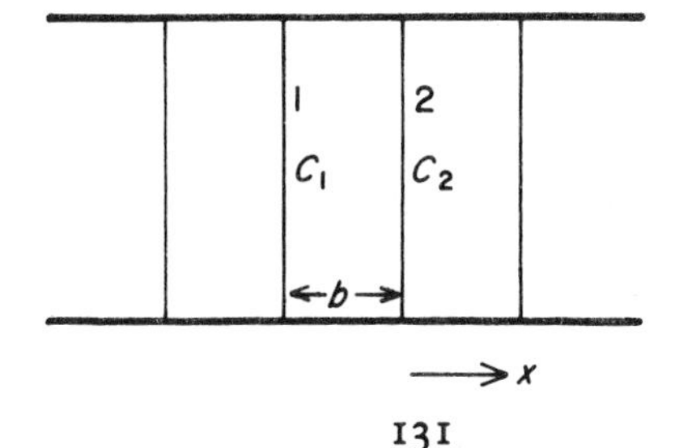

Fig. 5.45 Atomic planes 1 and 2 with atomic concentrations C_1 and C_2 respectively. b is the interplanar spacing.

either direction equally probable. For simplicity we will restrict ourselves to a two-dimensional situation. If a concentration gradient exists normal to the planes we can write that the concentration of atoms in plane 1 is C_1, and that in plane 2 is C_2 with $C_1 > C_2$. C is a volumetric concentration (number of atoms per cm^3). The number of atoms in plane 1 is therefore $n_1 = C_1 b$, and in plane 2, $n_2 = C_2 b$. In a small increment of time δt, the number of atoms leaving plane 1 is $n_1 \Gamma \delta t$. On average one half of these jumps are in each direction so that the number of atoms jumping from plane 1 to plane 2 is $\frac{1}{2} n_1 \Gamma \delta t$. Similarly the number of atoms jumping from plane 2 to plane 1 is $\frac{1}{2} n_2 \Gamma \delta t$. The flux J is the net flow of atoms which pass through a unit area in unit time, and is therefore given by

$$J = \tfrac{1}{2}(n_1 - n_2)\Gamma$$

Thus if $n_1 = n_2$ there is no net flux and the system is homogenous. It is only when some inhomogenity, such as a nucleus, is stabilized that a net flux is established.

In terms of concentration

$$J = \tfrac{1}{2} b (C_1 - C_2)\Gamma$$

In practice the value of C for a given plane is not measurable, but if the concentration is measured periodically through a sample and is assumed to vary continuously with distance x, we can write that the concentration gradient

$$\frac{\partial C}{\partial x} = -\frac{(C_1 - C_2)}{b}$$

$$\therefore \qquad J = -\tfrac{1}{2} b^2 \Gamma . \frac{\partial C}{\partial x}$$

The flux J is now related to the experimentally measurable concentration gradient. The ratio of $-J/\dfrac{\partial C}{\partial x}$ is defined as the *diffusion coefficient* D (with units cm^2 s^{-1})

$$D \equiv -\frac{J}{\partial C/\partial x} (= \tfrac{1}{2} b^2 \Gamma)$$

or

$$J = -D\left(\frac{\partial C}{\partial x}\right) \qquad (5.6)$$

Equation 5.6 is known as *Fick's first law* and simply states that the rate of flow is proportional to the concentration gradient. The minus sign is required because atoms flow toward the lower concentrations. In an isotropic mineral the diffusion rate will be independent of direction and there will be only one value of D. For an anisotropic material the diffusion rate may vary by many orders of magnitude in different directions.

Fick's first law describes a steady-state situation in which the concentration at each point does not change as a function of time. More commonly we are concerned with cases where concentration does change with time. As a simple illustration we can consider the growth of a lamella rich in B atoms forming on the cooling of an A–B solid solution. As the precipitate has more B atoms in it than its surroundings, its formation will have depleted the immediate surroundings in B. Thus atoms of B will diffuse from further away to make up this depletion only to be absorbed by the growing lamella. Thus the concentration of B around the lamella decreases continuously with time until at equilibrium the gradient is zero throughout the matrix. Fig. 5.46 shows the general form of the diffusion profiles around a lamella at times t_1, t_2 and t_3 during isothermal growth.

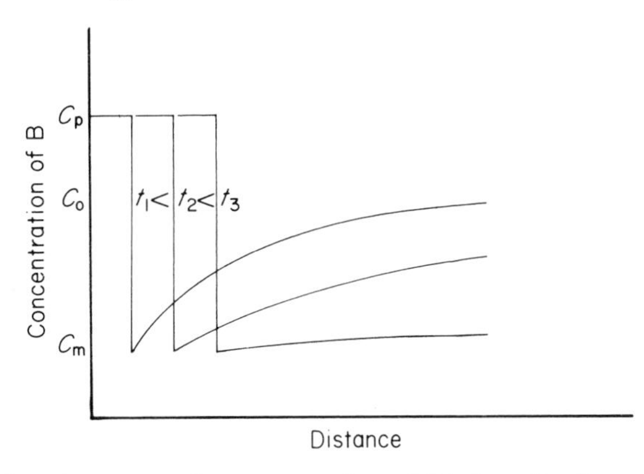

Fig. 5.46 A sequence of composition profiles around a lamella at times $t_1 < t_2 < t_3$. C_p and C_m are the equilibrium concentrations of element B in the lamella and matrix respectively; C_0 is the bulk composition of the original homogeneous solid solution.

To treat this more general situation we need a new equation which describes how C varies with both position and time. If we take a small part of the diffusion profile of Fig. 5.46 and consider two points on it x_1 and x_2 a distance Δx apart, as shown in Fig. 5.47(a), we can consider the flux due to the changing concentration gradient along the profile. The flux J_1 at x_1 coming out of the region, is higher than the flux

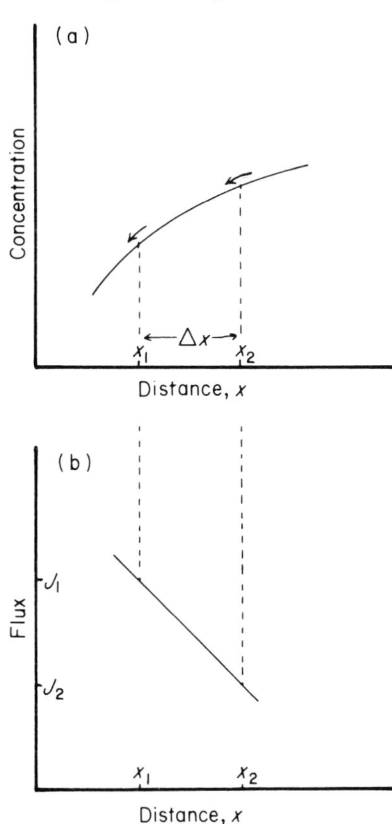

Fig. 5.47 (a) Part of the diffusion profile shown in Fig. 5.46, the arrows indicating diffusion direction of B atoms. (b) The resulting atomic flux as a function of distance x.

J_2 going into the region. This is illustrated in Fig. 5.47(b). If there are more atoms of B going out of the region than are coming in, the concentration of B between x_1 and x_2 must decrease. The decrease in the number of atoms of B in this volume in a time increment δt is

$$(J_1 - J_2)A\delta t = A\Delta x \delta C$$

where A is the cross sectional area through which the flux is flowing and hence $A\Delta x$ is the volume of the region considered. Since J varies continuously with x, J_1 and J_2 are related by

$$J_2 = J_1 + \left(\frac{\delta J}{\delta x}\right) \Delta x \quad \text{for small values of } \Delta x.$$

In the limit as $\delta t \to 0$, we can combine these two equations to give

$$\left(\frac{\partial C}{\partial t}\right) = -\left(\frac{J}{\partial x}\right)$$

If we substitute the equation for Fick's first law (equation 5.6) into this equation we obtain the relationship for the change in concentration with time as a function of the concentration gradient.

$$\frac{\partial C}{\partial t} = \frac{\partial}{\partial x}\left(D\frac{\partial C}{\partial x}\right) \tag{5.7}$$

If D is assumed to be independent of the concentration and hence x we can write

$$\frac{\partial C}{\partial t} = D\frac{\partial^2 C}{\partial x^2} \tag{5.8}$$

This equation is known as *Fick's second law*. The physical interpretation of this equation is quite straightforward. The term $\partial^2 C/\partial x^2$ is the curvature of C vs x. Therefore whenever the concentration profile is concave downward in a given region, the concentration in this region decreases with time. This is the situation described by the set of diffusion profiles in Fig. 5.46.

5.6.2 SOLUTIONS TO FICK'S LAWS AND THE DETERMINATION OF D

In order to apply Fick's laws to any problem, the equations must be solved for the experimental conditions in question. The mathematical theory of processes such as spinodal decomposition and nucleation and growth consist of deriving and solving a diffusion equation modified by the thermodynamic requirements of the process (i.e. the free-energy changes involved in atomic migration). The equations can also be modified to take into account elastic and interface parameters, and in metallic systems considerable progress has been made on these lines. Once the appropriate diffusion and thermodynamic parameters are known it is often possible to predict the scale and nature of the microstructures developed by certain thermal treatments and conversely, to define the thermal history from a study of the microstructures.

In minerals the study of diffusion lags behind the qualitative aspects of transformation process theory, partly because of the extra complexity of the structures, the very slow diffusion rates and other experimental problems. A knowledge of the appropriate diffusion coefficients is often the missing link in an attempt to quantify thermal histories of minerals from their microstructures.

Here we will outline the form of the solution to Fick's equations for one important case which forms the basis for most experimental determinations of the diffusion coefficient D. This consists of setting up a diffusion couple in which a specimen with a certain elemental concentration C_1 is physically joined to another in which the concentration of this element is C_2. The nature of this diffusion couple is important in distinguishing between several different diffusion coefficients. For example, if we make a couple by joining together olivine of composition Mg_2SiO_4 with olivine Fe_2SiO_4, Mg will diffuse in one direction and Fe in the other until equilibrium is reached. One type of atom cannot move independently of the other and the movement of both species contributes to the rate of homogenization and hence to the value of D obtained from Fick's second law. A value of D obtained in this way is called the interdiffusion coefficient.

In order to determine the diffusion coefficient of Mg in pure Mg_2SiO_4 (the self-diffusion coefficient) it is necessary to set up a couple between two samples of Mg_2SiO_4 in which one of the samples has been prepared with an excess of the stable isotope Mg–25. Thus both samples are identical except for this isotope difference. The diffusion of the isotope into the specimen of lower concentration is then used to measure the diffusion coefficient. Strictly speaking this is the tracer diffusion coefficient (i.e. the diffusion of the isotope Mg–25) but in practice it is virtually identical to the self-diffusion coefficient.

Fig. 5.47 shows the general form of the diffusion profiles obtained in such a case. As expected from Fick's second law, the concentration rises steadily in those parts where the curvature of the profile is concave upwards ($\delta^2C/\delta x^2$ is positive), falls in those parts where the curvature is concave downwards ($\delta^2C/\delta x^2$ is negative), and remains constant where the curvature is zero. Experimentally we measure the composition profile at various times and so obtain the concentration $C(x, t)$ as a function of both the distance from the boundary x, and the time t. The solution to the diffusion equation which fits the experimental conditions and describes these profiles is

$$C(x, t) = \frac{(C_1+C_2)}{2} + \frac{(C_1-C_2)}{2}\,\mathrm{erf}\left(\frac{x}{2\sqrt{Dt}}\right) \tag{5.9}$$

where 'erf' is short for 'error function' values of which are looked up in mathematical tables in the same way as 'sin' or 'log'. We are not concerned here with the details of this equation, except to point out one very important result. By calculating the distance x from the boundary at which various concentrations C will be found, we obtain the general relation that

$$x \approx (Dt)^{\frac{1}{2}}$$

This is a very significant result because it allows us, in the absence of a proper solution to Fick's second law, to estimate the distance over which diffusion is likely to operate in a given time. As the scale of the microstructures associated with exsolution processes is controlled by diffusion and cooling rate, this relationship enables simple quantitative estimates of D to be made in experimental systems where times are known and the separation of experimentally produced lamellae can be measured.

5.6.3 ATOMIC MECHANISMS OF DIFFUSION

We have already noted (Fig. 1.9) that atoms may diffuse along a number of pathways through a mineral. In the case of volume diffusion atomic jumps occur as a result of particularly violent oscillations of atoms. Here we describe a number of mechanisms by which an atom may move from one site to another in the structure. Four general types of volume diffusion mechanisms, illustrated in Fig. 5.48, have been proposed.

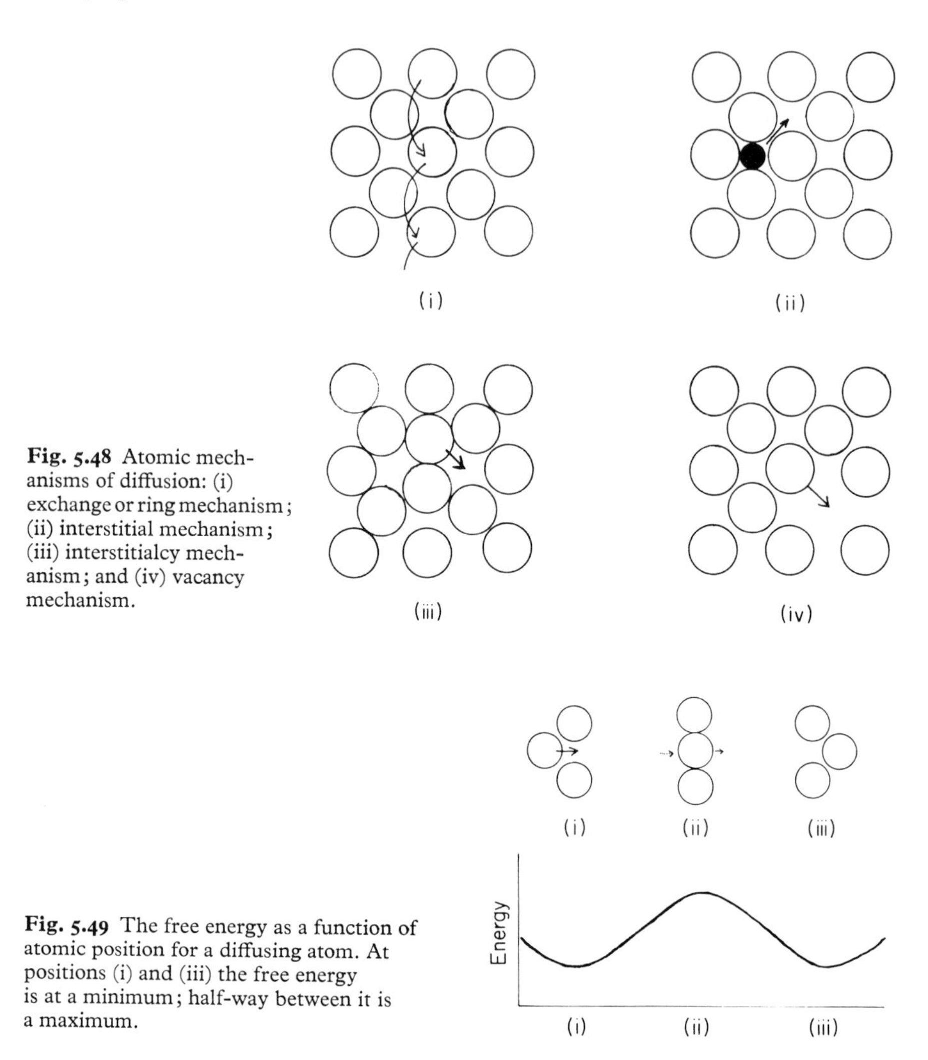

Fig. 5.48 Atomic mechanisms of diffusion: (i) exchange or ring mechanism; (ii) interstitial mechanism; (iii) interstitialcy mechanism; and (iv) vacancy mechanism.

Fig. 5.49 The free energy as a function of atomic position for a diffusing atom. At positions (i) and (iii) the free energy is at a minimum; half-way between it is a maximum.

1 The *exchange mechanism* is simply an interchange of position between two neighbours. Variations on this scheme include ring mechanisms in which three or more atoms move simultaneously on a ring and so jump to the next site on the ring. Any such mechanisms will cause large local distortions in the structure as atoms squeeze past one another. At the peak of this distortion the local energy of the structure must necessarily increase. This requires an activation energy, schematically illustrated in Fig. 5.49. A mechanism such as this would have a large activation

energy and hence would be unlikely to occur, particularly in tightly bonded structures.

2 The preceding mechanism may operate in perfect crystals. When there are imperfections other mechanisms requiring considerably less activation energy can operate. One type of imperfection is the presence of interstitial atoms. In an *interstitial mechanism*, small atoms can jump directly from one interstitial site to another without introducing much local strain. In a mineral structure where the atoms occupy specific sites and all such sites are filled, the number of possible interstitial sites may be small for all but the smallest impurity cations. If however there are more possible interstitial sites in a structure than there are interstitial atoms, appreciable diffusion may be possible by this mechanism. In many mineral structures where the anion effectively determines the volume (as in close-packed structures) cation diffusion may take place in this way.

3 The *interstitialcy mechanism* is basically similar to the previous mechanism and may occur when the diffusing atom is of a similar size to the lattice atoms. The interstitial atom moves by pushing a normal lattice atom into an interstitial site and moving into the lattice site itself. The region centred on the interstitial atom is called the interstitialcy, and during a single jump in this mechanism may move twice as far as either of the individual atoms themselves.

4 The *vacancy mechanism* is a very important type of diffusion mechanism especially in metals. It depends on the presence of point defects or vacant sites in the crystal. At temperatures above absolute zero any real crystal contains an equilibrium concentration of such defects. Although their presence always increases the internal energy of the crystal, the associated entropy increase lowers the free energy when the defect concentration is small. This is analogous to the situation in dilute binary solid solutions discussed in Chapter 4. Atoms neighbouring a vacancy can diffuse by jumping into the vacant site, thereby leaving another vacant site behind. A second atom may then jump into this vacant site, and the combined effect of such jumps results in a net diffusion of atoms. The distortion of the surrounding lattice would be relatively small, although the effectiveness of this mechanism obviously depends on the vacancy concentration.

Very little is known about diffusion mechanisms in minerals. The mechanism which operates in any particular case will depend on the type of crystal structure and the size of the migrating atom. In general a smaller ion will diffuse faster than a larger ion in the same structure. For example the diffusion of K^+ (radius 1.33 Å) in alkali feldspar is slower by about two orders of magnitude than that of Na^+ (radius 0.98 Å) at the same temperature.

Most of the research on diffusion in solids has been concerned with metals, but there is no reason to believe that the general theories will not apply also to non-metals. There are however a number of features of ionic compounds which are not found in metals. Electrostatic effects rule out a number of diffusion mechanisms which may apply to metals. An exchange or ring mechanism for example would not only involve local strains, but large electrostatic forces when positively and negatively charged atoms approached one another. Similarly, and probably more importantly, the formation of defect structures is not as simple in ionic compounds as it is in metals. Electrostatic imbalances arising from a large concentration of vacancies in one type of ion would effectively keep such concentrations low.

Whichever mechanism operates, the activation energy for volume diffusion is typically about twice that for grain boundary diffusion, which in turn may be about twice that for surface diffusion. This may be qualitatively related to the decreasing constraining influence of neighbouring atoms on the jump frequency. At low temperatures the process with the smallest activation energy is most likely to operate and therefore grain boundary diffusion will be more important than volume diffusion in a polycrystalline material under these conditions. Unless the grain size is very fine, this mechanism will only involve a small proportion of the atoms. As the temperature is increased volume diffusion becomes rapidly more important until at high temperatures it is the dominant process as it involves the bulk of the atoms in the material.

5.6.4 THE VARIATION OF D WITH TEMPERATURE

In the first part of this chapter we obtained the relationship $D = \frac{1}{2}b^2\Gamma$, where Γ was the jump frequency, and in the previous section we discussed the likelihood of any particular mechanism operating in terms of the free activation energy required to jump from one atomic site to the next. The diffusion rate is thus related to the probability that a certain number of atoms will have an energy sufficient to surmount this energy barrier ΔG_a, and make the chosen jump. Problems of this kind are solved by statistical thermodynamics. The value of ΔG_a will be much larger than the mean thermal energy of the atoms which may be expressed as RT (where R is the gas constant). Therefore the probability that such a large fluctuation in energy occurs will depend on the ratio $\Delta G_a/RT$. Boltzmann has shown that the probability of finding a given atom with an energy ΔG_a is proportional to $\exp\left(-\frac{\Delta G_a}{RT}\right)$. If the mean vibrational frequency of the atoms is ν, then the jump frequency Γ may be expressed as

$$\Gamma = \nu \exp\left(-\frac{\Delta G_a}{RT}\right)$$

$$\therefore \quad D = \tfrac{1}{2}b^2\nu \exp\left(-\frac{\Delta G_a}{RT}\right)$$

or

$$D = \text{Constant} \times \exp\left(-\frac{\Delta G_a}{RT}\right) \tag{5.10}$$

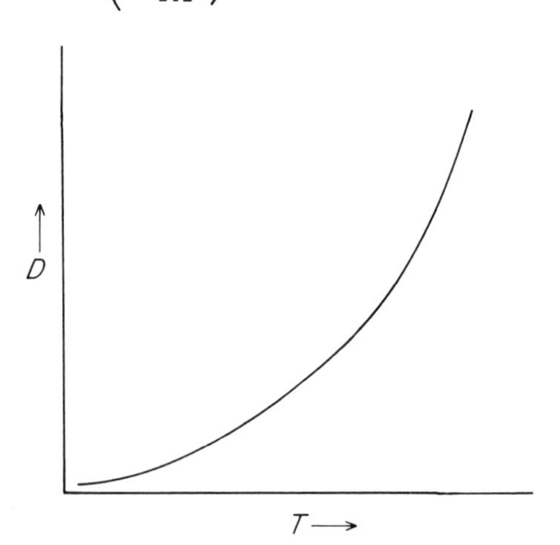

Fig. 5.50 Variation of the diffusion coefficient D with temperature T.

This equation shows the very strong temperature dependence of D, also shown schematically in Fig. 5.50. This exponential relationship has a profound effect on the rates of diffusion controlled processes as the temperature decreases, a theme which will be explored in the next chapter.

References and additional reading

Chadwick, G.A. (1972) *Metallography of Phase Transformations.* Butterworths.

Champness, P.E. and Lorimer, G.W. (1974) A direct lattice resolution study of precipitation (exsolution) in orthopyroxene. *Phil. Mag.* **30**, 357.

Champness, P.E. and Lorimer, G.W. (1976) Exsolution in Silicates. In *Electron Microscopy in Mineralogy.* (Wenk, H.R., Ed.) Springer Verlag.

Heuer, A.H. and Nord, G.L. Jnr (1976) Polymorphic phase transitions in Minerals. In *Electron Microscopy in Mineralogy.* (Wenk, H.R., Ed.) Springer Verlag.

Manning, J.R. (1974) Diffusion Kinetics and Mechanisms in simple crystals. In *Geochemical Transport and Kinetics.* Carnegie Inst., Washington.

Shewmon, P.G. (1969) *Transformations in Metals.* McGraw-Hill.

van Tendeloo, G., van Landuyt, J. and Amelinckx, S. (1976). The α–β-phase transition in quartz and $AlPO_4$ as studied by electron microscopy and diffraction. *Phys. Stat. Solidi* (*a*) **33**, 723.

Verma, A.R. and Krishna, P. (1966) Polymorphism and polytypism in crystals. Wiley and Sons.

Yund, R.A. and McCallister, R.H. (1970) Kinetics and mechanisms of exsolution. *Chem. Geol.* **6**, 5.

6

Kinetics

The basis on which the kinetics of processes can be discussed has been outlined in the last two chapters. The three most important aspects of any process from the point of view of kinetics are: the driving force for the transformation, i.e. the difference in free energy between the parent and product phases; the activation energy barrier for the transformation; and the possibility of alternative behaviour leading to the formation of metastable phases. Each of these aspects has already been discussed, and the importance of kinetics has been an underlying factor throughout. This chapter is a synthesis of many of these concepts expressed in terms of rates of transformations with emphasis on the interpretation of mineral microstructures.

Firstly we will consolidate a number of our ideas regarding activation energy and the activated state, a concept on which kinetic theory is based.

6.1 The activated state

Equilibrium thermodynamics is applicable to the beginning and the end states of a transformation and can tell us nothing about the intermediate states through which the system has to pass, nor therefore about the kinetics of the process. To overcome this difficulty we use what is termed a 'quasi-equilibrium approach', i.e. it is assumed that the intermediate configuration between the beginning and the end states can also be treated as if it has unique values of the thermodynamic functions. This intermediate configuration is called the activated state. There is nothing new in this although we are now expressing it in slightly more formal terms. Fig. 6.1 illustrates the changes in free energy involved in a transformation of this kind. The shape of

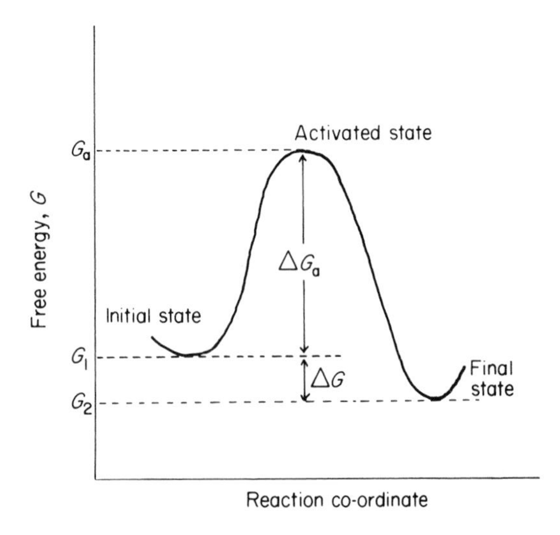

Fig. 6.1 The change in free energy involved in transforming from an initial state of free energy G_1 to a final state with free energy G_2. ΔG is the driving force for the transformation and ΔG_a is the free energy of activation.

the curve is a simple consequence of the fact that if the beginning and end states are at equilibrium (stable or metastable), their free energies must be minima, and any pathway from one to another must pass through a maximum. The height of this maximum is the free energy of activation, which is the term we have used in all the discussions involving energy barriers to processes.

The free energy of activation ΔG_a depends on the reaction pathway which is independent of the thermodynamics of the initial and final states. The pathway with the lowest ΔG_a will be the one taken. For example, the various diffusion mechanisms described in section 5.6.3 will have different values of ΔG_a; a more dramatic example is the use of catalysts to provide reaction paths of lower free activation energy. There is no direct link between the driving force and the activation free energy, although in nucleation processes they are both related to the extent of undercooling as explained in section 5.2. A greater degree of undercooling reduces the activation energy barrier and increases the driving force for nucleation. Note however that a greater undercooling also reduces the thermal energy of the atoms, which may make climbing even a reduced energy barrier a more difficult proposition. The question of the optimum undercooling for a transformation will be discussed later in this chapter.

6.1.1 ENTHALPY AND ENTROPY OF ACTIVATION

Just as we may define the enthalpy H and the entropy S and hence the Gibbs free energy $G(= H-TS)$ for the initial and final states, we can also apply the same thermodynamic functions to the activated state. The *enthalpy of activation*, ΔH_a, is therefore the difference between the enthalpy of the activated state and the enthalpy of the initial state. The curve for the enthalpy change during a reaction is much the same as that for the free energy (Fig. 6.1) except for one important difference. The enthalpy of the final state may be smaller or greater than that of the initial state, whereas the free energy of the final state must always be smaller. If the enthalpy of the final state is lower the reaction is exothermic; if it is higher the reaction is endothermic.

The *entropy of activation*, ΔS_a, is similarly the difference in entropy between the activated state and the initial state. Physically, the entropy of the activated state is associated with the number of configurations of the activated state, and all of the entropy terms described in Chapter 2 are applicable. It can also be described as the number of ways of passing from the initial to the final state. In practice there will always be more than one set of configurational changes capable of producing a given transformation.

The free energy of activation may therefore be written

$$\Delta G_a = \Delta H_a - T\Delta S_a$$

Therefore the activation free-energy barrier will decrease if the entropy of activation increases, i.e. the larger the number of reaction paths, the faster the process. An example of this is in order–disorder transformations. During the ordering process there are relatively few paths by which a disordered state may transform to an ordered state, compared to the large number of ways possible for the reverse reaction. This is one of the reasons why disordering is a faster process than ordering. Similarly, evaporation is usually faster than condensation, and crystallization is slower than melting.

Equation 5.10 which describes the temperature dependence of the diffusion coefficient is a general one for a thermally activated process, in that it is derived by considering the probability of a particular atom achieving sufficient energy to overcome an activation free-energy barrier. We may therefore write, that for any thermally activated process

$$\text{Rate of reaction} = \text{Const} \cdot e^{-G_a/RT}$$

We may substitute $G_a = H_a - TS_a$

$$\therefore \qquad \text{Rate of reaction} = C \cdot e^{S_a/R} \cdot e^{-H_a/RT}$$

The temperature independent constant $C \cdot e^{S_a/R}$ is called the frequency factor and is denoted by A. It depends on the frequency with which the reacting atom enters the activated state, and broadly speaking therefore depends on the mechanism of the process.

$$\text{Rate of reaction} = A \cdot e^{-H_a/RT} \tag{6.1}$$

This is often called the Arrhenius equation.

When we consider that the rate of ordering or exsolution processes in natural minerals as they cool, is controlled by an exponential function of this kind it is hardly surprising that metastable states are so commonly encountered. For example, the activation energy for Al, Si-ordering in albite is around 60 kcal mol^{-1}. Taking R to be 2 cal mol^{-1} $°C^{-1}$, we find that at 1000 K

$$e^{-H_a/RT} = e^{-\frac{60000}{2000}} \approx 10^{-13}$$

while at 300 K

$$e^{-H_a/RT} = e^{-\frac{60000}{600}} \approx 10^{-44}$$

Therefore the rate of the process is 10^{31} times faster at 1000 K than it is at room temperature! As the activation energy decreases the variation in the rate with temperature is not as marked. This is best seen on a plot of reaction rate against temperature:

$$\text{Rate} = A \cdot e^{-H_a/RT}$$

$$\log_{10} \text{rate} = \log A - \frac{H_a}{2.303R}\left(\frac{1}{T}\right)$$

Thus a plot of log reaction rate against $(1/T)$ will be linear with a gradient equal to $-H_a/2.303R$, and an intercept on the rate axis which is dependent on the reaction mechanism. Such a plot is known as an Arrhenius plot (Fig. 6.2). Experimentally determined reaction rates plotted in this way provide evidence for a thermally activated mechanism (if the graph is linear), and also provide values for H_a and A.

Figs 6.3(a) and (b) show the effect of high and low values of the activation energy on the slope of the Arrhenius plot. In Fig. 6.3(c) Arrhenius plots for two reactions with the same activation energy but with markedly different rates illustrate the effect of the frequency factor. This is the same as saying that the reaction with the highest entropy of activation will therefore have the lowest free energy of activation and hence proceed the fastest.

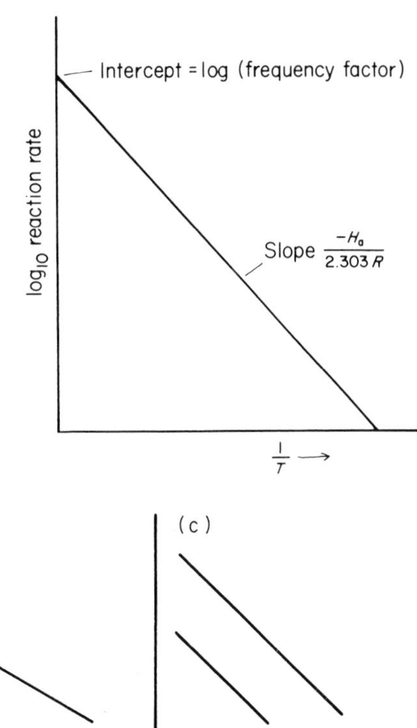

Fig. 6.2 An Arrhenius plot with log of the reaction rate plotted against $1/T$ (K^{-1}).

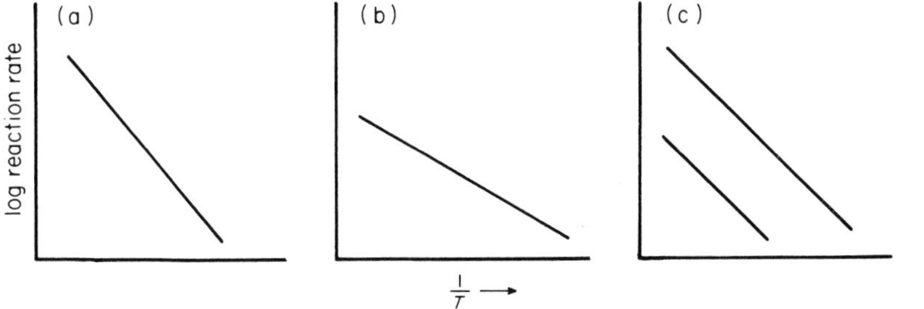

Fig. 6.3 Arrhenius plots illustrating (a) high activation energy (b) lower activation energy and (c) the same activation energy but different frequency factors leading to different reaction rates.

When applied to diffusion rates an experimental Arrhenius plot may be used to identify the temperature dependence of different diffusion mechanisms. For example, we would expect that at high temperatures volume diffusion would be the dominant mechanism, whereas at lower temperatures a mechanism with a lower activation energy (such as grain boundary diffusion) would predominate. A schematic plot of this kind is shown in Fig. 6.4.

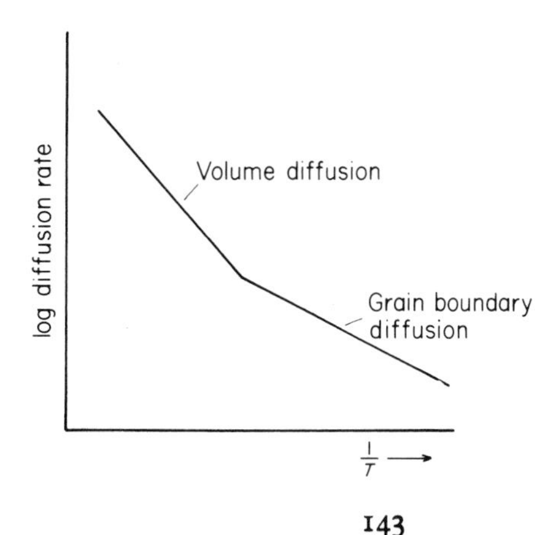

Fig. 6.4 A hypothetical Arrhenius plot showing different diffusion mechanisms (and hence different rates) at different temperatures.

6.2 The kinetics of nucleation and growth

In this section many of the points raised in sections 5.2 and 5.6 will be restated in terms of the rates of the processes. Up to this stage we have discussed nucleation of a new phase from a solid solution solely in terms of the free energy of formation of a critical nucleus, ΔG^*. The rate of nucleation is however also dependent on the rate at which material can diffuse through the matrix to form this nucleus. The first may be called the thermodynamic factor, and the second the kinetic factor. Both of these terms are temperature dependent and hence the rate will be temperature dependent. The free energy of activation ΔG^* depends on the supersaturation which is directly related to the temperature (see Fig. 5.11). At the solvus equilibrium temperature T_c, ΔG^* is infinite, but falls progressively to values approaching zero as the degree of undercooling increases. The rate equation for the formation of critical nuclei, taking only the thermodynamic factor into account is

$$R = \mathrm{c} \cdot \mathrm{e}^{-\Delta G^*/RT} \tag{6.2}$$

At the transformation temperature $\mathrm{e}^{-\Delta G^*/RT}$ is zero and increases continuously as the temperature decreases. This is shown schematically in Fig. 6.5(a).

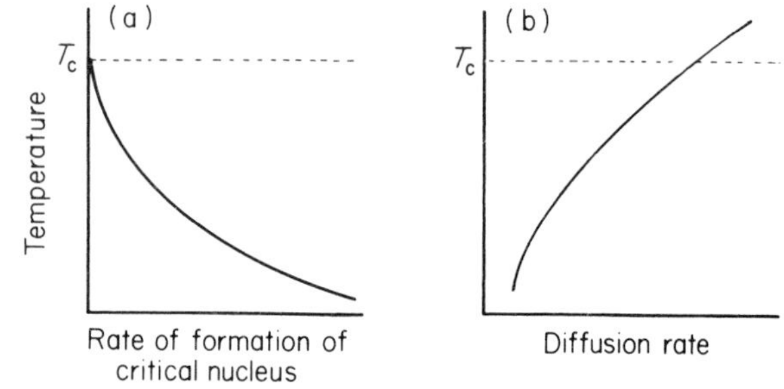

Fig. 6.5 (a) Rate of formation of a critical nucleus and (b) the diffusion rate as a function of temperature. T_c is the equilibrium transformation temperature.

To determine the nucleation rate however, this thermodynamic factor must be modified by the rate at which atoms can diffuse to the nucleation site. The temperature dependence of the diffusion rate is

$$D = D_0\, \mathrm{e}^{-H_a/RT} \tag{6.3}$$

(This equation should be compared with equation 5.10—the pre-exponential term D_0 now includes the entropy of activation term and hence H_a and not G_a is used.) The term $\mathrm{e}^{-H_a/RT}$ decreases rapidly with decreasing T, since H_a is constant for a given diffusion mechanism. This is shown in Fig. 6.5(b).

The rate of nucleation I at any temperature is the product of the thermodynamic term and the diffusion term.

$$\text{Rate of nucleation } I = \mathrm{K} \cdot \mathrm{e}^{-\Delta G^*/RT} \cdot \mathrm{e}^{-H_a/RT} \tag{6.4}$$

At temperatures near the transformation temperature T_c the diffusion rate may be sufficiently high, but the rate of nucleation is low because of the high value of ΔG^*. Thus at near equilibrium temperatures the rate of nucleation is controlled by the thermodynamics and not by the diffusion kinetics. At low temperatures, the activation free energy barrier ΔG^* is negligible and a large driving force for nucleation exists, but the low diffusion rates again result in a low nucleation rate. The rate is now kinetically controlled. At some intermediate temperature the nucleation rate increases to a maximum, essentially a compromise between the thermodynamics and the kinetics. The general form of the nucleation rate curve as a function of temperature is shown in Fig. 6.6(a). In practice it is more convenient to plot the time required for nucleation rather than the nucleation rate along the horizontal axis of such a diagram. The resulting C-shaped curve [Fig. 6.6(b)] is characteristic for all diffusion controlled nucleation processes in solids.

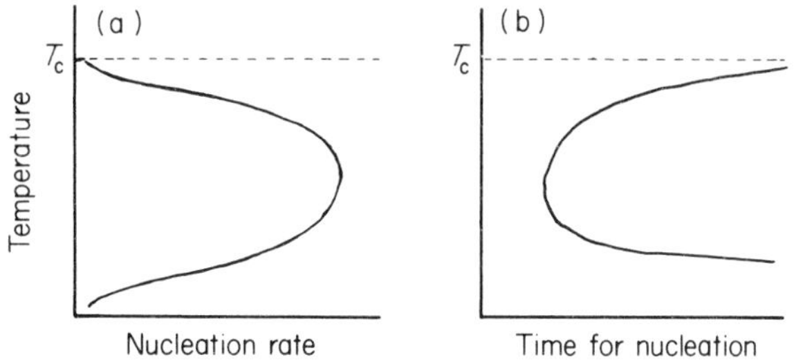

Fig. 6.6 (a) Nucleation rate and (b) the time for nucleation as a function of undercooling below the equilibrium temperature T_c.

The next stage to consider is the kinetics of the growth of the precipitates. A full analysis of growth rates is a particularly complex problem because of the number of variables involved, but it is usual to assume that the rate is mainly controlled by only one or two factors. The first is the rate at which atoms are brought to or removed from the growing particle by *diffusion*, and the second is the rate at which they can *cross the interface*. The effect of the nature of the interface on the growth has been discussed in section 5.2.4.

In the early stages interface controlled growth is likely to be the most important, as the small particles will have a considerable concentration gradient around them and so diffusion distances will be small. As the particle grows and drains solute atoms from a larger surrounding volume, the concentration gradient decreases and the driving force for diffusion is reduced. The interfacial area however, and therefore the flux of atoms across it, is meanwhile increasing. The net growth rate will be some combination of these two factors.

6.3 The Time–Temperature–Transformation (TTT) diagram

The overall rate of a phase transformation is usually displayed on a TTT diagram. This diagram describes the progress of a transformation on a graph of temperature T against time t. A logarithmic time axis is generally used because of the wide variation in times which may be involved. The shape of the curve (Fig. 6.7) is essentially the same as the C-curve in Fig. 6.6(b). In Fig. 6.7 the first curve represents the time and the temperature at which the product phase is first detectable, and the second curve the time at which the reaction is virtually complete. In many minerals where transformation times are very long, the reactions may never go to completion and only the position of the first curve is experimentally determined.

The theory on which the curve is based has been discussed in the previous section. Here we will rearrange the basic rate equation (equation 6.4) in order to show how an experimentally determined TTT curve can be used to determine both the free energy of activation for nucleation and the activation energy for diffusion.

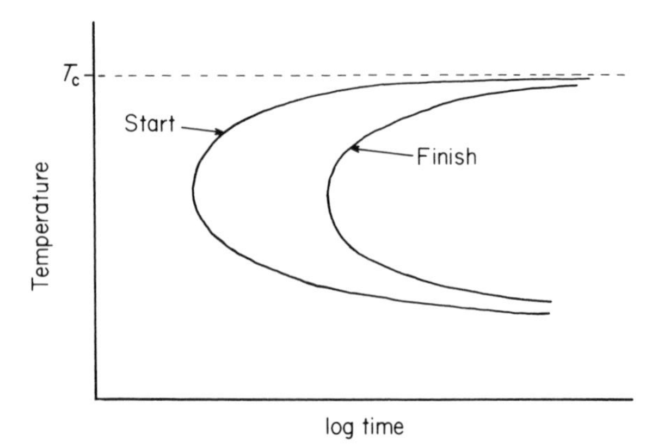

Fig. 6.7 Typical form of a time–temperature–transformation (TTT) plot for a transformation.

The first value dictates the rate of nucleation at near equilibrium temperatures, whereas the second determines the way in which this rate changes as the temperature falls.

Nucleation rate
$$I = K \cdot \exp\left(-\frac{\Delta G^*}{RT}\right)\exp\left(-\frac{H_a}{RT}\right) \tag{6.5}$$

$$\therefore \quad \ln I = \ln K - \left(\frac{\Delta G^*}{RT}\right) - \left(\frac{H_a}{RT}\right)$$

On the assumption that the time t for a given fraction to transform is inversely proportional to the nucleation rate, we may write

$$\ln t = \left(\frac{\Delta G^*}{RT}\right) + \left(\frac{H_a}{RT}\right) - \ln K' \tag{6.6}$$

Differentiating
$$\frac{d(\ln t)}{d(1/T)} = \frac{\Delta G^*}{R} + \frac{H_a}{R} + \frac{1}{RT}\left[\frac{d(\Delta G^*)}{d(1/T)}\right]$$

For all low T, $\Delta G^* \approx 0$, and

$$R\left[\frac{d(\ln t)}{d(1/T)}\right] = H_a \tag{6.7}$$

Thus if we plot ln t against $(1/T)$ as shown in Fig. 6.8, the activation energy for diffusion may be found from the slope of the linear part of the curve at lower T.

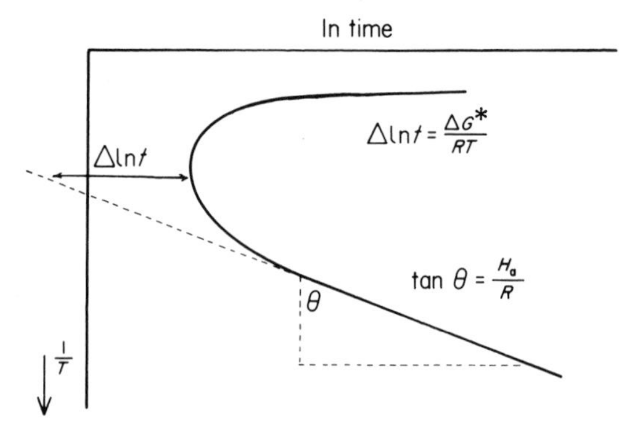

Fig. 6.8 Method of finding the activation energy for diffusion H_a and the activation energy for nucleation from a plot of ln time against $1/T$ (K^{-1}).

The linear part of this curve has the equation:

$$\ln t = \frac{H_a}{RT} - \ln K' \tag{6.8}$$

From equations 6.6 and 6.8

$$\Delta G^* = RT(\Delta \ln t).$$

This enables the free-energy barrier for nucleation to be determined at any temperature T, from the distance ($\Delta \ln t$) between the extrapolated straight line and the curve, as shown in Fig. 6.8. Note that where the curve becomes linear, the energy barrier falls to zero.

This type of situation in which the rates at near equilibrium temperatures are dominated by the thermodynamics, and at low temperatures by the kinetics, is typical of nucleation processes whether or not they involve a change in local chemical composition. Although the diffusion distances for an ordering transformation are much smaller than for exsolution, and hence the position of the curves on the time axis may differ, the overall shape will be the same.

6.3.1 TTT CURVES FOR HOMOGENEOUS AND HETEROGENEOUS NUCLEATION

In section 5.2 the differences between heterogeneous and homogeneous nucleation were discussed. Heterogeneous nucleation involves a lower free-energy barrier and so will be able to take place at relatively small values of undercooling below the equilibrium chemical solvus temperature. Homogeneous nucleation on the other hand, is highly improbable at temperatures above the coherent solvus (Fig. 5.17) due to the increased surface- and strain-energy terms. The temperature difference between the coherent and chemical spinodal is a measure of the magnitude of these terms. The TTT curves for the two nucleation mechanisms will therefore be asymptotic to different maximum temperatures. The slopes of the lower part of the two curves are likely to be similar if the bulk of the diffusion takes place by a volume diffusion mechanism in both cases. At any temperature homogeneous nucleation will be a slower process than heterogeneous nucleation, although this may be modified by the availability of suitable nucleation sites for heterogeneous nucleation. The net result of all of these factors defines the relative positions of the curves for the start of heterogeneous and homogeneous nucleation, as schematically shown in Fig. 6.9.

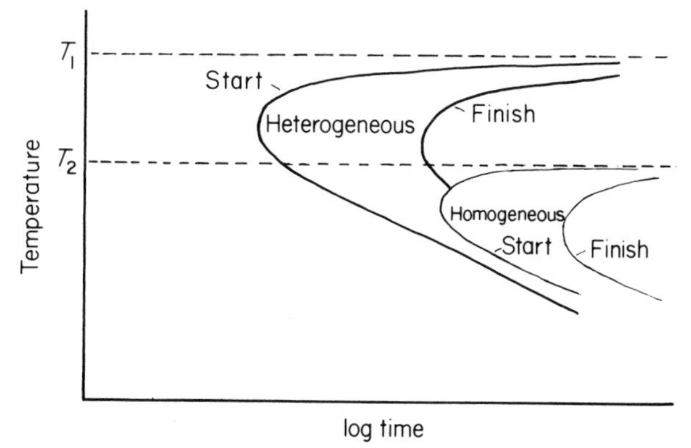

Fig. 6.9 A set of TTT curves showing the relative positions of start and finish curves for heterogeneous and homogeneous nucleation T_1 and T_2 are the equilibrium solvus and coherent solvus temperatures respectively.

It should be noted that many of the factors which define the activation free energy for heterogeneous nucleation are not entirely predictable and will depend a great deal on the precise nature of the mineral specimen (e.g. grain boundary and dislocation distributions, impurities etc.). Experiments involving the onset of heterogeneous nucleation are therefore not always reproducible. Homogeneous nucleation, while less dependent on extrinsic properties will also depend on the atomic defects which control volume diffusion and may vary with the thermal history of the specimen.

6.3.2 TTT CURVES FOR THE FORMATION OF TRANSITIONAL PHASES

By their nature (see section 5.4) the formation of transitional phases involves a greater degree of structural similarity to the parent phase and a smaller reduction in free energy, compared with the formation of the equilibrium phase. The relative free energies of GP zones, the semi-coherent second transitional phase B′, and the stable equilibrium phase A (Fig. 5.26), means that the degree of supersaturation required to form each phase will become progressively greater from A → B′ → GP zones (Fig. 5.25). Thus on the TTT plot, the upper temperature to which each curve is asymptotic decreases in the same order (Fig. 6.10).

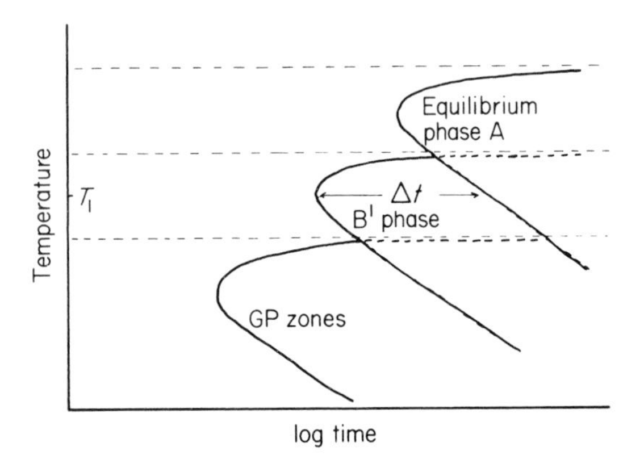

Fig. 6.10 Schematic TTT plots for the formation of equilibrium phase A, transitional phase B′ and GP zones.

At low temperatures, the free energy of activation for the nucleation of the phases decreases such that ΔG^* for A > ΔG^* for B′ > ΔG^* for GP zones. The time to start nucleating each phase therefore also decreases in this order, and is reflected by the relative positions of the noses of the TTT curves. The slopes of the low temperature part of the curves, dependent on the diffusion through the matrix, are all equal.

The schematic TTT plots in Fig. 6.10 illustrate the temperatures and times for which each phase will form. During an isothermal annealing period at temperature T_1 for example, the B′ phase will form from the solid solution first, followed after a specific time interval Δt, by the equilibrium phase A.

6.3.3 THE TTT CURVE FOR SPINODAL PROCESSES

As we have seen in section 5.3, the development of spinodal fluctuations depends on three factors. Firstly, the chemical composition must lie between the inflexion points of the G curve (i.e. where $\delta^2G/\delta C^2$ is negative). Secondly, the amplitude of the fluctuations is limited by the strain-energy term for the coexistence of regions of different composition, and finally the gradient energy term defines the limiting wavelength for which certain amplitude fluctuations are energetically possible.

The influence of all three factors combine to dictate the rate of a spinodal process as a function of the undercooling (ΔT) and the time. At temperatures just below the spinodal curve (see Fig. 5.23) only extremely long wavelength and small amplitude fluctuations are possible. These can only form very slowly due to the long diffusion distances involved. As the temperature is decreased, the increasing driving force provides the extra energy to balance the gradient-energy term (section 5.3.4) and shorter wavelength fluctuations become possible. The decreased diffusion distance increases the rate at which these fluctuations can form. At much lower temperatures the growth of even short wavelength fluctuations becomes slow due to the influence of falling temperature on the diffusion rate.

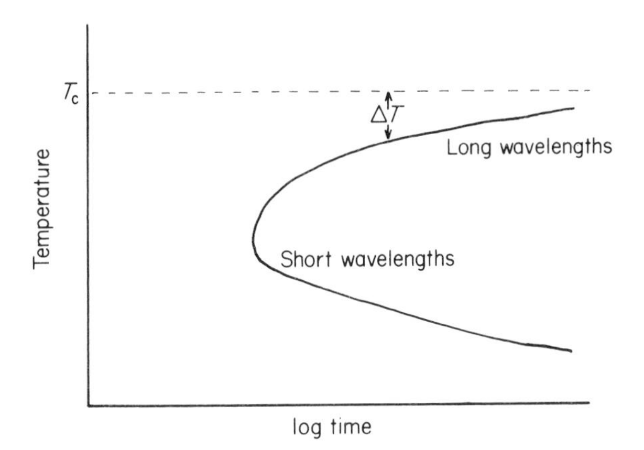

Fig. 6.11 TTT diagram for spinodal decomposition.

The general characteristics of the TTT diagram for spinodal behaviour (Fig. 6.11) are therefore similar to that for nucleation, with the gradient-energy term for the spinodal acting in a manner analogous to the activation free-energy term in a nucleation process.

6.3.4 TTT CURVES AND COOLING RATES

Experimentally, TTT curves are determined by carrying out a series of isothermal heating experiments on a solid solution or a disordered phase for various periods of time at different temperatures. By characterizing the products of each experimental run, data points are obtained which define the positions of the curves. In most geological systems of interest however, transformations do not take place isothermally but during continuous cooling, and the nature of the product formed is a function of the cooling rate. To a first approximation the TTT curve derived under continuous cooling conditions would be the same as that for isothermal conditions, but shifted to lower temperatures and to longer times. In practice, a set of TTT curves can be used to predict qualitatively the nature of the phases formed during different cooling rates by superimposing cooling curves on the TTT diagram.

This will be illustrated by describing the possible cooling behaviour of a solid solution of composition X in Fig. 6.12. The schematic phase diagram shows the curves for the chemical solvus, the coherent solvus and the coherent spinodal, thus outlining the regions in which heterogeneous nucleation, homogeneous nucleation and spinodal decomposition can occur. From the nature of these processes we can draw the schematic TTT curves for each, as shown in Fig. 6.13. Note the relative rate of the continuous spinodal process compared to the nucleation processes.

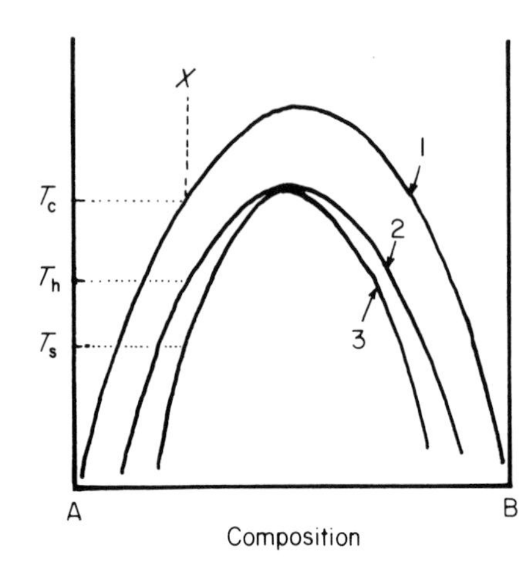

Fig. 6.12 Equilibrium phase diagram showing (1) the equilibrium solvus, (2) the coherent solvus and (3) the coherent spinodal.

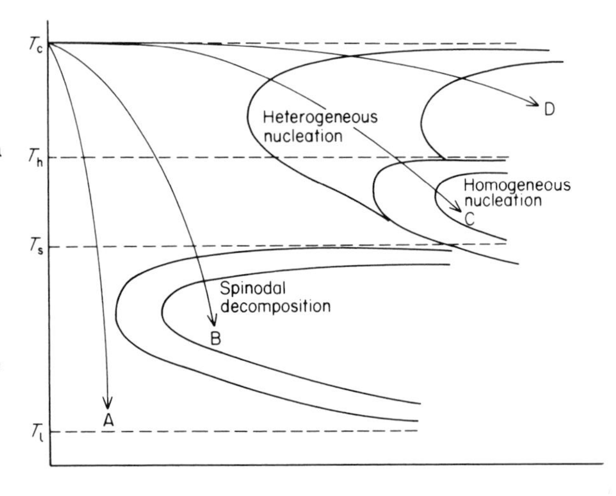

Fig. 6.13 TTT diagram showing the relative position of start and finish curves for heterogeneous nucleation, homogeneous nucleation and spinodal decomposition for the phase X in Fig. 6.12. The temperatures T_c, T_h and T_s are the temperatures at which the composition X intersects the equilibrium solvus, the coherent solvus and the coherent spinodal curves, respectively.

Four cooling curves have been superimposed on the TTT diagram. The fastest cooling rate, curve A, does not cross any of the rate curves, and hence the solid solution is quenched without decomposition being initiated. At any temperature below T_l it would remain in this state indefinitely, as T_l is the kinetic cut-off temperature for any decomposition. At a slightly slower cooling rate B, spinodal decomposition will be the only process able to operate. Spinodal modulations of this kind are not uncommonly preserved in minerals from many terrestrial and lunar rocks. At a cooling rate C, heterogeneous nucleation of the equilibrium phase will begin at relatively high temperatures, but the process will not have gone to completion before the curve for homogeneous nucleation is intersected. The rest of the exsolution will then take place by this mechanism. For very slow cooling rates, D, exsolution will be entirely by the heterogeneous nucleation of the equilibrium, phase.

6.3.5 TTT CURVES WITH RISING TEMPERATURE

TTT data are not restricted to situations involving the formation of phases on cooling. Where a polymorphic transformation to a high-temperature form takes place with rising temperature, a TTT diagram can also be drawn. For small values of ΔT above the transformation temperature, T_c, the rates of transformation will again be dominated by the overall change in the free energy for the process. Thus, near T_c the driving force will be very low and the activation energy for nucleation will be very high. Consequently, transformation rates will be very sluggish.

With increasing temperature, however, the rate of the transformation will increase and will continue to increase as the diffusion rate also increases with temperature. The general characteristics of the TTT diagram associated with rising temperature are shown in Fig. 6.14. Thus in a heating and cooling cycle over a relatively short period (as shown by the dotted line), no evidence of transformation will be observed at temperatures above T_c. Only in a long-term heating and cooling cycle is it likely that the transformation will take place under near-equilibrium conditions. This type of diagram is relevant to the behaviour of minerals during thermal and regional metamorphism.

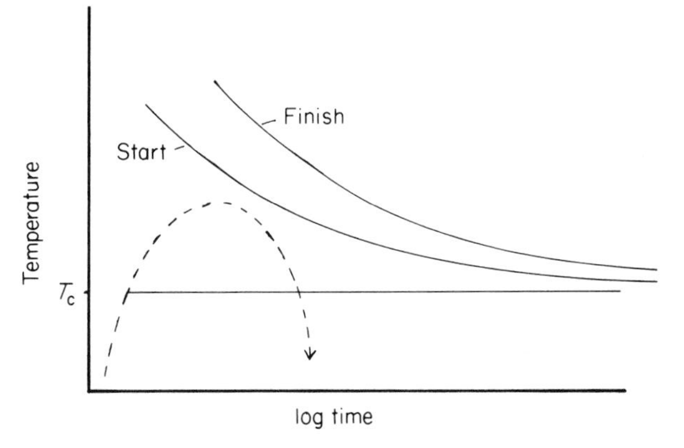

Fig. 6.14 TTT curves for the start and finish of a transformation which takes place with rising temperature. T_c is the equilibrium transformation temperature. The dotted curve is a heating and cooling cycle for which no transformation would be observed.

The kinetics of many metamorphic reactions may also be illustrated in this way. As an example we will consider the results obtained by Greenwood on the breakdown of the layer silicate talc, $Mg_3Si_4O_{10}(OH)_2$ on heating. Although not originally formulated in TTT terms, this work demonstrates the importance of kinetics in reactions, and the misleading conclusions which could be drawn if this was not recognized.

At high temperatures talc is not stable relative to the assemblage enstatite + quartz + vapour, but this decomposition reaction usually takes place via the intermediate formation of the amphibole anthophyllite $Mg_7Si_8O_{22}(OH)_2$. Over longer periods of heating, the anthophyllite ultimately breaks down to enstatite + quartz. Above about 750°C at 1 kbar pressure the stable reaction is

$$\text{talc} \longrightarrow \text{enstatite} + \text{quartz} + \text{vapour} \qquad (1)$$

The intermediate reactions are:

$$\text{talc} \longrightarrow \text{anthophyllite} + \text{quartz} + \text{vapour} \qquad (2)$$

and

$$\text{anthophyllite} \longrightarrow \text{enstatite} + \text{quartz} + \text{vapour} \qquad (3)$$

Greenwood's data for the time taken for the reactions (2) and (3) to be 50% completed is shown plotted on the TTT diagram in Fig. 6.15. Several important points are shown here. Firstly, for the reactions to occur at equilibrium temperatures

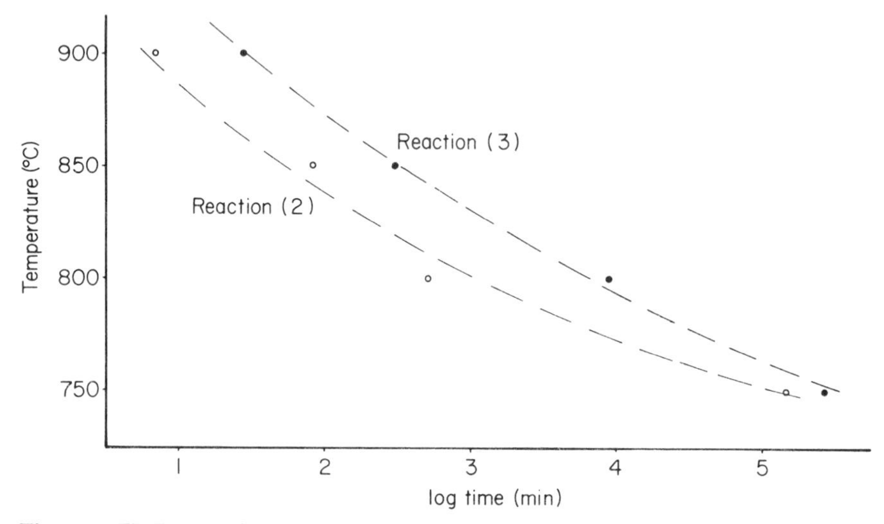

Fig. 6.15 TTT curves for the reactions:
2 talc → anthophyllite + quartz + vapour;
3 anthophyllite → enstatite + quartz + vapour.
(Data from Greenwood, 1963.)

(to which the curves are asymptotic) requires very long experimental reaction times. Nearly six months are required before reaction (3) is 50% completed at 750°C. Secondly, at temperatures above about 750°C where enstatite + quartz + vapour is clearly the more stable assemblage, anthophyllite is the first phase to form in the breakdown of talc. Short-term experiments would lead to the wrong conclusion that anthophyllite is a stable phase under these conditions. Finally, the stability field of anthophyllite should be defined by the temperature interval between the two curves at the equilibrium temperatures for the reactions. The curves in Fig. 6.15 are not conclusive on this point but clearly the stability of anthophyllite under these conditions can only be very limited. The kinetics of forming anthophyllite within any stability field it may possess would obviously be very sluggish. The problem of the stability of anthophyllite has yet to be solved.

The reason for the kinetically favourable metastable formation of anthophyllite from talc can be seen by considering the relationship between the two structures. The structure of talc is made up of sheets of silica tetrahedra which can be converted into strips of the double-chain anthophyllite structure by simply breaking some of the Si–O bonds. This is crystallographically a simpler process, and hence kinetically more favourable than the direct formation of the single-chain enstatite structure. The principles involved in these reactions are the same as those already described for the formation of metastable phases: under equilibrium conditions, where a number of possible reactions can take place, the fastest one which will usually involve a smaller reduction in free energy, will take place first.

6.3.6 THE IMPORTANCE OF THE TTT DESCRIPTION OF PROCESSES

Traditionally, the method of determining the changes of structure that occur in minerals with changes in temperature, pressure, composition etc. has been to experimentally derive an equilibrium phase diagram on which these phase relationships are displayed. Basically, the method involves making up a sample of the appropriate composition, holding it at the known temperature and/or pressure conditions for a certain period of time and then quenching to retain the mineral structure or assemblage formed The pioneering experimental work of Bowen and his colleagues on the crystallization of silicates from the melt inspired a great deal of research in the determination of phase diagrams in geologically important systems.

In the first instance most of the work was carried out at high temperatures in systems containing a liquid melt phase. Later, similar methods were applied to solid state systems with the result that phase diagrams have been published for a large number of systems. One fact which often has not been sufficiently appreciated is the vast gap that separates liquid–solid and solid–solid transformations. In general, liquid to solid transformations follow the phase changes displayed on an equilibrium phase diagram fairly closely due to the rapidity of diffusion in the molten state—but even here the glassy state rather than the crystalline state can be formed if the cooling rate is sufficiently fast (e.g. in volcanic glasses), or else if a crystalline phase forms it may be compositionally zoned if the rate of growth is faster than the supply of components from the melt by diffusion. Such departures from stable equilibrium are generally fairly obvious and easily recognizable.

For phase changes occurring in the solid state, the rate of the transformation is dependent on solid-state diffusion which may be many orders of magnitude slower than that in liquids. For solid-state transformations, therefore, *time* is a very important variable to consider. Departures from stable equilibrium are not only much more likely, but in complex minerals may be very much more subtle and difficult to recognize without an understanding of the behaviour. Alternative behaviour may produce metastable equilibrium phases which are not the stable phases appropriate to an equilibrium phase diagram, and under certain conditions, these metastable phases may be produced in preference to the stable state. Thus it is possible to obtain quite different products in any given starting composition, resulting from different transformation mechanisms, depending on the actual temperature. An illustration of this is shown in Fig. 6.10 which shows the general form of the temperature and time dependence of the formation of transitional phases.

Time-dependent transformations are best displayed on a TTT diagram, which may be thought of as a non-equilibrium phase diagram in that it represents phase changes as a function of the degree of undercooling and the time, i.e. as a function of the thermal history. Ideally, for a solid-state transformation we should consider both the equilibrium phase diagram, which indicates how a transformation or reaction will tend to proceed given an infinite period of time, and the TTT diagram which indicates how it does proceed in practice. The importance of TTT diagrams in understanding solid-state transformations cannot be overemphasized, but unfortunately, at the present time very few published TTT diagrams exist compared to the number of equilibrium phase diagrams. It is worth noting that a word of caution is often necessary when interpreting the data on so called 'equilibrium' phase diagrams. Even when equilibrium is known to have been established, the possibility that it may be metastable equilibrium cannot always be ruled out. A study of the role of kinetics in the transformation behaviour of the phases involved can ascertain the true nature of the phase relationships.

6.4 The effect of cooling rate on the scale of exsolution textures

In most minerals exsolution textures occur as the result of a single-phase solid solution breaking down due to the presence of a low-temperature solvus as shown in Fig. 6.12. The TTT curves for the possible processes which may operate allow us to define broad categories of cooling rates (Fig. 6.13) in terms of the sequence of processes taking place. In this section we will look a little closer at a single one of these categories, shown by curve D in Fig. 6.13 in which the only process is one of heterogeneous nucleation, to see the likely effect of a range of cooling rates within this category on the scale of the exsolution texture.

One of the common characteristics of exsolution textures in minerals is that in any single specimen the exsolution lamellae are of approximately equal thickness and are generally uniformly distributed throughout the grain. In slowly cooled rocks the scale of this exsolution is coarser than in rapidly cooled rocks and this fact has for a long time been used by petrologists to qualitatively describe the cooling rate. Equipped with some knowledge of nucleation and diffusion and their temperature dependence, we can now investigate how these textures could be used to quantify cooling rates.

All of the cooling rates within category D of Fig. 6.13 could be described as near-equilibrium in that they involve a relatively small amount of undercooling. At such slow cooling rates the initial nucleation event is determined very sensitively by this degree of undercooling. As we have seen in section 6.6 nucleation will not take place at any significant rate until after a certain degree of supersaturation is reached. At this stage a number of isolated nuclei will form. Around each nucleus we can define a 'sphere of influence' or diffusion distance, which is the region over which there is some depletion of solute by this nucleus. On very slow cooling and with a relatively high solvus temperature this sphere of influence [with radius $(Dt)^{\frac{1}{2}}$] will be large, and within it the concentrations of the solute will have been reduced to a value less than that for critical supersaturation. Therefore within this

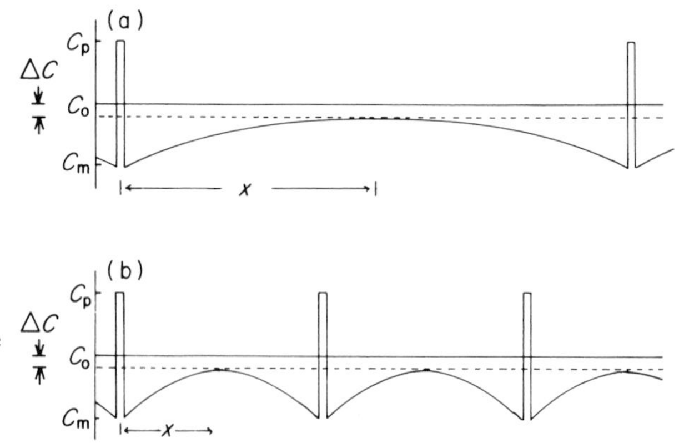

Fig. 6.16 Concentration profiles around lamellae formed during (a) slow cooling and (b) more rapid cooling. C_p and C_m are the equilibrium concentrations of the lamella and matrix, respectively, C_0 is the bulk composition of the initial solid solution, and ΔC is the critical amount of supersaturation required for nucleation.

distance no other nucleation event can take place unless the supersaturation increases. Beyond the sphere of influence other nuclei may form and will rapidly establish their own depleted region around them. The fact that no nuclei will form in the depleted region because of the sensitivity of the nucleation rate to supersaturation at small values of ΔT effectively determines the equilibrium spacing of nuclei in the very earliest stages of the exsolution process.

If the temperature is decreased slowly, the diffusion rate down the concentration gradients will be sufficiently rapid to maintain a depleted zone around each nucleus so that the supersaturation between the nuclei will not increase above the critical value. Hence no further nucleation will occur. Steady growth of the early-formed nuclei will continue under near-equilibrium conditions until the equilibrium concentrations of precipitate and matrix are reached. Fig. 6.16(a) illustrates this situation diagrammatically. ΔC is the value of the solute concentration in the matrix which corresponds to the critical degree of supersaturation required for nucleation. C_0 is the original matrix bulk composition prior to nucleation, and C_p and C_m are the equilibrium concentrations of precipitate and matrix, respectively. The fact that the values of ΔC, C_p and C_m may change with temperature does not materially affect the argument at this stage.

If the initial cooling rate were faster, the time scale for nucleation would be less and hence the diffusion distance or sphere of influence around the intitial nuclei would be smaller. Under near-equilibrium conditions the same argument would apply as in the former case and no further nucleation would occur between the early-formed nuclei. In this case, however, the separation of the nuclei would now be smaller, as shown in Fig. 6.16(b).

Where nucleation operates in this way under near-equilibrium conditions it is possible to use the scale of the initial nucleation process itself as a measure of the rate of cooling, provided that the operative diffusion constants are known. The variation in the scale of exsolution textures in the pyroxenes illustrates this point qualitatively as shown in the series of micrographs in Fig. 6.17. Under extremely slow cooling rates, relatively few widely spaced lamellae are present, whereas in rapidly cooled pyroxenes the lamellae are on an electron-microscopic scale. Although the simple relationship $x \approx (Dt)^{\frac{1}{2}}$ can be used to say that a change in one order of magnitude in the scale of the microstructure is equivalent to a change in cooling rate by two orders of magnitude, we have insufficient data on diffusion rates to make quantitative estimates of cooling rates at present.

6.4.1 STRANDED DIFFUSION PROFILES

In the above examples we assumed that after the initial near-equilibrium nucleation event, the diffusion rate was sufficiently rapid for solute atoms to be drained from the whole region between adjacent lamellae as the temperature was decreased, i.e. that the diffusion distance did not significantly change. In practice such a situation is only possible while temperatures are maintained near equilibrium. As the temperature falls below these values the diffusion distance must necessarily decrease due to the strong temperature dependence of the diffusion coefficients, particularly when the activation energy is large. As the diffusion distance becomes appreciably smaller than the initial separation of the nuclei, the solute concentration in the regions between lamellae becomes fixed and a stranded diffusion profile results.

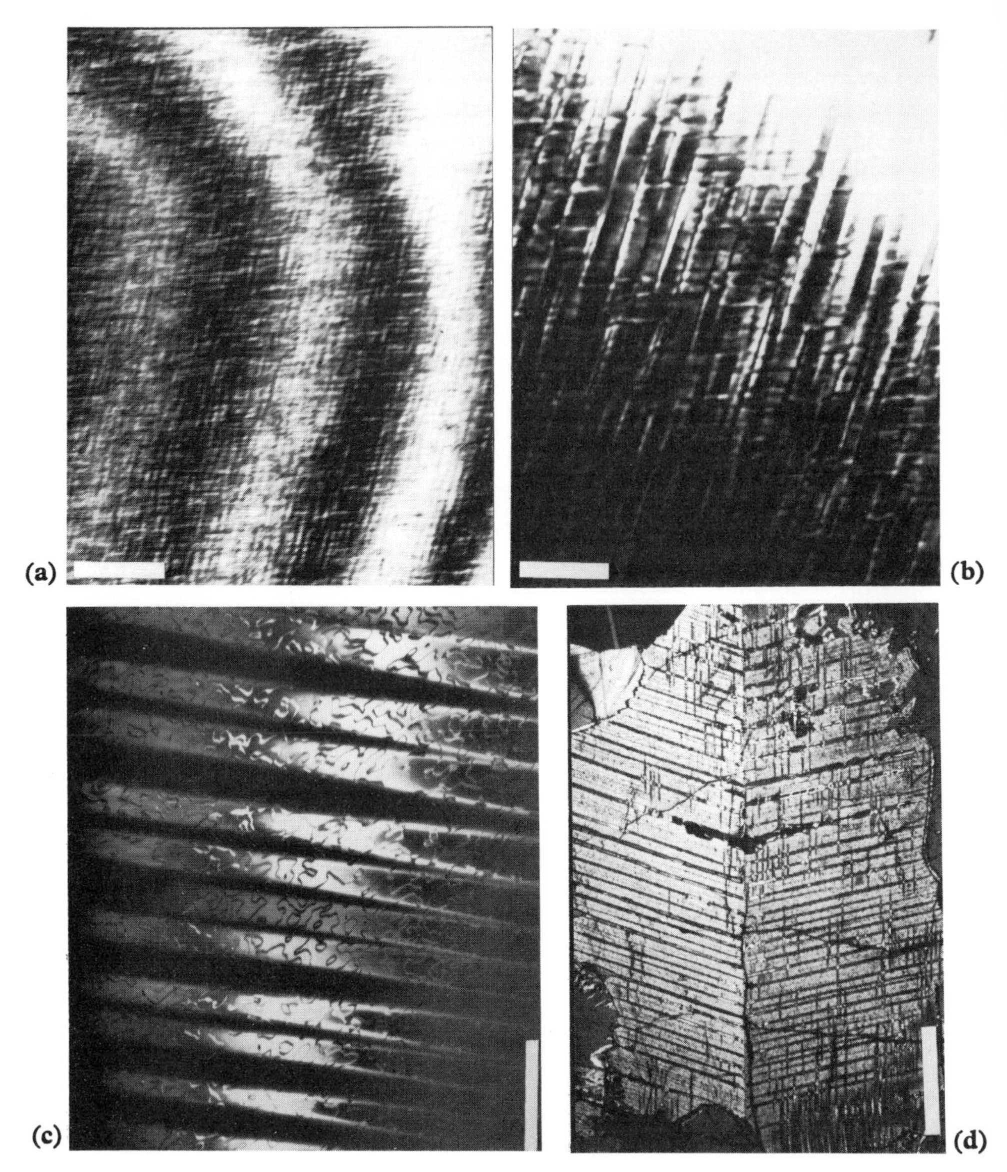

Fig. 6.17 Series of micrographs illustrating the range in the scale of exsolution from the augite–pigeonite solid solution. The scale of the exsolved lamellae will depend on both the cooling rate and the initial composition of the solid solution. (a) Fine scale spinodal 'tweed' microstructure in a lunar augite. The length of the scale bar is 0.1 μm. (b) Coarser pigeonite forming two sets of lamellae within an augite matrix. The length of the scale bar is 0.1 μm. (c) Dark field electron micrograph of pigeonite lamellae within an augite matrix. The pigeonite lamellae show antiphase domains due to the high–low transformation. The length of the scale bar is 0.5 μm. (Photos courtesy of G. L. Nord Jnr.) (d) Optical micrograph of a twinned augite crystal with exsolution lamellae of pigeonite. The length of the scale bar is 0.2 mm.

With falling temperature the regions within the stranded profile may become supersaturated with respect to exsolution, and eventually a second generation of nuclei appears. Where no defects exist this second generation of nuclei may occur by homogeneous nucleation at a greater undercooling. In cases where nucleation of

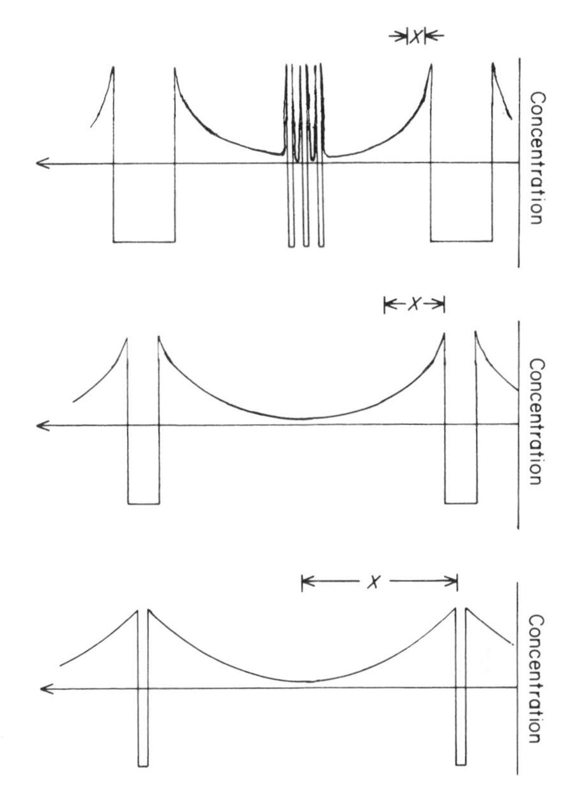

Fig. 6.18 The decreasing diffusion distance x as the temperature falls leads to a stranded diffusion profile. If the solute concentration between the lamellae exceeds a critical value for supersaturation second generation lamellae may form.

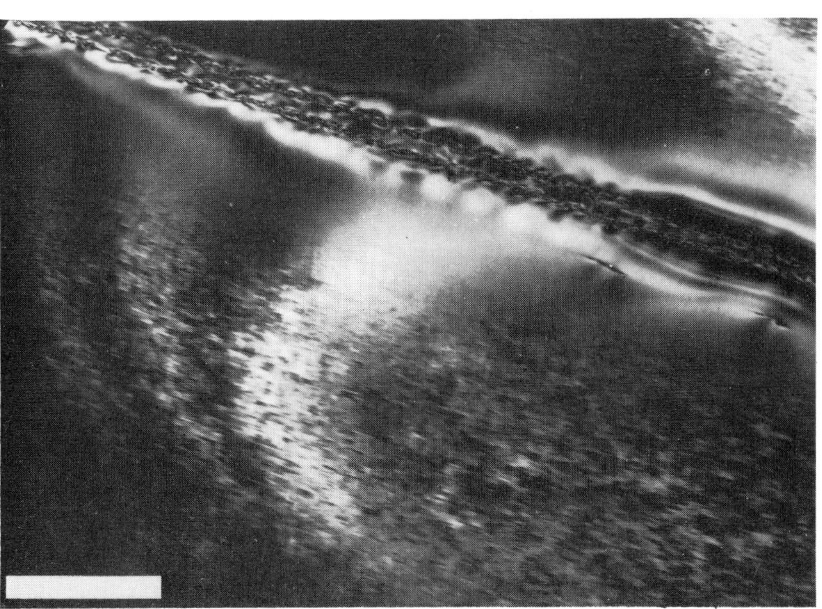

Fig. 6.19 Electron micrograph of a lamella of augite exsolved from a matrix of orthopyroxene. Subsequently, second generation precipitation has formed the fine GP zones. The precipitate free zone (PFZ) around the first formed lamella shows the region over which the Ca concentration has been depleted to a value below that required for the nucleation of the second generation precipitates. The length of the scale bar is 1.0 μm. (Photo courtesy of P. Champness.)

the stable phase is kinetically impeded, GP zones may form within the stranded profile. Typically, the second generation lamellae are separated from the initial lamellae by a precipitate free zone in which the supersaturation was insufficient for nucleation. Fig. 6.18 shows this development schematically, and Fig. 6.19 is an electron micrograph showing second generation lamellae and a precipitate free zone around exsolution lamellae of augite in orthopyroxene.

As outlined above, it is possible to use the primary scale of the microstructure, as defined by the initial nucleation event to estimate the cooling rate when the diffusion coefficients are known. The shape of the stranded diffusion profile, which develops in the later stages of cooling, can also be used to obtain information on cooling rates. The model example of this type of determination has been carried out on exsolution textures in Fe–Ni meteorites, and will be described briefly in the following section.

6.4.2 EXSOLUTION TEXTURES AND COOLING RATES IN Fe–Ni METEORITES

(a) The Fe–Ni phase diagram

Fig. 6.20 shows the subsolidus phase diagram for part of the Fe–Ni system. Above 900°C a single homogeneous iron-nickel alloy called taenite, which has a face centred cubic structure, is stable. In pure iron there is a phase transformation below this temperature to a body-centred structure, in which the solubility of nickel is considerably reduced. This body-centred cubic phase is known as kamacite. The result of the transformation is the existence of an inversion interval in the phase diagram, as we have described in section 4.5. The wide two-phase kamacite+taenite field exists over the bulk compositions of most iron meteorites.

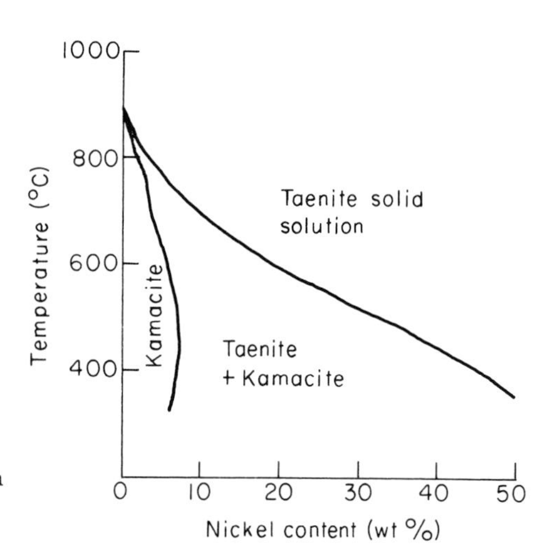

Fig. 6.20 The equilibrium phase diagram for the Fe-rich part of the Fe–Ni system below 1000°C.

Consider the behaviour of an iron meteorite with 10% Ni cooling from about 900°C. At 700°C the meteorite would enter the two-phase kamacite plus taenite field, and if cooling was sufficiently slow, kamacite would begin to exsolve from the parent taenite. The energetically most favourable orientational relationship between the phases (providing the best fit for the two structures) is when kamacite occurs as lamellae on the {111} lattice planes of the taenite. The four sets of kamacite plates

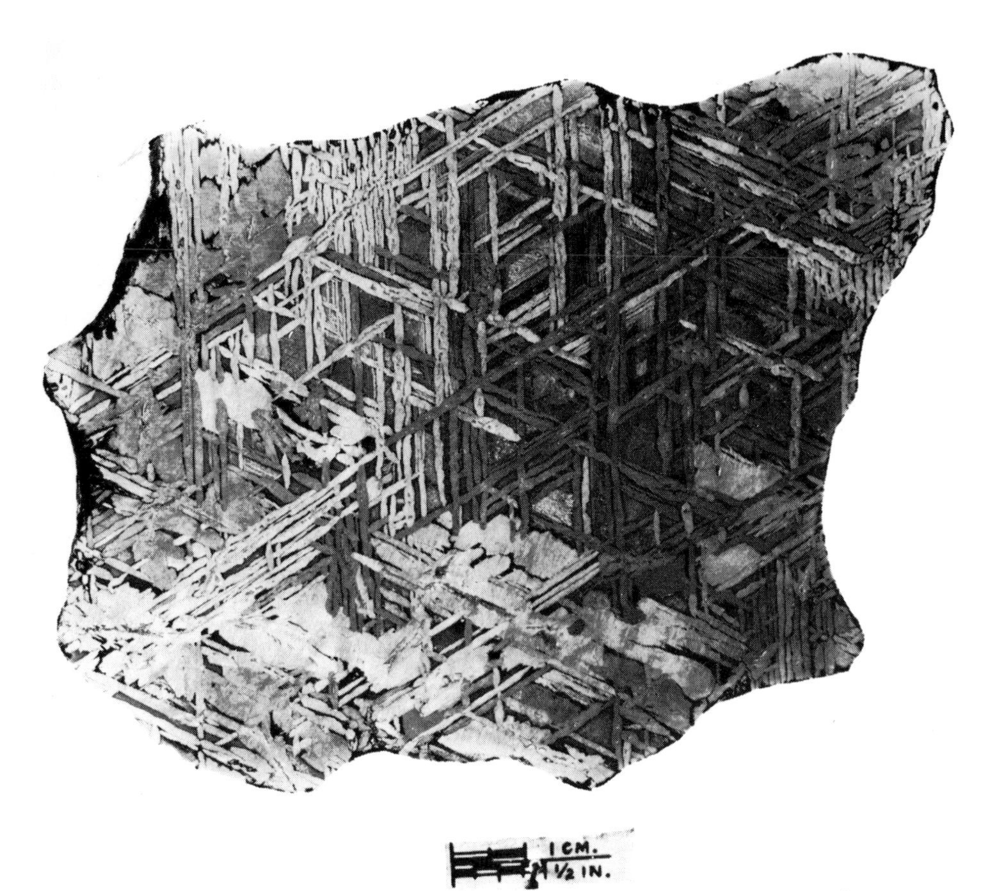

Fig. 6.21 Polish and etched section of the Waingaromia, New Zealand iron meteorite showing the Widmanstätten pattern produced by the exsolution of kamacite lamellae on {111} planes of the host taenite phase. (Courtesy of the Smithsonian Institution, Washington.)

form an octahedral pattern which is particularly striking on a polished and etched surface of an iron meteorite (Fig. 6.21). Such a structure is referred to as a Widmanstätten structure, after Count Alois de Widmanstätten who observed it in the early nineteenth century, and meteorites which show this structure are called octahedrites. Such structures cannot be reproduced experimentally due to the extremely slow cooling rates required to maintain equilibrium.

As the 10% Ni meteorite cools, the Ni content of both the kamacite and the taenite increases (Fig. 6.20) which means that the kamacite must grow at the expense of the taenite to maintain the bulk composition. The lever rule (section 4.4) can be used to estimate the equilibrium proportions of the two phases. As more taenite is converted to kamacite Ni must diffuse out into the remaining taenite matrix as required by the phase diagram. This process continues to about 500°C when the Ni content of the kamacite begins to decrease. The result of this diffusion process is a characteristic diffusion profile which becomes stranded as the temperature falls.

(*b*) *The diffusion profiles*

At high temperatures the diffusion distance for Ni is sufficiently large to maintain the equilibrium Ni concentrations in both the taenite and the kamacite regions. However as the temperature and hence the diffusion distances decrease the Ni can no longer diffuse to the centres of the taenite bands and the Ni concentration increases at the edges. As the temperature falls even further, the thickness of the equilibrated edge between the phases becomes thinner leading to the typical M-shaped diffusion

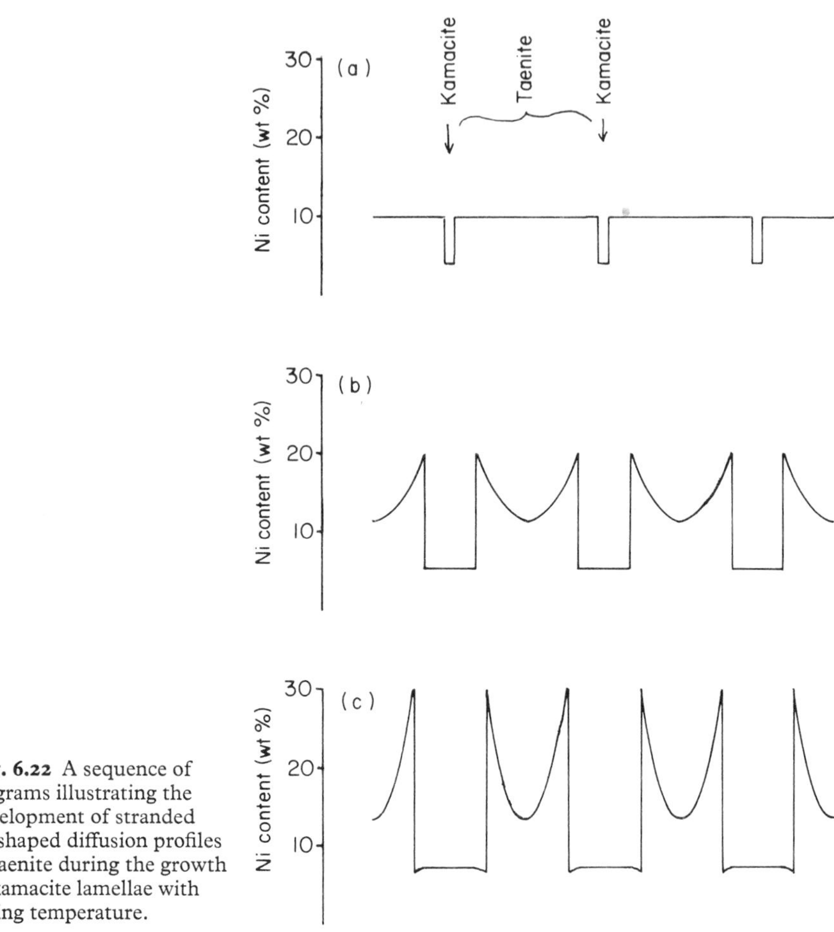

Fig. 6.22 A sequence of diagrams illustrating the development of stranded M-shaped diffusion profiles in taenite during the growth of kamacite lamellae with falling temperature.

profile within the taenite (Fig. 6.22). The very low Ni interiors of the taenite which are badly out of equilibrium at low temperatures, may eventually become sufficiently unstable to break up into a fine-grained intergrowth of kamacite + taenite termed plessite. There are a number of different morphologies possible in plessite, and it is probable that under non-equilibrium conditions different mechanisms, which do not result in the Widmanstätten pattern characteristic of nucleation and growth, may operate.

Fig. 6.22 shows a series of schematic diffusion profiles illustrating their development with falling temperature as the kamacite lamellae grow. Equilibrium is maintained within the kamacite down to relatively low temperatures as Ni diffusion is much faster in kamacite than in taenite. Below about 500°C the kamacite boundary on the phase diagram changes slope and the Ni content of the kamacite must therefore decrease. The edges of the kamacite depleted in Ni indicate that at temperatures below 500°C diffusion within the kamacite also failed to keep pace with the cooling.

The thickness of the kamacite lamellae in iron meteorites is mainly dependent on three factors. Firstly, the Ni concentration determines the amount of kamacite at any equilibration temperature, as shown by the phase diagram. Secondly, the nucleation temperature has a strong effect on the diffusion coefficients and hence on the extent to which equilibrium is maintained. For a given Ni content, material which nucleates at higher temperatures (i.e. nearer to the critical temperature) will produce coarser structures than that which nucleates at lower temperatures. This point is related to the third factor which is the cooling rate. A meteorite which spends a longer time within a certain temperature range will approach equilibrium more closely and will have thicker lamellae of kamacite than a meteorite with the same Ni content which cooled more rapidly. We have already discussed this point in section 6.4.

(*c*) *Cooling rates*

The determination of cooling rates from the microstructure of iron meteorites can be made either by an analysis of the stranded diffusion profiles or from the final kamacite plate thickness, as both are ultimately related to the same factors.

When the diffusion coefficients and their temperature and composition dependence are known, it is possible to compute the shape of stranded diffusion profiles associated with different cooling rates and nucleation temperatures. These computed profiles may then be compared with profiles measured by direct analysis of meteorites until a good match is obtained. This method has been applied to a large number of meteorites (see Wasson, 1974) and it has been found that octahedrites must have cooled very slowly, losing only 1 to 10°C every million years. Such slow cooling rates indicate that meteorites must have cooled within a shield of poorly conducting

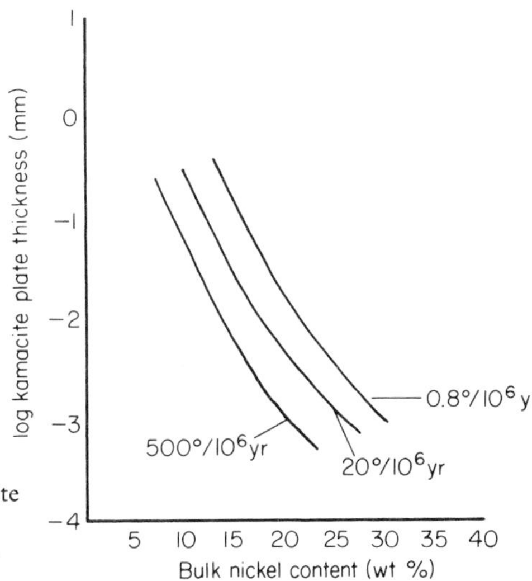

Fig. 6.23 The plate thickness of kamacite lamellae as a function of bulk nickel content for three different cooling rates. (After Goldstein and Short, 1967.)

material, presumed to be the silicate crust of larger parent bodies. Estimates, based on cooling rates and conductivities, of the size of such parent bodies vary from 70 to 200 km, which is further evidence that octahedrites may be derived from the small planets of the asteroid belt.

More rapid methods of cooling rate determination are based on the final kamacite plate thickness in a meteorite of given nickel concentration. The data obtained from the previous method can be plotted as shown in Fig. 6.23 and the cooling rate directly read from the graph. Although this method does not take into account the effect of variable nucleation temperatures on the plate thickness, its general application to problems in terrestrial minerals must be appreciated. It is to be anticipated that as more data on diffusion rates in minerals becomes available similar methods may be devised to obtain quantitative cooling rate information for microstructures such as those shown in Fig. 6.17.

References and additional reading

Burke, J. (1965) *The kinetics of phase transformations in metals.* Pergamon Press, Oxford.

Chadwick, G.A. (1972) *Metallography of phase transformations.* Butterworths.

Goldstein, J.I. and Short, J.M. (1967) The iron meteorites, their thermal history and parent bodies. *Geochim. Cosmochim. Acta* **31**, 1733.

Greenwood, H.J. (1963) The synthesis and stability of anthophyllite. *Journ. Petrology* **4**, 317.

McConnell, J.D.C. (1975) Microstructures of minerals as petrogenetic indicators. *Annual Review of Earth and Planetary Sciences* **3**, 129.

Wasson, J.T. (1974) *Meteorites.* Springer-Verlag, pp. 54–60.

7

Mineral Transformations I—Polymorphic Transformations

In the previous chapters we have discussed the thermodynamics, the mechanisms and the kinetics of the types of transformations most commonly encountered in minerals. We are now in a position to look at a few mineral transformations in more detail, both to re-emphasize some of the points already made, and to illustrate their application to mineral behaviour. No attempt will be made to cover any mineral system in full and the examples selected are those which have been fairly well studied and which serve to illustrate some particular process. The same general principles may be subsequently applied to any mineral.

To appreciate the structural and chemical background of the mineral systems we will discuss here it may be necessary to refer back to the appropriate sections in Chapters 2 and 3. The following chapters are in a sense a continuation of some of the mineral problems outlined in these earlier chapters, described within the framework of transformation process theory. In this chapter some examples of polymorphic transformations are discussed. The silica minerals have already been dealt with, and the emphasis here is on aspects of feldspar, pyroxene and sulphide mineralogy. In Chapter 8 examples of exsolution processes will be discussed, while in Chapter 9 we will describe some systems in which the behaviour is highly metastable and kinetics dominate the processes which take place. Under such conditions the distinction between polymorphic transformations and exsolution processes become blurred and the mechanism operating depends on the cooling rate.

7.1 Polymorphic transformations in the feldspar minerals

Polymorphic transformations in the end members albite $NaAlSi_3O_8$, sanidine $KAlSi_3O_8$ and anorthite $CaAl_2Si_2O_8$ are primarily involved with the ordering of Al and Si over the tetrahedra in the aluminosilicate framework. The behaviour in each case however is quite different, due both to the Si : Al ratio and the nature of the cation within the framework.

The feldspar structure has been described in section 3.6.2. The topologic symmetry (when the distortions are eliminated and the chemical contents of the atomic sites the same) of the three-dimensional network of linked tetrahedra is monoclinic at high temperatures and the tetrahedra form two sets related by symmetry. These may be labelled T_1 and T_2 as shown in Fig. 3.30. The Al, Si-distribution must under these circumstances be disordered. This structure is the starting point for the subsequent transformations which will take place on cooling.

7.1.1 TRANSFORMATIONS IN ALBITE, $NaAlSi_3O_8$

A key feature of the process of Al, Si-ordering in albite is that it takes place within a triclinic structure. Prior to ordering, a transformation from the monoclinic to the triclinic state takes place at about 1100°C. We will discuss briefly the nature and the consequences of this transformation before considering Al, Si-ordering.

(*a*) *Monoclinic–triclinic albite*

As noted in section 3.6.2 the effect of temperature on the feldspar structure is to expand the crankshaft-like chains and increase the size of the cation site. On cooling, the framework tends to collapse around the cation site. The presence of a large cation such as potassium is able to prevent this collapse and hold the framework apart in its monoclinic state. The sodium atom is too small to achieve this and below the critical temperature the framework suddenly becomes puckered around the sodium atoms whose thermal vibration is no longer sufficient to maintain the framework in its expanded state.

The vibration of the sodium atom in monoclinic albite has been described in terms of positional disorder of the type shown in Fig. 2.1. If this is so then domains with the Na atom in one of the four possible positions would exist throughout the crystal, although as yet no direct evidence for such a domain structure has been found. As a possible explanation for the existence of four possible Na positions, it has been suggested that the four kinds of domains are linked to some local order within the $AlSi_3$ tetrahedral units. Within each domain the Al may occupy a different one of the four possible tetrahedral sites and hence each Na position might then be a function of the immediate Al, Si-environment. It is worth noting that when we speak of disorder within the tetrahedral sites we do not mean a completely random 'peppering' of Al and Si atoms in the structure. In order to maintain local charge balance Al atoms are still likely to be surrounded by Si atoms, even in the disordered state. The order may exist over a very short range (down to unit cell size) with no correlation between different $AlSi_3$ units. In this sense the disordered state is different in feldspars from that in simple alloys, for example.

The transformation to the triclinic structure is a distortion which takes place too rapidly to be quenched. The temperature of the transformation seems to depend on the thermal history of the specimen studied, and hence on the degree of local Al, Si-order. The temperature falls rapidly with the substitution of K for Na, due to the fact that the larger cation hinders the structural collapse. With the addition of 15 mole % K-feldspar component the critical temperature falls from over 1000°C to around 700°C. In the monoclinic state there is a complete solid solution between $NaAlSi_3O_8$ and $KAlSi_3O_8$, although in the triclinic state where the cation site is smaller the solubility of potassium will obviously be greatly restricted.

One of the important features of the monoclinic to triclinic transformation is that, due to the symmetry change, there are four possible orientations of the distorted triclinic cell which can form from the monoclinic parent structure. The way this comes about is shown in Fig. 7.1. In the monoclinic cell the *c* and *b* axes are perpendicular and lie parallel to a mirror plane and a diad axis, respectively. In the transformation to the triclinic structure the cell distorts and both mirror plane and diad axis are lost. The distortion is equally likely to be either to the 'left' or the 'right'. During nucleation, therefore, regions of these two different configurations are produced and the boundary between them is a twin plane. Now there are two different ways of relating these two regions, i.e. two different types of twin planes. If the regions are related such that they share a common *c* direction, the twin plane being the old mirror plane of the monoclinic structure, the twin is termed an albite twin. If the *b* direction is common in the two regions and the twin plane is parallel to the old diad axis, a pericline twin is formed. As can be seen from Fig. 7.1 a slight rotation is necessary to relate the regions in a pericline twin to those in the albite twin, and hence there are a total of four different orientations possible.

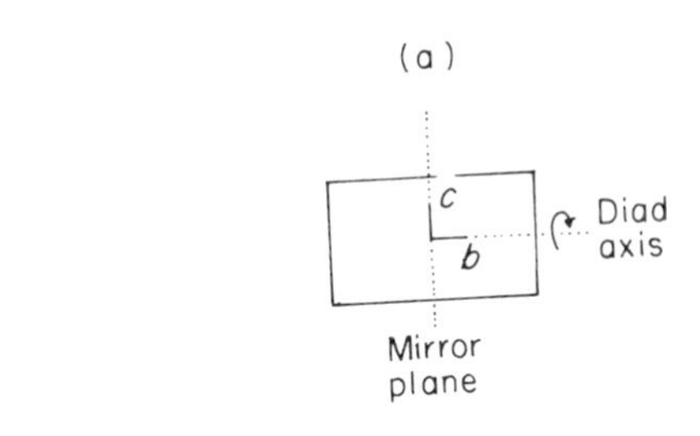

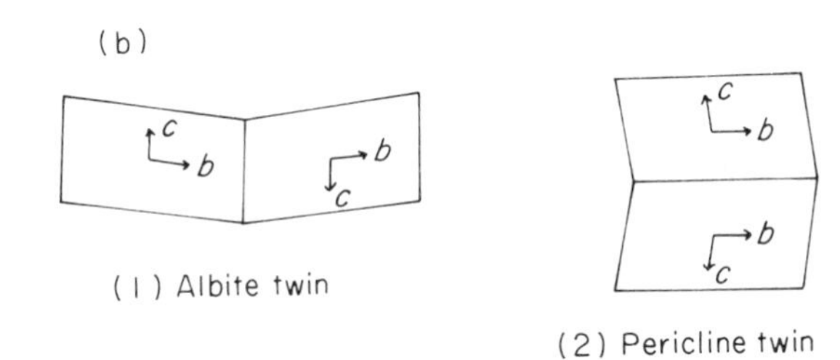

Fig. 7.1 (a) The unit cell of monoclinic albite showing the mirror plane and diad axis. (b) Two different ways of relating the two equally likely distortions to the triclinic cell: (1) with the *c* axes parallel, leading to an albite twin, and (2) with *b* axes parallel leading to a pericline twin.

During nucleation of the triclinic structure, domains of each of the four orientations will form and a finely domained structure results. This twinned structure can only coarsen by a reversal of one of the domains, a process which is difficult to achieve in practice. The albite twin plane has a lower energy than the pericline twin and hence may be expected to be the more favourable. Such twinning is very common in albite and may also form during growth.

(*b*) *Al, Si-ordering in albite*

In triclinic albite there are four distinct sets of tetrahedral sites in the structure, designated T_1o, T_1m, T_2o, T_2m as illustrated in Fig. 3.33 and described in section 3.6.2. With an Al : Si ratio of 1 : 3 it is possible to form a completely ordered structure by segregating the Al atoms into one set of sites. In albite, aluminium diffuses into the T_1o sites as the oxygen atoms of this tetrahedron are more closely bonded to the Na atom and hence the lower valency Al is electrostatically more favourably located in this site than Si. Fig. 7.2 shows the Al distribution over the four sites as a function of temperature. In the disordered state the Al occupancy is 0.25 for all four sites while in the completely ordered state all of the Al is in T_1o.

Disordered triclinic albite is known as high albite and the low-temperature ordered form is low albite. At any intermediate temperature there appears to be an 'equilibrium' value for the state of order which is achieved after long annealing periods. These intermediate states of order are usually referred to as intermediate

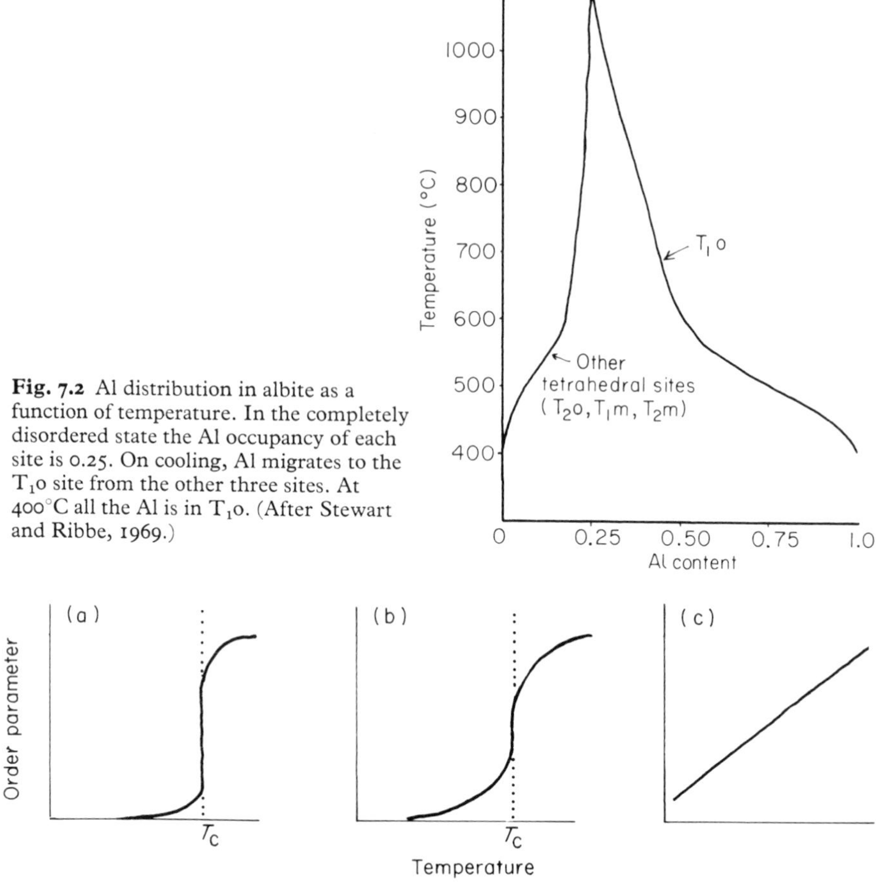

Fig. 7.2 Al distribution in albite as a function of temperature. In the completely disordered state the Al occupancy of each site is 0.25. On cooling, Al migrates to the T_1o site from the other three sites. At 400°C all the Al is in T_1o. (After Stewart and Ribbe, 1969.)

Fig. 7.3 The change in order parameter as a function of temperature for (a) a first-order transformation, (b) a second-order transformation, and (c) a continuous exchange between the sites where a certain degree of order is associated with each temperature.

albites. As intermediate albite is not very common it has been suggested that the high–low albite transformation may not be continuous, but may involve a discontinuous change in the order parameter at a certain critical temperature.

Although there is still considerable uncertainty as to the type of transformation which takes place we will discuss briefly the nature of the problem and one of the ways in which it has been tackled. Firstly, we need some way of determining the degree of order. During ordering there is an appreciable change in the cell dimensions, and some geometrical property based on this change is generally used. The problem then is to find how this property and hence the degree of order varies as a function of temperature. Three possibilities are shown in Fig. 7.3. These need not be strictly adhered to and intermediate possibilities may exist.

Under dry conditions the rate of ordering, which must involve Al, Si-exchange, is extremely sluggish, whereas in the presence of water vapour it is speeded up by a factor of around 10^7. This indicates a different diffusion mechanism in each case. Although these mechanisms are not known it has been suggested that the presence of protons (perhaps associated with vacant sites) may be utilized by the Al^{3+} to locally balance the charge with Si^{4+} and hence facilitate Al, Si-exchange in the structure.

Experiments to determine the degree of order as a function of temperature and time have been carried out under hydrothermal conditions by McKenzie and others. Between 980 and 700°C the degree of order asymptotes to a constant value after a period of time, this value being characteristic of the temperature. This implies a continuous ordering process. Below 700°C the situation is uncertain but there is evidence of a discontinuity with the ordering rate considerably speeded up at lower temperatures.

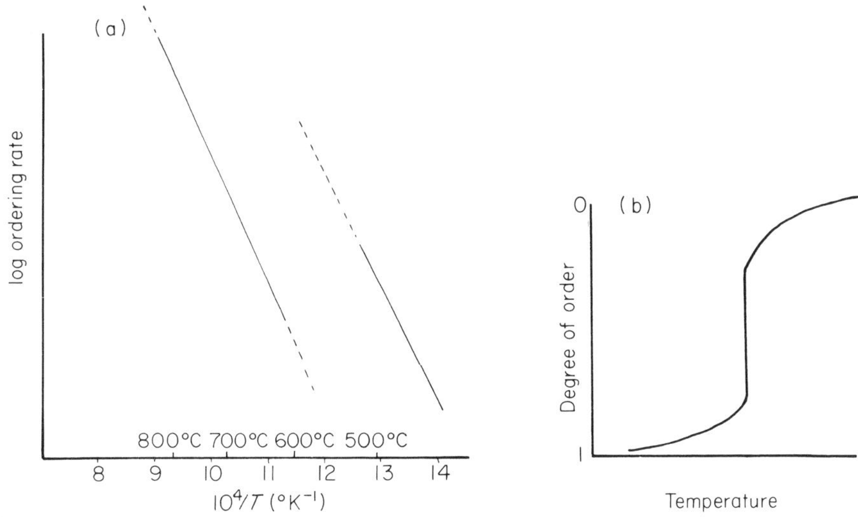

Fig. 7.4 (a) An Arrhenius plot for the rate of Al, Si-ordering in albite. (After McConnell and McKie, 1960.) (b) Variation in the degree of order with temperature for the situation shown in (a).

By studying the kinetics of the ordering process as a function of temperature some insights may be obtained into the nature of the transformation. McKenzie's data has been analysed in this way and the results plotted on an Arrhenius plot (section 6.1). The diagram obtained is shown in Fig. 7.4(a). A marked break in the plot occurs between 700 and 550°C and indicates the presence of a discontinuity. The low-temperature line corresponds to rates about 500 times faster than would be indicated by extrapolating the high-temperature line to lower temperatures. This situation corresponds to a transformation of the type shown in Fig. 7.4(b), with some of the characteristics of both first- and second-order transformations.

The activation energy for diffusion [given by the slopes of the lines in Fig. 7.4(a)] is similar in both cases: 61.5 kcal mol^{-1} at high T, and 56 kcal mol^{-1} at low T. The frequency factor and hence the entropy of activation is however higher for the low-temperature line. Thus at low temperatures the free energy of activation is lower and the process is more rapid. This increase in the pre-exponential part of the Arrhenius relationship may be associated with an appreciably larger free energy change for Al,Si-ordering below 600°C. This comes about if the stable state below this temperature has a substantially higher degree of Al,Si-order. There is also still some controversy over the interpretation of the experimental results in terms of two separate activation energies and a more detailed analysis may be found in the sources listed in the references.

7.1.2 ORDERING IN POTASSIUM FELDSPAR. SANIDINE–ORTHOCLASE–MICROCLINE

The disordered form of potassium feldspar which exists stably at high temperature is monoclinic with structure similar to that of monoclinic albite. However, the presence of the larger K ion in sanidine precludes the simple collapse of the structure to triclinic symmetry as observed in albite at high temperatures. While such a collapse may be implicated to a lesser extent, the principal cause of the inversion of sanidine to triclinic microcline at relatively low temperatures is the ordering of the Si and Al ions on tetrahedral sites. Ideally, the monoclinic–triclinic inversion from sanidine to microcline should nevertheless closely parallel the monoclinic–triclinic inversion in albite. That this is not so is due to the fact that the ordering process, on which the transformation depends, requires the inter-diffusion of Si and Al ions on the tetrahedral sites. This is an extremely sluggish process and has a time scale which must be defined on a geological rather than a laboratory time scale.

In monoclinic sanidine there are only two sets of tetrahedral sites (Fig. 3.32) and hence it is not possible to form a fully ordered state with the Al : Si ratio of 1 : 3. Any tendency to order Al into a particular site will distinguish it from the

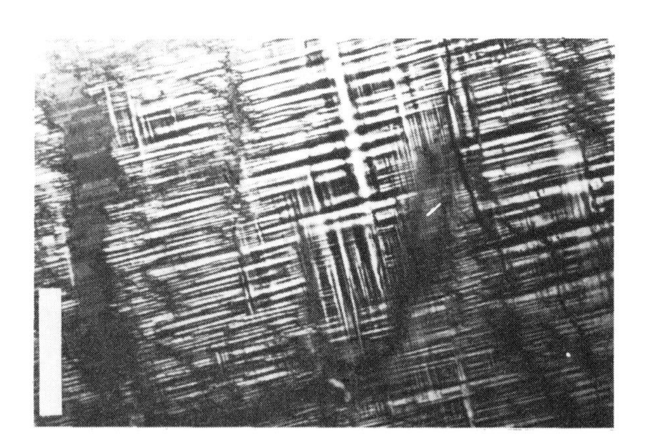

Fig. 7.5 Cross-hatched twinning in microcline in which two sets of twin planes form domains related either by the albite twin law or by the pericline twin law. The length of the scale bar is 0.1 mm. (Photo courtesy of R.H. Colston.)

others in the set, i.e. a region in which the Al ion is systematically on one of the T_1 sites while Si is on the other three sites will become triclinic due to a local loss of both mirror plane and diad axis (Fig. 3.33). In the sanidine structure there are two possible sites either of which may be favoured locally for occupation by the Al ion. This two-fold possibility leads to the observed twinning in microcline where we may consider separately the possibilities of pericline and albite twinning, as has already been noted in Fig. 7.1. Thus fine, complex twinned microcline (Fig. 7.5) contains four sub-individuals, two related by the albite law (characteristic of the loss of the mirror plane) and two related by the pericline law (characteristic of the loss of the original diad axis in the parent sanidine structure).

Were it not for the extreme sluggishness of the Al, Si-diffusion in potassium feldspar the monoclinic–triclinic inversion would be quite straightforward. In the circumstances the situation is more complex since it is necessary to consider that the simple transformation between sanidine and microcline may be precluded by its kinetics. Where this constraint is imposed, the only possible behaviour within the parent sanidine is for it to order as far as possible on the basis of its monoclinic symmetry and in practice this means that there are very small local regions in which the alternative choices of Al site are made. This local ordering does not involve the need to nucleate the triclinic phase. Electron-microscope studies indicate that in this alternative structure the local regions are of the order of 100 Å apart and that, unlike the case in microcline, there are no twin type boundaries between the regions. The regions could be described as incipient triclinic domains. An electron micrograph of this microstructure is shown in Fig. 7.6(a) and a schematic diagram illustrating the pattern of distortions is in Fig. 7.6(b). It is important to note that this fine cross-hatched structure is not the same as fine-scaled microcline twinning. Firstly, there are no internal phase boundaries formed nor any change in symmetry. Furthermore, the mechanism by which this modulated structure forms is quite different.

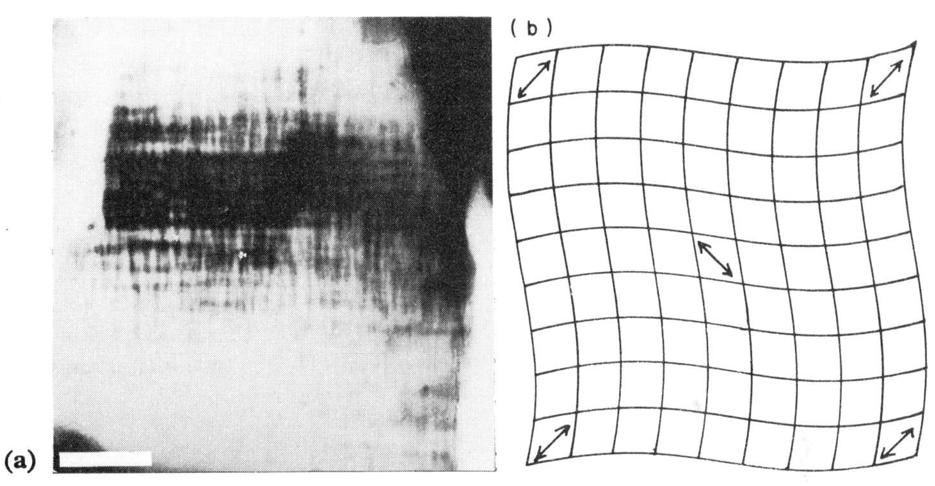

Fig. 7.6 (a) Transmission electron micrograph of adularia showing the typical modulated microstructure. The directions of the two modulations correspond to the composition planes of albite and pericline twins in cross-hatched microcline. The length of the scale bar is 0.05 μm. (b) A model illustrating the pattern of distortions in adularia which account for the microstructure observed in (a). The individual regions are strained due to local Al, Si-ordering, although the overall symmetry remains monoclinic. (After McConnell, 1971.)

This alternative monoclinic structure is particularly well observed in adularia, a potassium feldspar phase which crystallized with monoclinic symmetry well below the ordering temperature. An entirely similar explanation may be presented in the case of the metastable phase orthoclase. Here again the fine-scale local ordering is present and the material is stranded kinetically in relation to the possible inversion to the triclinic phase microcline. Electron-microscope study of a typical orthoclase from a slowly cooled granodiorite shows that the process of inversion from orthoclase to microcline involves the local growth of the regions of differing Al site preference (Fig. 7.7) and the process has close similarities with a normal nucleation process.

It has also been found that at sufficiently low temperatures this process of direct nucleation of microcline proper may itself be inhibited and that in this case the process of inversion to microcline takes place by a homogeneous process in which the relative proportions of the two local ordered alternatives becomes unbalanced. In this way the crystal moves slowly and uniformly from a state where

Fig. 7.7 Electron micrograph of orthoclase from the Nevada granodiorite showing the preferential growth of one of the alternative regions of Al order, from within the modulated orthoclase structure. The homogeneous region has nucleated and grown from the orthoclase and marks the first stage in the formation of microcline. The length of the scale bar is 0.2 μm.

the lattice parameters are monoclinic to a state with the final parameters of microcline. In other words the result of such an imbalance in the distribution of strain in the crystal is that these coarsened regions, in which some degree of Al, Si-order has developed, are able to gradually 'tilt over' from the monoclinic state to the triclinic state through a continuous series of intermediate stages. Thus in these low-temperature intermediate microclines, as they have been termed, various 'degrees of triclinicity' corresponding to the extent of order are possible. We will refer to this alternative mechanism of the formation of microcline from the modulated structure as the intermediate microcline route.

We can summarize the various possible modes of behaviour of potassium feldspar on cooling by a schematic route map (Fig. 7.8) which shows the relative kinetics and temperatures of the processes involved. Other aspects of the behaviour of alkali feldspars will be dealt with in the next chapter where we will discuss the processes which take place when intermediate compositions in the disordered monoclinic solid solutions are cooled.

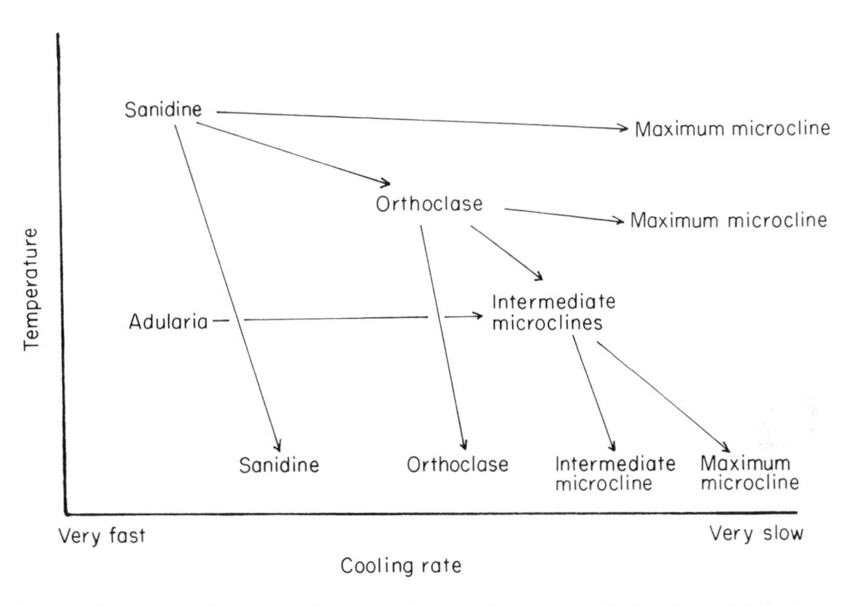

Fig. 7.8 A schematic 'route map' showing the possible modes of behaviour of K feldspar as a function of temperature and time.

7.1.3 TRANSFORMATIONS IN ANORTHITE, $CaAl_2Si_2O_8$

Anorthite differs from the alkali feldspar end members in a number of important respects. Although the basis framework structure is similar, the Al, Si-distribution is determined by the presence of the divalent Ca^{2+} ion within the framework and hence the Al : Si ratio of 1 : 1 which is required to maintain charge balance. The consequences of this are that anorthite has a much stronger tendency to form an ordered Al, Si-distribution than the alkali feldspars, and we may assume that pure anorthite An_{100} remains fully ordered virtually right up to the melting point.

(a) *Al, Si-ordering in anorthite*

As shown schematically in Fig. 3.34 alternating Al and Si atoms cause the *c* axis of the unit cell to be doubled relative to that in high albite. The cell is triclinic. We will refer to such Al, Si-ordering as the anorthite ordering scheme. A fully ordered structure is only possible in pure anorthite—addition of the albite component involves the coupled substitution:

$$Na^{+} + Si^{4+} \leftrightharpoons Ca^{2+} + Al^{3+}$$

and necessarily introduces some degree of disorder. At high temperatures it results in the formation of a disordered solid solution with the triclinic high-albite end member. The field of this disordered solid solution is terminated toward the anorthite composition by the formation of the doubled cell due to the strong tendency to order on the anorthite scheme.

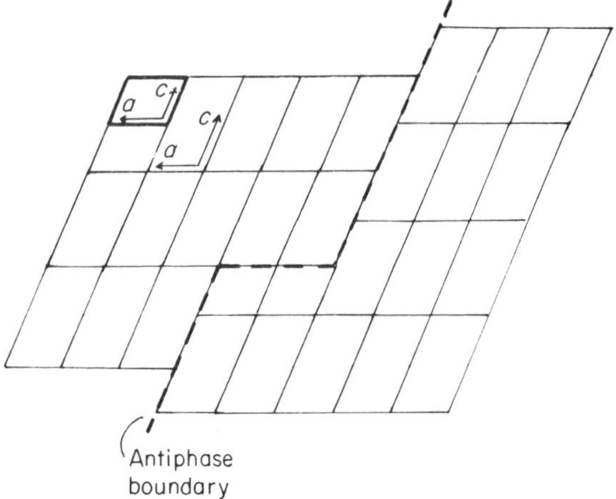

Fig. 7.9 An antiphase boundary produced by the doubling of the *c* axis in the ordered anorthite structure. The unit cell in heavy outline is that of the disordered state.

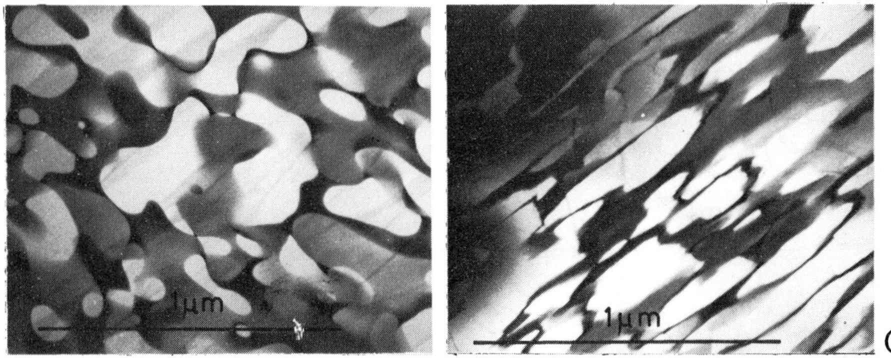

Fig. 7.10 (a) Dark-field electron micrograph showing the antiphase domains formed as a result of doubling the *c* axis during Al, Si-ordering in anorthite. (b) Dark-field electron micrograph of the same part of the crystal as in (a) but imaged with a different diffraction maximum to show the antiphase domains formed as a result of the loss of symmetry during the positional ordering of Ca atoms in anorthite. (Photos courtesy of A. C. McLaren.)

In plagioclases of composition between about An_{70} and An_{90} a transformation from the high-albite disordered solid solution to the anorthite structure occurs on cooling. As this involves doubling the cell size there is the possibility of the formation of translational variants of the anorthite structure as described in section 5.5.2 and illustrated schematically in Fig. 7.9. The boundaries between these regions are antiphase domains. In the electron microscope these domains can be observed by forming a dark-field image (section 1.2) using the extra spots which appear in the electron diffraction pattern when the unit cell is doubled. Such an image is shown in Fig. 7.10(a).

Observations of such domains in Ca-rich plagioclases from a number of geological environments suggests that their size and morphology depend on both composition and thermal history. They may thus be potentially useful in determining geological cooling rates. The compositional dependence arises from the fact that the temperature of the transformation to the anorthite ordering scheme decreases as the albite content increases. This is because an increase in the albite content necessarily reduces the degree of Al, Si-ordering possible and allows the inversion from the ordered to the disordered state to occur at a lower temperature. Therefore for a given cooling rate, coarser domains will be formed in more Ca-rich plagioclases which transform at a higher temperature where Al, Si-diffusion is more rapid.

The shape of the antiphase domains also appears to depend on cooling rate. We would expect that boundaries in some orientations would be energetically more favourable than others, and hence in a slowly cooled specimen the migration of boundaries to low energy planes could take place.

The qualitative determination of cooling rates from the antiphase domain microstructure requires the kinetic parameters for domain formation and subsequent coarsening to be known. At present there are no such data available, nor have the various factors which may influence their size been evaluated. It has also been noted that not all Ca-rich plagioclases contain such domains, presumably because they have crystallized below the transformation temperature. There is also the possibility that domains could be formed during growth under such circumstances, further complicating interpretations in terms of cooling history.

(*b*) *Domains of Ca ordering*

In pure anorthite above about 240°C, the relatively small Ca atom is able to 'rattle about' within the ordered aluminosilicate framework and is in this sense disordered. Below this temperature the framework twists and collapses around the cation which as a result may be displaced in one of two different ways. The change may be thought of in terms of a change from short- to long-range order of the Ca atoms. The translational symmetry of the triclinic structure is decreased by this transformation and therefore again we have the possibility of an antiphase domain structure being formed. This may be distinguished from that formed during Al, Si-ordering as it is imaged using different electron diffraction spots. Fig. 7.10(b) shows this domain structure in the same grain as that in Fig. 7.10(a), indicating that both Al, Si-ordering and Ca-positional ordering have taken place.

The size of the domains is dependent on the cooling rate but the situation becomes more complex with the addition of albite component and hence an increase in Al, Si-disorder. As with many aspects of plagioclase mineralogy, the interpretation of the nature and origin of these domains is controversial. Some sources for further reading are given in the reference section.

7.1.4 SUMMARY

In our brief treatment of the transformation behaviour of the feldspar end members a number of points should have become apparent.

1 In the alkali feldspars the cation size difference determines whether Al, Si-ordering can take place within a triclinic structure where no further symmetry changes are required for complete ordering, or within a monoclinic structure where the Al, Si-ordering process is responsible for a change to triclinic symmetry. In the

first case (albite) where the triclinic domains have already been formed by the framework collapse around the Na atom, ordering although sluggish takes place without structural complications. In the second case (sanidine) two possibilities ensue: either a fully ordered triclinic structure (microcline) is formed by a nucleation process, or alternatively an order and hence distortion modulated structure (orthoclase) may form by a continuous process.

2 There is a complete solid solution between monoclinic albite and sanidine, but the change to triclinic symmetry in albite severely restricts the amount of solid solution possible. The disparate Na and K cation sizes will therefore lead to a solvus at lower temperatures.

3 In the plagioclase feldspars a solid solution exists at high temperatures between disordered triclinic high albite and a composition of about An_{90} beyond which an ordered structure is formed with the anorthite ordering scheme.

4 The Al, Si-ordering scheme in albite is fundamentally different from that in anorthite and they are not compatible in that it is not possible to mix them in a single homogeneous feldspar framework. On this basis we would expect that at low temperatures no intermediate phases should exist between ordered albite and ordered anorthite. However, as we shall see in a later chapter, the possibility of alternative metastable behaviour leads to complex microstructures in the intermediate plagioclases.

7.2 Polymorphic transformations in the pyroxenes

When considering the behaviour of the pyroxenes within the pyroxene quadrilateral (Fig. 3.22) it is convenient to do so in terms of two basic silicate chain configurations. As we have already outlined in section 3.4.1 the type structure for pyroxenes is that of diopside (Fig. 3.20). The silicate chains are practically straight and are symmetrically equivalent (chains with tetrahedral apices to the 'right' are related by a diad axis to those with apices to the 'left'). The resulting structure is monoclinic and will be referred to as the diopside or clinopyroxene structure. In the end member diopside $CaMgSi_2O_6$, Ca^{2+} occupies the larger M2 site and Mg^{2+} the smaller M1 site. (Note that the cation sites in the pyroxenes are considerably smaller than in the feldspars and Ca^{2+} is regarded as a large cation within this structure.) The diopside structure, with its straight chains and eight-fold co-ordination of the M2 site, forms when the M2 site is mainly occupied by a large cation such as Ca^{2+}.

At the other side of the quadrilateral where there is practically no calcium and the cations Fe^{2+} and Mg^{2+} are considerably smaller, a different silicate chain arrangement becomes more stable. By rearranging the chains relative to one another the co-ordination of the M2 sites is reduced to six, the same as for M1 thus being more compatible with the smaller cations. This structure is orthorhombic and hence referred to as orthopyroxene.

The relationship between orthopyroxene and clinopyroxene is shown in Fig. 7.11 by comparing their unit cells. The orthopyroxene could be regarded in terms of clinopyroxene which has been twinned on a unit-cell scale. This results in an orthorhombic unit cell repeat with $c_{\text{ortho}} = c_{\text{clino}}$, and $a_{\text{ortho}} = 2a_{\text{clino}} \sin \beta$ (Fig. 7.11). The relationship between these structures will be one of the problems we will discuss later in this section.

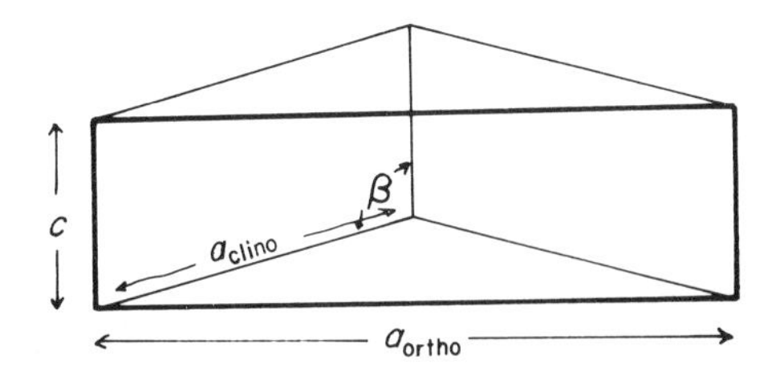

Fig. 7.11 The relationship between the unit cell of orthopyroxene (heavy line) and two twin-related unit cells of clinopyroxene (light line).

The immediate question which we must ask next is what happens in the compositions between the two opposite sides of the quadrilateral, i.e. at intermediate calcium contents. As the orthopyroxene structure is essentially a response to the increasing content of small cations we might expect that below a certain calcium content the orthorhombic structure becomes more stable.

At high temperatures the diopside structure is stable over a considerable range of Ca contents, almost up to the Ca-free end of the quadrilateral. As we have already outlined in section 3.4.1, lattice vibrations at high temperatures may keep the silicate chains extended even as the content of the M2 sites decreases in calcium. The high-temperature structure of pigeonite is similar to that of diopside and augite and considerable solid solution exists between them.

At lower temperatures however orthopyroxene becomes the stable phase at Ca-poor compositions. The temperature of the transition between the diopside structure and the orthopyroxene structure decreases rapidly as the calcium content increases. In other words, the stability field of the diopside structure narrows with falling calcium content and may actually 'pinch out' at high temperatures at the Ca-free end of the quadrilateral. On this basis the augite composition retains the diopside structure from the melting point down to room temperature. A composition such as pigeonite should have the diopside structure at high temperature and the orthopyroxene structure at lower temperatures while in the almost Ca-free pyroxene termed hypersthene (an enstatite-ferrosilite solid solution with a composition between En_{50} and En_{70}) the orthorhombic structure may well be stable right up to the melting point.

The situation at the end-member composition enstatite $MgSiO_3$ is further complicated by the fact that above 1000°C a third major pyroxene-structure type, termed protoenstatite becomes stable. We shall not consider the complex transformation behaviour in this corner, but direct our attention to the more common pyroxene compositions around pigeonite.

7.2.1 THE BEHAVIOUR OF PIGEONITE ON COOLING

(a) *The low-temperature modification of the clinopyroxene structure*

As we have seen, at high temperatures pigeonite has a clinopyroxene structure similar to that of diopside, whereas at low temperatures the stable form has the orthopyroxene structure. The transition from clino- to orthopyroxene requires a reorganization of the silicate chains and is thus a reconstructive transformation.

Again we have the situation where the ideal behaviour on cooling involves a process requiring a considerable structural change and a relatively large activation energy. Kinetically the transformation will be sluggish and only expected to take place under near-equilibrium conditions in slowly cooled deep-seated rocks. Note however that the twinning relationship between the two structures suggests that applied stress may affect the mechanism, a point which we will take up later.

In more rapidly cooled rocks, as the temperature falls to below the equilibrium clino to ortho transition temperature, we again enter the regime where we must consider other alternatives by which the high-temperature structure might reduce its free energy. The diopside structure in pigeonite is unstable at lower temperatures where the lattice vibrations can no longer maintain the practically straight silicate chain framework which is too large for the cations present. If the chain rearrangement to the more economical orthopyroxene structure does not take place on cooling, a temperature is reached where the chains kink due to a structural collapse around the M2 site. This is a simple displacive transformation which is rapid enough to take place on quenching. The temperature of the transformation is strongly composition dependent, decreasing rapidly from around 1000°C for Mg-rich pigeonites as either the Ca or the Fe content increases.

In the diopside structure the silicate chains are all equivalent, related by a diad axis which passes through the cation positions (Fig. 3.20). During the transformation the chains do not kink in the same way and the equivalence between them is lost, with layers of kinked chains alternating with layers of still extended chains. Although the diad axis is lost, the structure remains monoclinic and will be referred to as low clinopyroxene. Diopside can therefore be said to have the high clinopyroxene structure.

The high–low clinopyroxene transformation can be regarded as alternative metastable behaviour taking place under non-equilibrium conditions when the formation of the stable orthopyroxene structure is kinetically impeded.

As the transformation involves the formation of two non-equivalent sets of silicate chains, two different possibilities arise. A chain in the high-temperature

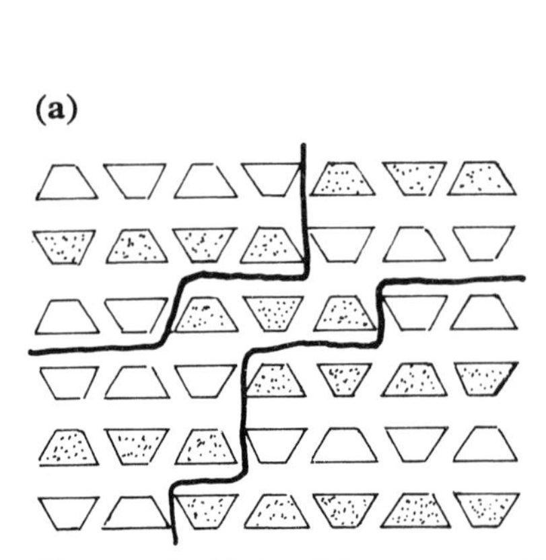

Fig. 7.12 (a) Model of the structure of low pigeonite with the chains viewed from the end. The stippled chains are those which are kinked and the heavy lines are antiphase boundaries. (b) Electron micrograph of antiphase domains formed as a result of the high–low transformation in pigeonite. The length of the scale bar is 0.2 μm. (Photo courtesy of M. A. Carpenter.)

structure may either remain extended or become kinked. If the process occurs by random nucleation and growth, as appears to be the case, antiphase boundaries will form where two domains with reversed sequences of kinked and extended chains meet. This is shown schematically in Fig. 7.12(a) with an electron micrograph of antiphase domains in pigeonite [Fig. 7.12(b)].

The collapse of the high clino structure further reduces the solubility of calcium in the pigeonite and some expulsion of a Ca-rich component might be expected to follow. It has been shown that some of this calcium migrates to the domain boundary. The local crystal structure at an antiphase boundary depends on its orientation. In some orientations the boundary has a structure similar to that of diopside, with a relatively large M2 site associated with extended parts of the silicate chain. The extent to which the calcium may diffuse to these favourable cation sites at the boundaries depends on the cooling rate.

In relatively rapidly cooled pigeonites the antiphase domains are ragged and irregular (Fig. 7.12). On slower cooling reorganization of the boundaries takes place and straighter, more regular domains are formed, the boundaries adjusting their orientation to provide the maximum number of suitable sites for the local Ca concentration. Where a Ca-concentration gradient exists (e.g. in the stranded diffusion profile around a previously exsolved augite lamella) the antiphase boundaries continuously alter their orientation as shown in Fig. 7.13.

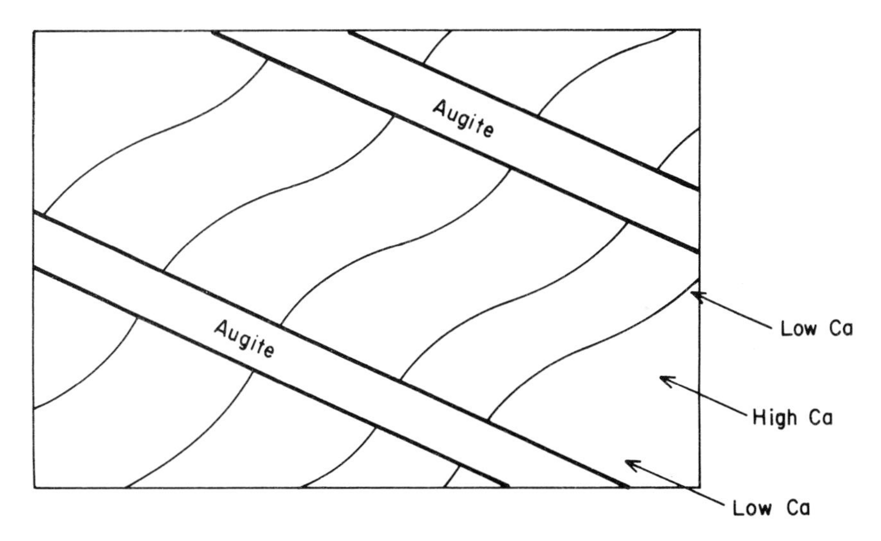

Fig. 7.13 Ca-concentration gradients around pre-existing augite lamellae result in continuously altering orientations of the antiphase boundaries (curved lines) in the host pigeonite. (Drawn from an electron micrograph taken by M. A. Carpenter.)

As the solubility of calcium in pigeonite decreases with temperature, pigeonite from slowly cooled intrusions invariably contains exsolution lamellae of a Ca-rich augite phase. The relationships between exsolution lamellae and antiphase boundaries will be discussed in a later section on exsolution processes, but it is as well to point out here that Ca-enriched antiphase boundaries may subsequently provide highly suitable sites for exsolution having both a structure and composition tending towards that of augite.

As the high–low clino transformation in pigeonite appears to occur by random nucleation, the size of the antiphase domains should be related to the cooling rate through the transition temperature. On slower cooling fewer nuclei with more time to grow are formed and a coarser microstructure is produced. Where the transformation occurs by homogeneous nucleation within a single-phase pigeonite the domain size may be used as an indicator of cooling rate. Most pigeonites however contain exsolved phases and the interphase boundaries provide suitable sites for the enhanced formation of the low clino structure by heterogeneous nucleation. This results in deceptively large antiphase domains. Compositional gradients which may also exist in such cases would locally change the transformation temperature and hence the domain sizes. Their use in estimating cooling rates may therefore in many cases be limited.

(*b*) *The pigeonite to orthopyroxene transformation*

In deep-seated igneous intrusions pigeonite eventually transforms to orthopyroxene. The mechanism of the transformation appears to be sensitive to external stresa conditions. In fact the relationship between clino and orthopyroxenes at low-Cs compositions is still controversial and seriously complicated by the effect of stress on the transformation. It will be assumed here that in pigeonitic compositions, orthopyroxene is the most stable phase at low temperatures, in the absence of stress,

In orthopyroxenes, which have inverted from pigeonite on slow cooling, the structural orientation is generally random with respect to the original pigeonite structure, despite the close structural similarity of the two phases. This implies a reconstructive transformation in which a randomly oriented nucleus of orthopyroxene forms and grows at the expense of the pigeonite, maintaining an incoherent boundary with the parent phase and involving a minimum amount of diffusion. As the orthopyroxene structure can accept even less calcium than the low clino pigeonite structure, further cooling usually involves the exsolution of small amounts of a Ca-rich phase such as augite from the orthopyroxene. The sequence of transformations from high clino to low clino to orthopyroxene involves a reduction of calcium solubility at each step and hence a number of episodes of augite exsolution are possible.

When pigeonite is in the form of lamellae within augite (having exsolved from a Ca-rich solid solution) the strain associated with the existence of a relatively thin lamella within a monoclinic structure will in many cases prevent the transition to orthopyroxene. The monoclinic phase will be 'stabilized' by the augite due to the similarity of their structures. If the transition does occur under these conditions the mechanism of inversion appears to be different. As the structures of the two polymorphs differ in the relative positions of adjacent silicate chains, a shift of the chains parallel to their length is the basis for a possible transformation mechanism. Strain may cause such a change in the stacking of silicate chains which can be described in terms of a 'unit cell twinning' mechanism as suggested by Fig. 7.11. Fig. 7.14 is an electron micrograph of augite containing a lamella of pigeonite which shows numerous 'stacking faults' in the silicate chain arrangement representing the first stages in the transformation to orthopyroxene. Orthopyroxene formed in this way inherits its structural orientation from the parent clinopyroxene and may be coherent with it.

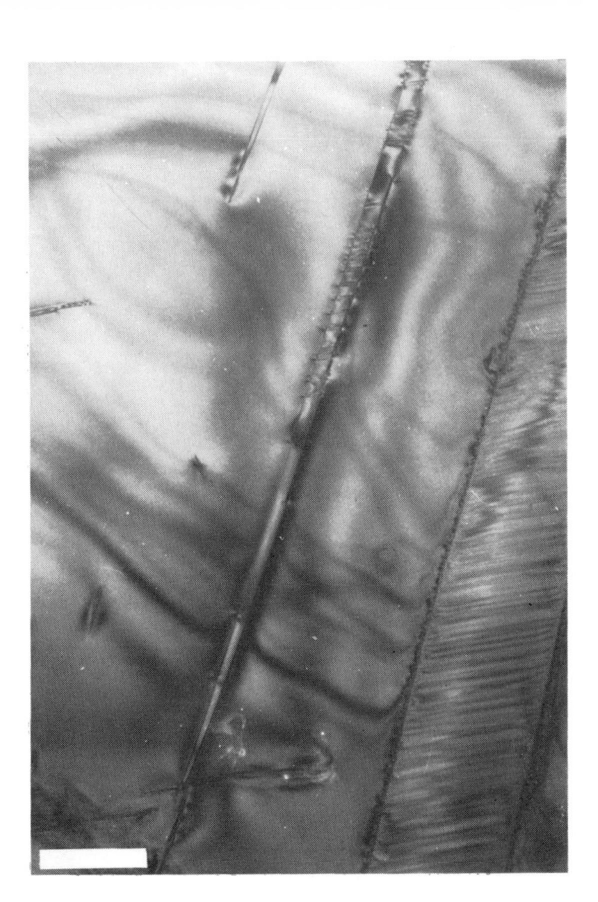

Fig. 7.14 Electron micrograph of augite containing lamellae of exsolved pigeonite. The stacking faults in the pigeonite lamellae may be precursors to the formation of orthopyroxene, the stable low-temperature phase at low temperatures. The length of the scale bar is 1.0 μm. (From Champness and Copley, 1976. Photo courtesy of P. E. Champness.)

The transformation from low clinopyroxene to orthopyroxene in the absence of stress illustrates some of the problems encountered when the structural similarity is close yet a reconstructive transformation is required from one to the other. The free-energy differences between the phases will be small so that neither is appreciably more stable than the other. The combination of a small free-energy drive but a large activation energy for nucleation results in a very sluggish transformation and the probability of non-equilibrium behaviour if the mechanism is one of random nucleation. The close relationship between the phases however suggests that the presence of any shear stress (even that generated by heating) opens the possibility for a different transformation mechanism and a reduced activation energy. Stress-induced coherent intergrowths between the phases become common and phase relations difficult to unravel. The thermodynamic effect of the stress terms must be recognized although the distinction between those factors that affect the mechanism and those that affect the thermodynamics becomes a difficult one to draw. Thus phase relations in the Ca-free phases are still the subject of debate.

7.3 Transformations in the amphiboles

The structural analogies between pyroxenes and amphiboles have already been outlined (section 3.4.2) and transformations within the amphiboles will therefore have many parallels with the behaviour of the pyroxenes. The behaviour again depends mainly on the distribution of the relatively large calcium atom. In the monoclinic amphiboles which have the high clino structure at all temperatures, the M4 site (cf. M2 in pyroxene) has a high-Ca content which prevents the structural collapse of the silicate double chains on cooling. Cummingtonite is the analogue of pigeonite and has the high-clinoamphibole structure at high temperatures and the low-clino structure at lower temperatures. In the Ca-poor compositions a transformation to the orthoamphibole structure takes place on further cooling.

As the general features of the behaviour of the amphiboles are similar to the pyroxenes these transformations will not be described here. In this section we will discuss ordering of Fe^{2+} and Mg^{2+} in an orthoamphibole. A direct analogy exists with ordering in orthopyroxene but we will describe the amphibole study here as the kinetics have been determined and hence the geological applications are more apparent.

7.3.1 THE KINETICS OF THE Fe^{2+}–Mg^{2+} EXCHANGE REACTION IN ANTHOPHYLLITE

We refer to Fe^{2+}, Mg-ordering in anthophyllite as an exchange reaction because no phase transformation takes place as ordering proceeds. As outlined in section 5.5.1, when ordering takes place by a partitioning between two or more topologically non-equivalent sites no symmetry changes are necessary for the change from a disordered to an ordered state. The symmetry is already determined from the structural topology and is independent of the cation distribution. Partitioning of the cations between sites can take place continuously as a function of temperature and there is an equilibrium distribution characteristic of any particular temperature.

In the (Mg, Fe) orthopyroxenes, both Mg and Fe^{2+} are distributed over M1 and M2 sites, although as the temperature decreases Fe^{2+} shows a marked preference for the larger M2 site. In the orthoamphibole anthophyllite $(Mg, Fe)_7Si_8O_{22}(OH)_2$ the Fe^{2+} similarly shows a preference for the M4 site as ordering proceeds, with only a small fraction in the M1, 2 and 3 sites in the ordered state. Even in the disordered state the distribution is far from random with Fe^{2+} still predominantly in the M4 site.

As no transformation takes place during ordering we are not concerned with the notions of critical temperature and undercooling when discussing the kinetics. However, it is a diffusion-controlled reaction and the kinetics will still be described by similar rate laws and TTT curves to those already discussed. There are a number of features of the TTT curves for a continuous ordering reaction which differ from those described in section 6.3 and these will be dealt with before proceeding to discuss anthophyllite in more detail.

(a) *TTT curves for continuous cation exchange reactions*

TTT curves may be drawn for the different degrees of order achieved as a function of temperature and time. An equilibrium distribution corresponds to the horizontal part of the curve (i.e. no further ordering takes place even after infinite time), and there is a different equilibrium distribution at each temperature. Successive curves showing increasing degrees of order will asymptote to lower temperatures. This is shown in the schematic TTT curves in Fig. 7.15. The equilibrium distribution is attained after different times for different degrees of order, higher degrees of order

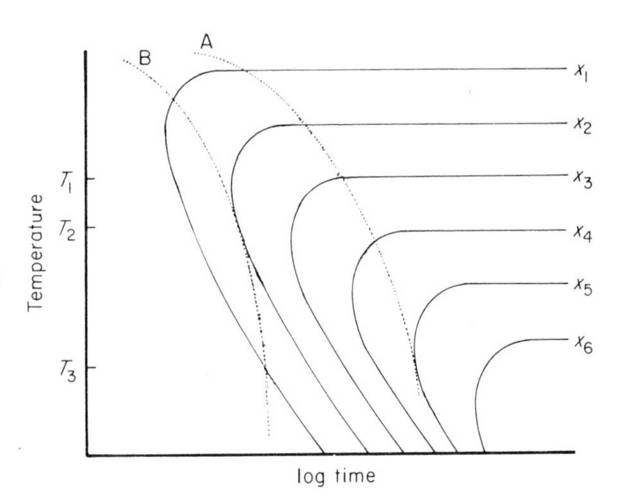

Fig. 7.15 TTT curves drawn for different degrees of order ($x_1 \rightarrow x_6$) achieved as a function of temperature and time in continuous exchange reactions. For cooling curve A equilibrium is maintained down to temperature T_1. Disequilibrium ordering continues down to T_3 where the cooling curve is tangential to a TTT curve. For cooling curve B, equilibrium is not maintained during cooling and ordering stops at T_2 where the degree of order is given by x_2. (After Seifert, 1977.)

taking longer times to achieve at lower temperatures. In the figure, six curves are drawn varying from complete order x_6 to almost complete disorder x_1. Under isothermal conditions the curves indicate the times required for a disordered material to reach the various degrees of order, until ultimately the equilibrium distribution for that particular temperature is reached.

The effect of a continuous cooling rate on the ordering process can be seen by superposition of a cooling curve on the TTT diagram. Ordering can only proceed when curves of higher order parameter are intersected as the temperature falls. If the horizontal parts of the curves are crossed then equilibrium will be achieved by the mineral at this temperature. If we consider the slow cooling curve A in Fig. 7.15, it can be seen that equilibrium is maintained down to temperature T_1 where the degree of order is given by line x_3. Below this temperature ordering proceeds but equilibrium is no longer maintained. Below a temperature T_3 no further ordering can take place at this cooling rate, and the final state has a degree of order x_5. In the more rapidly cooled sample (curve B) only a degree of order x_2 can be achieved and ordering stops at temperature T_2. In the fast cooling case the divergence from equilibrium as well as the kinetic cut-off for further ordering occurs at a higher temperature.

The maximum degree of order will be attained at the kinetic cut-off temperature which may be defined as the temperature at which the cooling curve is tangential to the TTT curve for that particular degree of order. Below the cut-off temperature the cooling rate of the mineral exceeds the Fe^{2+}–Mg exchange rate and no further ordering can take place. If the exchange rate at the cut-off temperature is known the cooling rate at this temperature can be calculated.

(*b*) *TTT curves for ordering in anthophyllite*

The kinetics of Fe^{2+}–Mg exchange in anthophyllite are such that equilibrium can only be achieved experimentally in reasonable times at high temperatures. Therefore the set of TTT curves cannot be experimentally determined nor the cut-off temperatures directly evaluated.

A solution to this problem has been presented by Mueller (1969) who has derived a theoretical model for the kinetics and thermodynamics of such exchange reactions in simple systems. This model can be applied to phases which are part of an ideal binary solid solution, i.e. only two cations will be involved in the exchange process, with the added prerequisites that no phase transformations should occur on cooling and that the energy differences between the cation sites be large enough to produce a marked temperature dependence of the cation distribution. These criteria rule out most silicates apart from orthopyroxenes and amphiboles.

Mueller's model has been successfully applied to an anthophyllite by Seifert and Virgo, who found that experimental data obtained at high temperatures fitted well to the predictions of the model. They studied the exchange of Fe^{2+} between the M4 sites and the M1, 2 and 3 sites. Fe^{2+} site occupancies can be determined by Mossbauer spectroscopy. The application of the model requires that the following parameters be experimentally determined:

1 the bulk Fe^{2+} content of the sample;

2 the initial concentration of Fe^{2+} on the M4 site;

3 the change of Fe^{2+} concentration on the M4 site as a function of time at a given temperature;

4 the equilibrium cation distribution at this temperature. The model then enables the calculation of rate constants for ordering and disordering from which the TTT curves can be derived.

Fig. 7.16 shows TTT curves obtained in this way for an anthophyllite of composition $Na_{0.05}Ca_{0.09}Fe^{2+}{}_{1.17}Mg_{5.79}Si_{7.87}Al_{0.18}O_{22}(OH)_2$, i.e. with some Na on the A sites and a small amount of Al^{3+} replacing Si^{4+} in the framework. As we

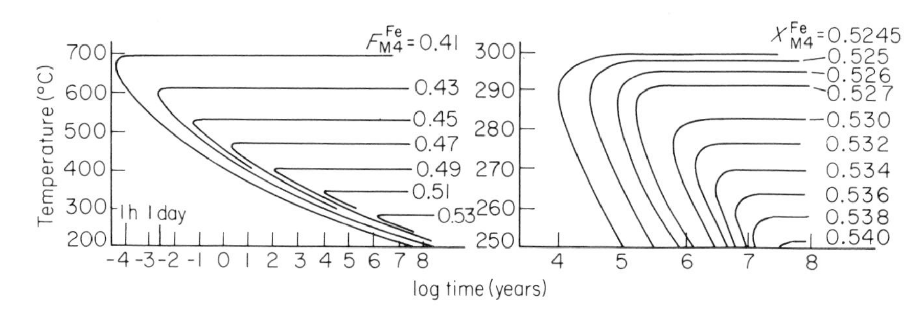

Fig. 7.16 TTT curves plotted for various values of X_{M4}^{Fe} which is the atomic fraction $\frac{Fe^{2+}}{Fe^{2+} + Mg}$ on the M4 sites. (a) For anthophyllite crystallized at 720°C. (b) For anthophyllite crystallized at 300°C. (After Seifert and Virgo, 1975.)

have mentioned above, one of the parameters required is the initial concentration of Fe^{2+} on the M4 site, i.e. the degree of order at the crystallization temperature assuming that equilibrium is attained. In Fig. 7.16 TTT curves are shown for anthophyllite crystallized at 720 and 300°C. The degree of order is measured by the parameter X_{M4}^{Fe} which is the atomic fraction $Fe^{2+}/Fe^{2+}+Mg$ on the M4 site.

In practice, such curves can be used to determine cooling rates (assumed linear) when the crystallization temperatures are known and the degree of order in the natural specimen measured. The effect of a change in the bulk composition on the kinetics must also be known. Using these curves Seifert and Virgo found that the maximum cooling rate of their anthophyllite specimen was about 1×10^{-4} °C per year for all values of the crystallization temperature consistent with this composition.

This method of determining cooling rates is free of many of the uncertain effects of heterogeneous processes where the nucleation of a new phase is involved, and is independent of the other methods described previously.

7.4 Cation ordering at compositions around chalcopyrite, $CuFeS_2$

Cation ordering in sulphides differs from that in our previous example, anthophyllite, in a number of important respects. In anthophyllite the cations occupy topologically distinct sites and no symmetry change takes place as ordering proceeds. In most sulphides, on the other hand, cations occupy sites which are equivalent in the disordered state but lose this equivalence on ordering. As described in section 5.5.1 this results in the formation of a superstructure. A symmetry change is involved in the process and there is a critical temperature above which the disordered form is stable and below which the ordered form is stable.

The second difference is one of kinetics. Ordering in sulphides is very rapid compared with that in silicates and often the disordered state is not quenchable. Although rapid kinetics may seem to suggest that metastability of the sort common in silicates is unlikely, the reverse is often the case. When a disordered state is cooled below its stability limit, the process which operates will be that which is most favourable kinetically, not necessarily that which is the most stable. Thus ordering processes may often mask the ideal behaviour which might involve longer range diffusion and hence be slower.

In our first example of sulphide behaviour we will describe a situation in which ordering occurs, but various phases which are ordered to different extents are formed in response to different cooling rates.

In stoichiometric chalcopyrite $CuFeS_2$ at low temperatures, copper and iron atoms form an ordered distribution over one half of the tetrahedral sites in a cubic close-packed sulphur structure. The unit cell is therefore doubled compared with that of a disordered distribution (Fig. 2.4) .In the disordered state a solid solution with this cubic structure exists over a fairly wide range of cation compositions (Fig. 3.12) and in this section we enquire into the cooling behaviour of one such composition of the general form $Cu_{1+x}Fe_{1+x}S_2 (x \approx 0.125)$, i.e. a disordered cation enriched chalcopyrite.

In the disordered state the extra cations will be distributed more or less randomly over the vacant tetrahedral sites, so that there are two aspects of any ordering process to consider. Firstly, which are to be the vacant sites and which the occupied sites, and secondly, how are the two cations Cu and Fe distributed over the occupied sites ? Thus we have the possibility of a number of stages of partial order.

Experiments on such solid solutions can be carried out by observing the behaviour of a sample *in situ* in an electron microscope during heating and cooling cycles. The rapid kinetics of the transformations involved enable this to be done. The results show that for a composition near $Cu_9Fe_9S_{16}$ three different superstructure phases could be formed depending on the cooling rate. These are labelled

I, II, III on the semi-quantitative TTT diagram in Fig. 7.17. The first point to note is that the only stable phase on this diagram is that labelled III. It is formed at relatively slow cooling rates. At more rapid cooling rates metastable states are formed.

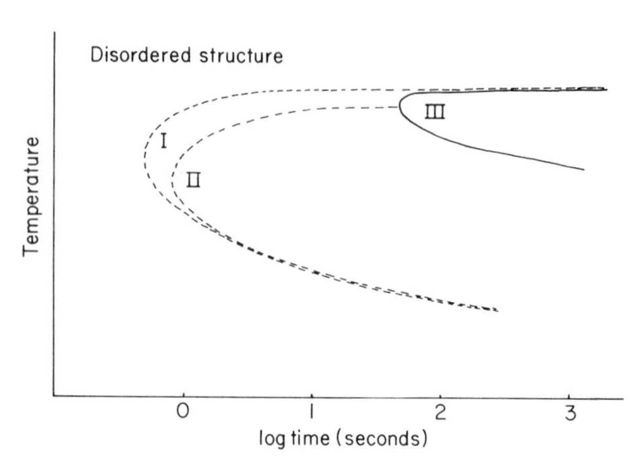

Fig. 7.17 TTT diagram for cation ordering transformations in metal-enriched chalcopyrite. Phases I and II are metastable phases relative to phase III, and are formed during more rapid cooling.

At the fastest cooling rates, the disordered structure may be quenched in. At slightly slower cooling rates phase I is formed. This retains the cubic cell but the repeat distances in the lattice are doubled due to some ordering process. Slower cooling results in the formation of phase II from phase I. Phase II is still cubic but involves yet a further doubling of the unit-cell repeat. This larger superstructure implies that some further ordering process has taken place. The TTT diagram shows that at low temperatures this phase may be retained metastably for very long periods, as the kinetic cut-off for the formation of the stable phase III is at a higher temperature.

The formation of phase III involves a greater structural change than the previous two steps. Phase III is a tetragonal superstructure of the cubic cell and hence a change in crystal system is required. It can be shown theoretically (see refs) that the symmetry changes involved imply a substantial degree of ordering but a more difficult process kinetically. Over a range of cooling rates it is therefore a kinetically inaccessible phase and alternative metastable behaviour leads to the formation of partially ordered phases I and II. Phase I is concerned with ordering of vacant and occupied sites, while in phase II some further ordering occurs within the cation sites.

The fact that phase II is not yet fully ordered is demonstrated by its microstructure and its subsequent behaviour on annealing at higher temperatures. It has a perturbed or cross-hatched microstructure suggesting that there is some attempt on a local scale to transform to another phase. Annealing confirms that it is attempting to transform to phase III and that at higher temperatures where the kinetics are more favourable it may achieve this by one of two ways. Firstly, twin-related lamellae of phase III may form by nucleation and growth from the perturbed structure (Fig. 7.18). The second mechanism involves progressive coarsening of the cross-hatched texture by the preferential growth of one or other of the modulations to eventually produce a twinned intergrowth of phase III (Fig. 7.19).

This behaviour is highly reminiscent of disordered sanidine which under equilibrium conditions should transform to microcline, but on more rapid cooling

Fig. 7.18 (a) Transmission electron micrograph of the perturbed (mottled) microstructure of phase II in metal-enriched chalcopyrite. (b) The same grain as in (a) after experimental annealing, showing the formation of lamellae of the more stable phase III by nucleation from phase II. The length of the scale bar is 0.2 μm.

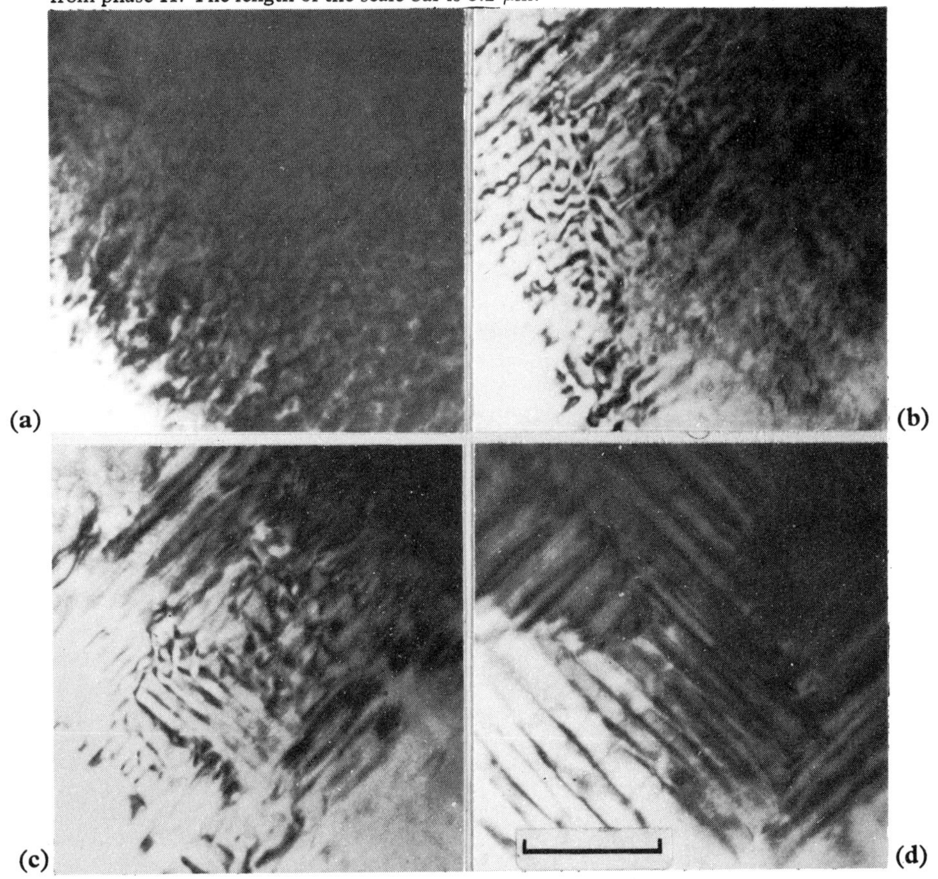

Fig. 7.19 A sequence of transmission electron micrographs of metal-enriched chalcopyrite showing the formation of phase III (d) from phase II (a) by the preferential growth of one or other of the modulations. The mechanism should be compared with that in Fig. 7.18. The length of the scale bar is 0.2 μm.

forms a modulated structure (orthoclase) which may then ultimately form microcline by direct nucleation, or by a coarsening of the modulation. Although the details of the processes are not completely understood, the principles involved appear to be the same in each case.

This treatment of ordering behaviour in $Cu_{1+x}Fe_{1+x}S_2$ illustrates some of the problems encountered in sorting out mineral assemblages when the kinetics of transformations are not taken into account. In this compositional region there are two natural mineral phases known. One named talnakhite appears to have a composition around $Cu_9Fe_8S_{16}$ and the other, mooihoekite around $Cu_9Fe_9S_{16}$. They have been assumed to be separate and stable phases, yet their structures are the same as the phases II and III, respectively.

Electron-microscope examination of the natural minerals confirms that talnakhite has a perturbed structure similar to phase II, but that at a composition around $Cu_9Fe_8S_{16}$ the stable, ordered mooihoekite structure (III) is not accessible. At this composition a fully ordered structure is not possible geometrically and the metastable talnakhite structure remains stranded. Natural mooihoekite of composition around $Cu_9Fe_9S_{16}$ behaves in the same way as described in Fig. 7.17. Thus the minerals talnakhite and mooihoekite are related kinetically in a manner analogous to orthoclase and microcline, although in the former case a change in the kinetics of ordering with composition is partly responsible for the metastable persistence of talnakhite.

Although the minerals talnakhite and mooihoekite are relatively rare, the importance of their behaviour lies in the fact that the principles involved are applicable generally and that, in this case, the kinetics of the processes are sufficiently rapid for the transformations to be observed dynamically whereas in other analogous systems this may be impossible.

7.5 Polymorphism in Al_2SiO_5—andalusite, kyanite, sillimanite

In this section we will outline briefly some of the problems associated with this very well known but poorly understood polymorphic system. It has been discussed widely in books on metamorphic petrology (e.g. Turner, 1968), so that here we will only make a number of observations which illustrate the principles of their behaviour.

The structures of the three polymorphs have a number of features in common. In all three structures isolated SiO_4 tetrahedra are linked via Al atoms which form parallel chains of edge-sharing AlO_6 octahedra [Fig. 7.20(a)]. The octahedra in these chains contain half of the Al in the structural formula; the remaining Al atoms are in polyhedra which vary from structure to structure. The main differences in the structures can be described in terms of the ways in which the AlO_6 chains are oriented with respect to one another (Fig. 7.20). This results in the other half of the Al lying in four-fold coordination in sillimanite, five-fold in andalusite and six-fold in kyanite. The main point to be made by comparing these structures is that although they are very similar in many respects, any transformation from one to the other will involve breaking Si—O and Al—O bonds, i.e. a reconstructive transformation.

Phase equilibria between the minerals kyanite, andalusite and sillimanite have been studied widely by petrologists, as they appear to be a key group in interpreting pressure–temperature regimes in metamorphism. The general form of the

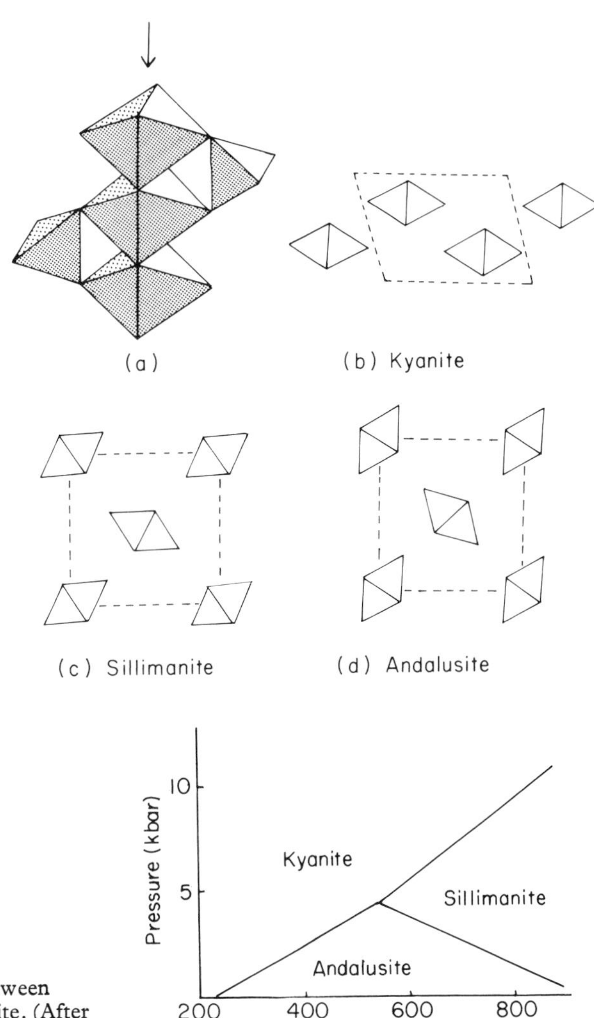

Fig. 7.20 (a) Linkage between AlO_6 octahedra and SiO_4 tetrahedra in the aluminosilicate structures. (b) The arrangement of the chains of octahedra [viewed in the direction of the arrow in (a)] in kyanite. (c) The arrangement of the chains of octahedra in sillimanite. (d) The arrangement of the chains of octahedra in andalusite.

Fig. 7.21 Stability relations between kyanite, andalusite and sillimanite. (After Richardson, Gilbert and Bell, 1969.)

phase diagram is shown in Fig. 7.21 but there have been very significant deviations in the details, particularly in the position of the invariant 'triple point'. This is usually attributed to difficulties inherent in the thermodynamics of the system.

When discussing the thermodynamics of this type of polymorphism it is very important to appreciate the difference between the thermodynamic parameters of the phases involved, and the thermodynamics of the process by which one is transformed to the other. The latter is expressed in terms of the thermodynamics of the activated state, a concept discussed in section 6.1. Confusion often arises when the difference between the terms free energy ΔG and free energy of activation ΔG^* is not made clear. The first term expresses the difference in free energy between the two phases at a certain temperature (or pressure) and being independent of the mechanism is only one of the terms on which the kinetics of the transformation depends. The second term which is the free-energy difference between the starting phase and the activated complex determines the kinetics of the transformation and is dependent on the mechanism.

(a) *Free-energy differences between the polymorphs*

At a given temperature each phase has a certain free energy given by $G = H - TS$, where H is the enthalpy and S the entropy of the phase. When polymorphism occurs there is a critical temperature and pressure at which the free energies of the two phases are equal, i.e. $\Delta G = 0$ and they are in equilibrium (Fig. 3.4). At this point the driving force for the transformation is zero. The extent to which the curves diverge as a function of temperature and pressure indicates the change in ΔG as a function of these variables. This in turn clearly depends on the slope of the G curve for each phase, i.e. on the entropy of each phase.

Entropy data have been estimated from high-temperature heat capacities and for the aluminium silicates the structural similarities lead to very similar entropy values. Over a range of temperature, ΔS for andalusite–sillimanite is only about 0.3 cal mol^{-1} deg C^{-1}, and for kyanite–andalusite and kyanite–sillimanite ΔS is around 2 cal mol^{-1} deg C^{-1}. Translated into free-energy terms this results in trivial free-energy differences between the polymorphs even at temperatures and pressures considerably above or below the critical equilibrium values.

Small values of ΔG need not affect the kinetics of phase transformations—we have described examples where very small values of ΔG may be sufficient to drive a transformation if the mechanism and kinetics are favourable. However, where a reconstructive nucleation process is the only possible mechanism for a transformation, as is the case here, the small values of ΔG greatly increase the size of the critical radius and the activation energy for nucleation as shown by equations 5.3 and 5.4 (section 5.2).

The consequences of the small ΔG values are threefold: they decrease the likelihood of the nucleation of one phase in the other taking place under equilibrium conditions; they make it extremely difficult to determine phase boundaries experimentally; and they introduce problems of metastable crystallization from parent reactants which themselves have a high free energy. Although these aspects are related we will comment on each in turn.

1 The small change in ΔG over a wide range of temperatures means that very large departures from equilibrium will be required to reduce the activation energy for nucleation and drive the reaction forward. Thus any evidence of a transformation from one phase to another in the solid state does not imply that the P,T conditions were anywhere near the equilibrium values, rather the opposite.

2 Phase-equilibrium determinations where a starting mix is raised to various P,T conditions and the nature of the products determined will be extremely prone to small effects, and strain energy, surface energy, the presence of impurities etc. will become critical. For example, if starting materials are finely ground, the surface-energy terms introduced would be of the order of 100 cal mol^{-1} and strain energies would be even higher. A change in ΔG of around 100 cal mol^{-1} would result in a shift of around 50°C in the position of the apparent equilibrium curve for kyanite–andalusite, and 300°C in the andalusite–sillimanite curve, thus making any petrological application meaningless. The possible thermodynamic effect of Fe^{3+} substitution for Al^{3+} and of Al, Si-disorder in sillimanite, must also be taken into account in this context.

Conversely, if we want to determine the phase boundaries with an uncertainty of ±50°C we must use procedures which will give a free energy with an accuracy of ± 15 cal mol^{-1} for andalusite–sillimanite, and ± 100 cal mol^{-1} for reactions involving kyanite. These are trivial quantities when compared to the internal lattice energies of these minerals which are around 10^5 cal mol^{-1}. Some of the methods which have been used to determine the equilibrium phase diagram have been summarized by Fyfe.

3 The third problem introduced by the small ΔG values is that of metastable crystallization from a starting phase or phases which have high free energies. If we start with a glass or an oxide mix of Al_2SiO_5 composition and attempt to synthesize any of the polymorphs around their equilibrium P,T conditions, the ΔG for the synthesis reaction will be very high compared to the ΔG between the polymorphs themselves. This has been illustrated in Fig. 5.8. Thus, energetically there is a very strong tendency to crystallize any of the polymorphs, and which of them actually crystallizes will depend less on the thermodynamics than on kinetic factors such as preferential nucleation etc. Experiments based on these methods must be regarded as suspect.

In metamorphic rocks, the aluminium silicates are usually formed by dehydration reactions such as:

$$\text{muscovite} + \text{quartz} \rightarrow Al_2SiO_5 + \text{sanidine} + H_2O$$

which have a large ΔG compared to the ΔG between the polymorphs. Whether the stable or metastable polymorph forms depends on their relative rates of nucleation and growth. In a heterogeneous system such as a rock, the role of preferential nucleation is difficult to evaluate, although the presence of numerous possible nucleation sites and a vapour phase does suggest that over geological time there would not be a marked preference for the formation of the metastable phase and simultaneous suppression of the stable phase, i.e. metastable growth may not be a very important factor. Over a fairly wide range of conditions both phases might be expected to occur, a situation borne out by natural assemblages.

The overall consistent pattern of Al_2SiO_5 occurrences in nature supports the argument that an equilibrium phase diagram may be used in a general way to divide low-pressure and intermediate-pressure metamorphic regimes, although any quantitative deductions must be considered speculative.

(b) *Transformations between the polymorphs*

The kinetics of the direct transformation from one polymorph to another as P,T conditions change will be governed by the overall free energy of activation of the process. This includes a number of separate activation energies and depends on the mechanism of the transformation. Nucleation in the dry state of one phase in another will involve Al—O and Si—O bond breaking and diffusion, both processes associated with large activation energies. The free energy of activation for the formation of a critical nucleus will also be great, as already mentioned. Thus, in the solid state the kinetics of such a transformation will be extremely sluggish even on a geological time scale and metastable persistence of these phases in dry systems would be expected to be common.

The overall free energy of activation is related to the enthalpy of activation and the entropy of activation in the usual way ($\Delta G_a = \Delta H_a - T\Delta S_a$) so that by providing a mechanism which increases the entropy term, the free energy of activation can be decreased and the transformation speeded up. (Note that this is independent of the thermodynamics of the phases themselves.) The presence of a fluid phase provides such mechanisms, and experiments have shown that the gradual dissolution of the unstable phase coupled with growth of the more stable phase would be expected to take place in geologically short times, although the deviations from equilibrium are not known.

In conclusion, therefore, the occurrence of the polymorphs in nature is determined both by the thermodynamics of the phases themselves (leading to the possibility of metastable growth) and by the mechanism of the transformation from one to the other (leading to metastable persistence of phases). The common coexistence of two and even all three polymorphs in nature suggests that some combination of the factors mentioned here is operating. However, in a complex rock it is also likely that the apparently simple polymorphic change from one phase to another takes place by a series of reactions involving other minerals. An evaluation of the extent of metastability must be made in each specific case. Of primary importance is the role of H_2O, which not only affects the kinetics but the phase chemistry, a topic discussed in many metamorphic petrology books.

It is worthwhile making a brief comparison at this stage of the relative merits of using mineral reactions in a rock and homogeneous processes such as cation ordering in a mineral, as potential indicators of the geological history of a rock. Many mineral reactions are beset by the same problems as described for the Al_2SiO_5 polymorphs, and kinetic studies are greatly hindered by the insufficiently understood effects of grain size, stress, catalytic effects of fluid phases, nucleation problems etc. The extent to which an experimental equilibrium reaction curve, if indeed it does represent equilibrium, may be applied to different natural rocks is questionable. Continuous ordering transformations, on the other hand, although not yet fully understood are largely exempt from these effects, and thus are potentially more likely to provide quantitative thermal history data.

7.6 High-pressure phase transformations

The changes which take place in a mineral as the temperature decreases involve a reduction in entropy and in internal energy. The internal energy is reduced because of the increased co-ordination of the atoms as cation sites contract in size. The response of cation polyhedra to changes in temperature has already been discussed, particularly in the case of the alkali feldspars and the pyroxenes.

An increase in pressure has, in many ways a similar effect as a decrease in temperature. We can see this from the simple thermodynamics of pressure transformations. For any phase

$$G = H - TS$$

When the pressure is not negligible the term PV must be considered in the enthalpy term,

$$G = E + PV - TS$$

If two polymorphs exist, then at equilibrium ΔG between them is zero

$$E_1 + PV_1 - TS_1 = E_2 + PV_2 - TS_2$$
$$\Delta E - T\Delta S + P\Delta V = 0$$

Since an increase in pressure results in a decrease in volume the $P\Delta V$ term will be negative. Therefore the polymorphic transformation on increased pressure must involve an increase in internal energy, a decreased entropy or both.

Thus, to a first approximation it might be expected that an increase in pressure or a decrease in temperature may produce similar structural changes in those minerals whose structures can be described in terms of simple cation polyhedra. In many silicates the tetrahedral framework remains inert while the cation sites expand or contract as a function of temperature and pressure. A temperature-induced transformation of this kind, which we have discussed already, is that from monoclinic to triclinic albite (section 7.1.1). Our first example of the effects of pressure on some minerals is the analogous structural collapse which can be brought about in alkali feldspars by an increase in pressure.

At higher pressures grosser transformations take place in many minerals producing denser, more highly co-ordinated structures. Such transformations have been studied widely in an attempt to explain density changes and discontinuities which are known, from seismic data, to occur in the earth's mantle. A large number of such transformations have been discovered, notably by Ringwood and coworkers, and described in his book (see refs). In this section we shall look at one such transformation: that from olivine to the spinel structure. As yet, the mechanisms of many of these high-pressure transformations are not known. The principles of behaviour and the range of transformation types apply equally well however to both pressure- and temperature-induced transformations.

7.6.1 MONOCLINIC–TRICLINIC TRANSFORMATIONS IN SANIDINE AT HIGH PRESSURE

In albite, below about 1100°C at atmospheric pressure, a structural collapse of the Al–Si framework takes place about the alkali site which is too small to support a framework with monoclinic symmetry. The transformation temperature decreases as the potassium content increases, due to the effective increase in the size of the cation site produced by this substitution. The cell dimensions of the feldspar at the transformation temperature are nearly constant over a range of compositions suggesting that there is some critical effective cation size below which the transformation will take place. In other words the transformation is a structurally controlled event.

The effective size of the cation site may change in three ways: by changing the temperature, the potassium content, or the pressure. On this basis Hazen predicted that potassium-rich alkali feldspars, which in the disordered state remain monoclinic at room temperature, should transform to the triclinic state at elevated pressures, with more potassic sanidines requiring greater pressures. The experiments which he carried out subsequently confirmed this prediction with a sanidine of composition Or_{67} transforming at 12 ± 1 kbar and Or_{82} at 18 ± 1 kbar, at room temperature. Thus in this case temperature, pressure and composition have behaved as structurally analogous variables, an increase in pressure being equivalent to a decrease in temperature or an increase in the Na : K ratio.

These results suggest that it is possible to define a certain pressure (P), temperature (T) and composition (X) at which the structure is at its critical geometry

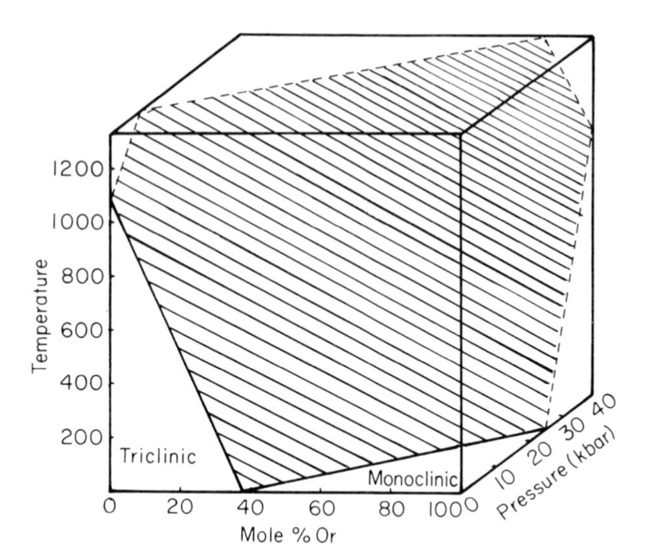

Fig. 7.22 Pressure–temperature–composition (P–T–X) surface for the monoclinic–triclinic transformation in disordered alkali feldspars. (After Hazen, 1976.)

and about to collapse to triclinic symmetry. This can be drawn as a P–T–X isostructural surface (Fig. 7.22). Where it can be demonstrated that a transformation is controlled in this way by the structural geometry, these P–T–X surfaces which can be calculated, may be closely related to phase boundaries. Hence basic structural data might be used to predict some phase equilibria.

This relatively new approach may provide data on the behaviour of a number of minerals whose stabilities are limited by geometrical limits in the relative sizes of adjacent polyhedra. Pyroxenes and amphiboles may also be suitable candidates for such treatment. The effect of pressure on the high–low pigeonite transformation, for example, should be to increase the temperature at which the structural collapse can take place. Pressure will therefore need to be considered in the interpretation of antiphase domain sizes in pigeonite.

7.6.2 THE OLIVINE–SPINEL TRANSFORMATION

It is generally agreed that olivine of approximate composition $(Mg_{0.9}Fe_{0.1})_2SiO_4$ is probably the major mineral in the earth's upper mantle. Structural changes which may occur in olivine as a function of increased pressure will involve the formation of denser, structurally more compact phases and will therefore have important consequences on the properties of the mantle. Seismic data indicate that changes in seismic velocity and hence density do occur at depth with a major discontinuity at around 400 km, corresponding to pressure of around 130 kbar and temperatures in the vicinity of 1500°C.

More recently it has been proposed that the increase in density brought about by phase transformations may be one of the mechanisms generating deep-focus earthquakes and may also provide a driving mechanism for the sinking oceanic crustal slabs in plate tectonics. Interpretations of this kind depend to a large extent on the kinetics of the phase transformations involved, which in turn depends on the mechanism. At the present time no electron-microscope studies have been made to determine such mechanisms, although the phase transformations can be carried out experimentally using high-pressure and temperature apparatus.

The olivine structure can be described in terms of hexagonal close-packed oxygen atoms with Si occupying one-eighth of the tetrahedral sites and Mg,Fe in one half of the octahedral sites. In the spinel structure the oxygens are approximately cubic close-packed with the cations in the same co-ordination as in olivine. The transition from olivine to spinel thus does not involve a change in the co-ordination number of the cations, which is unusual in high-pressure phase transformations which involve a large increase in density. The density increase of around 10% in transforming to the spinel structure is accomplished by changing the distribution of the cations in the available sites in such a way that the linkages between cation polyhedra form a more rigid and compact structure.

The phase diagram for the system Fe_2SiO_4–Mg_2SiO_4 at high pressure is shown in Fig. 7.23. At the forsterite end which is the most relevant to mantle studies, an intermediate β phase which has a 'modified-spinel' structure is formed prior to the formation of the spinel phase. The temperatures and pressures of the transformation to spinel are consistent with depths of around 400 km in the mantle, and have been invoked to partially explain the seismic velocity discontinuities which have been observed.

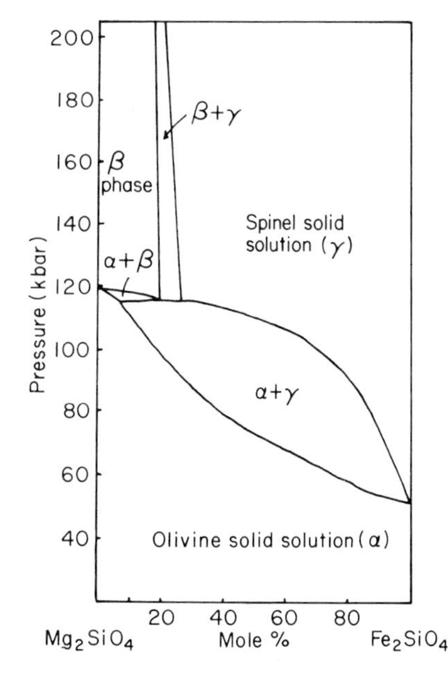

Fig. 7.23 Approximate phase diagram for the system forserite (Mg_2SiO_4)–fayalite (Fe_2SiO_4) at 1000°C (after Ringwood, 1975).

The structural changes involved in transforming olivine to spinel require a reconstructive process of bond breaking and reforming, and thus nucleation and growth is the likely mechanism. The factors important in the kinetics of such a transformation are similar to those already described for both homogeneous and heterogeneous nucleation, although the events are now both pressure and temperature controlled. Although the transformation is essentially a response to the increased pressure, it is a thermally activated process. The rate therefore increases exponentially with temperature as well as the overpressure (ΔP) beyond equilibrium. Just as in the previous polymorphic transformations that have been discussed, the temperature and the extent of overpressure required to achieve such a major transformation will depend on those factors which determine the size of the critical nucleus (i.e. strain- and surface-energy terms) and the diffusion rate across the interface.

In experimentally induced olivine–spinel transformations, the olivine is usually very finely ground and may have water added to it. The shear stresses generated in the experimental apparatus are much higher than those expected in the mantle. All of these factors affect the free energy of activation and hence the kinetics, so that transformation rates determined in this way will be orders of magnitude higher than the rate in the mantle. Mantle olivines will also probably have recrystallized at high temperature and so be large and relatively stress free. Therefore homogenous nucleation, which requires further penetration into the metastable region may be the dominant mechanism. The extent to which olivine in the mantle remains metastable relative to the transformation to spinel is an important question if phase transformations are to be held responsible for many of the properties of the mantle.

The temperature dependence of the kinetics implies that apart from the over-pressure required to achieve the transformation, the temperature must be high enough for atomic diffusion to be significant. Below a certain cut-off temperature the transformation may not be able to take place no matter how high the pressure. Thus if the mantle material is not hot enough the olivine may not transform to spinel. Sung and Burns have calculated that this cut-off temperature is around 700°C for Mg_2SiO_4. Above this temperature the transformation rate is dependent on both T and ΔP.

Figure 7.24 shows four schematic Time–Pressure–Transformation (TPT) curves for heterogeneous nucleation of spinel to illustrate the kinetics at four different temperatures. The following features should be noted.

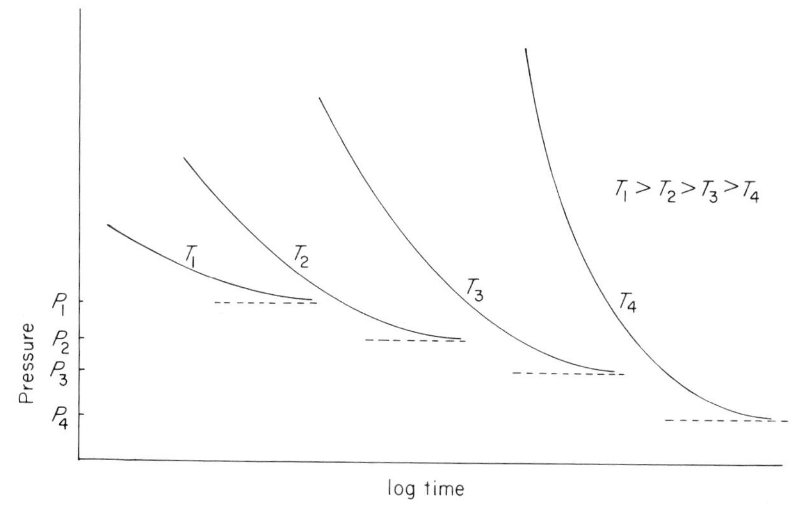

Fig. 7.24 Schematic Temperature–Pressure–Transformation (TPT) curves illustrating the kinetics of the olivine–spinel transition at four different temperatures $T_1 > T_2 > T_3 > T_4$.

1 The equilibrium pressure for the transformation (shown by $P_1 \rightarrow P_4$) decreases as the temperature decreases ($T_1 \rightarrow T_4$). The average gradient dP/dT for the transformation is around 30 bars °C^{-1}.

2 The rate of the transformation at a given pressure above the equilibrium value is strongly temperature dependent.

3 At lower temperatures the overpressure ΔP required to start the transformation in a given time increases markedly. Thus at high temperatures departures from equilibrium may be relatively slight while at lower temperatures the tendency for the metastable retention of olivine will be high.
4 At some lower value of temperature (not shown) the rate of the transformation will be negligible at any amount of overpressure.

One of the consequences of this behaviour which has been discussed by Sung and Burns concerns the fate of olivine in a cold downgoing slab of lithosphere being subducted into the mantle. If the plunging velocity of the slab is high the temperature at its cold centre may remain below the kinetic cut-off of 700°C even down to depths in excess of 600 km, where the overpressure ΔP will be very high and the olivine in a highly metastable state. As the slab heats up to above the cut-off temperature the rate of the transformation will be so rapid, and the change in free energy so great that the transformation becomes implosive. The energy, which is released as seismic waves, is more than sufficient to generate deep-focus earthquakes. The depth at which the transformation occurs will decrease with decreasing plunging velocity of the slab.

Other consequences of the olivine–spinel transformation on the dynamics of the mantle have been discussed in some depth by Sung and Burns and will not be treated further here. It is becoming clear however that an understanding of the thermodynamics and kinetics of mineral behaviour of this kind may be vital to an understanding of large-scale tectonic processes.

References and additional reading

7.1

McConnell, J.D.C. (1971) Electron optical study of phase transformations. *Mineral. Mag.* **38**, 1.

McConnell, J.D.C. and McKie, D. (1960) The kinetics of the ordering process in triclinic $NaAlSi_3O_8$. *Mineral. Mag.* **32**, 436.

MacKenzie, W. S. (1957) The crystalline modifications of $NaAlSi_3O_8$. *Am J. Sci.* **255** 481.

Ribbe, P.H. (Ed.) (1975) *Feldspar Mineralogy*. Mineral. Soc. America Short Course Notes.

Stewart, D.B. and Ribbe, P.H. (1969) Structural explanation for variation in cell parameters of alkali feldspar with Al/Si ordering. *Am. J. Sci.* **267A**, 144.

7.2

Burnham, C.W. (1973) Order-disorder relationships in some rock-forming silicate minerals. *Ann. Review Earth and Planetary Sciences* **1**, 313.

Champness, P.E. and Copley, P.: The transformation of pigeonite to orthopyroxene. In: Electron microscopy in Mineralogy. Wenk, H.U. (Ed.) Springer-Verlag 1976.

Papike, J.J. and Cameron, M. (1976) Crystal chemistry of silicate minerals of geophysical interest. *Reviews of Geophysics and Space Physics* **14**, 37.

7.3

Mueller, R.F. (1969) Kinetics and thermodynamics of intracrystalline distributions. *Miner. Soc. Am. Spec. Pap.* **2**, 83.

Seifert, F.A. (1977) Reconstruction of rock cooling paths from kinetic data on the Fe^{2+}-Mg exchange reaction in anthophyllite. *Phil. Trans. Roy. Soc. Lond. A* **286**, 303.

Seifert, F.A. and Virgo, D. (1975) Kinetics of the Fe^{2+}-Mg order-disorder reaction in anthophyllites: quantitative cooling rates. *Science* **188**, 1107.

7.4

Putnis, A. (1978) Talnakhite and mooihoekite: the accessibility of ordered structures in the metal-rich region around chalcopyrite. *Canad. Mineral.* **16**, 23.

Putnis, A. and McConnell, J.D.C. (1976) The transformation behaviour of metal-enriched chalcopyrite. *Contrib. Mineral. Petrology* **58**, 127.

7.5

Brown, G.C. and Fyfe, W.S. (1971) Kyanite-Andalusite equilibrium. *Contrib. Mineral. Petrol.* **33**, 227.

Fyfe, W.S. (1976) Stability of Al_2SiO_5 polymorphs. *Chem. Geol.* **2**, 67.
Richardson, S.W., Gilbert, M.C. and Bell, P.M. (1969) Experimental determination of kyanite-andalusite and andalusite-sillimanite equilibria; the aluminium silicate triple point. *Am. J. Sci.* **267**, 259.
Turner, F. J. (1968) *Metamorphic Petrology*. McGraw-Hill.
Vernon, R.H. (1976) *Metamorphic processes.* George Allen and Unwin.
7.6
Hazen, R.M. (1976) Sanidine: predicted and observed monoclinic to triclinic reversible transformation at high pressure. *Science* **194**, 105.
Hazen, R.M. (1977) Temperature, Pressure and Composition: Structurally analogous variables. *Phys. Chem. Minerals* **1**, 83.
Ringwood, A.E. (1975) *Composition and Petrology of the Earth's Mantle.* McGraw-Hill, Ch. 11.
Sung, C.M. and Burns, R.G. (1976) Kinetics of high-pressure phase transformations: implications to the evolution of the olivine–spinel transition in the downgoing lithosphere and its consequences on the dynamics of the mantle. *Tectonophysics* **31**, 1.

8

Mineral Transformations II—Transformations Involving Exsolution

In this chapter we will describe the behaviour of a number of mineral systems in which a solid solution exists at elevated temperatures, while exsolution into compositionally different phases tends to take place on cooling. In Chapter 4 the simple thermodynamics of such behaviour was discussed in terms of the effect of a positive ΔH of mixing on the phase diagram, and in Chapter 5 the possible mechanisms of phase separation were described. The aim here is to illustrate how some of these mechanisms operate in well-known minerals and how the relative kinetics of the processes may be used to interpret the microstructures and thermal histories of the minerals.

In many cases the microstructures which form are not only the result of exsolution processes but also of polymorphic transformations which may take place in one or both of the separated phases. The relative rates of the exsolution and polymorphic transformations, coupled with the cooling rate may produce markedly different mineral textures. This is illustrated by both the alkali feldspars and the pyroxenes.

Exsolution may also take place as a result of a change in the bulk composition of a mineral. Oxidation in minerals containing Fe^{2+} is the most common example and often results in a process termed oxidation exsolution. We will illustrate this with reference to the early stages of oxidation in olivines and spinels.

The final example in this chapter illustrates the point that solid solutions may form metastably in minerals which have crystallized at low temperatures. Below a certain temperature these solid solutions will persist indefinitely, but if the minerals are subsequently heated to a temperature above this kinetic cut-off, exsolution can take place. This method may be used to determine an upper limit for the crystallization temperature. Iron-bearing rutiles which have crystallized under hydrothermal conditions behave in this way.

8.1 The cooling behaviour of alkali feldspar solid solutions

We have already discussed the polymorphic transformations which take place in the two end members albite $NaAlSi_3O_8$ and sanidine $KAlSi_3O_8$ on cooling (section 7.1). We now consider the cooling behaviour of intermediate compositions, which at high temperatures belong to a continuous monoclinic solid solution between the two end members. As we have already anticipated on the basis of relative cation sizes (section 3.6.2) a solvus exists in this system at lower temperatures. The phase diagram is shown in Fig. 8.1.

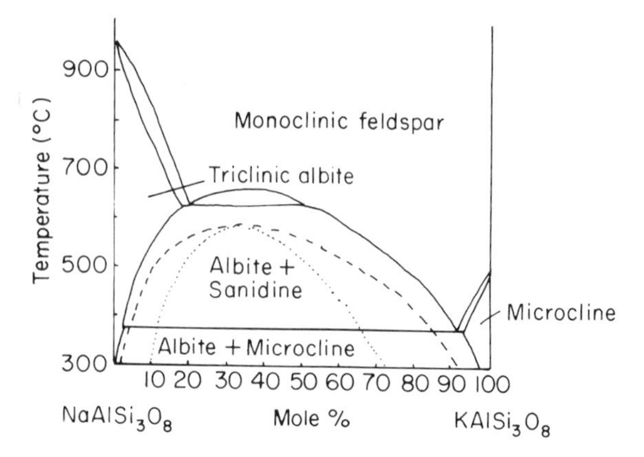

Fig. 8.1 Equilibrium phase diagram for the alkali-feldspar system. The dashed line is the coherent solvus, and the dotted line the coherent spinodal. (After Robin, 1974.)

The microstructures which may form on cooling involve three different processes:

1 the inversion from monoclinic to triclinic albite, which is instantaneous;
2 the segregation of sodium and potassium ions, which is relatively rapid;
3 the Al, Si-ordering in the K-rich phase which is extremely sluggish.

The final result therefore depends on the bulk composition and the cooling rate. The exsolution intergrowths formed are called perthites and may be on a scale from tens of Angstroms to several millimetres. Compared with other silicates slowly cooled alkali feldspars may form very coarse textures due to the relatively rapid diffusion rates of sodium and potassium and the fact that exsolution does not involve the migration of silicon or aluminium in the framework structure.

The fact that the aluminosilicate framework may remain essentially unaltered while sodium and potassium segregation takes place makes it possible for coherency to be maintained and processes such as coherent nucleation and spinodal decomposition are possible. In fact the similarity in cell dimensions and elastic constants of Na- and K-rich regions results in only a small depression ($\approx 80°C$) of the coherent solvus below the equilibrium solvus, so that coherent behaviour is likely to be important when cooling rates are relatively rapid. The coherent solvus and coherent spinodal are shown in Fig. 8.1 by the dashed and dotted lines, respectively.

Thus there are two different mechanisms by which exsolution can take place. Firstly the diffusion of Na and K can form compositionally different regions, while the aluminosilicate framework remains continuous across the interfaces, merely flexing to accommodate the different sizes of Na and K atoms. This coherent mechanism takes place under conditions where the second mechanism, heterogeneous nucleation, has been impeded kinetically. In this process nucleation at surfaces and defects takes place with the structure of the new phase being almost parallel to but structurally discontinuous from the framework in the matrix.

Here we will be concerned mainly with the first of these processes and with the early stages of exsolution in rapidly cooled alkali feldspars. These are reasonably well understood and have in some cases been reproduced experimentally. The microstructures described are therefore limited to the cryptoperthites (on a scale not resolvable in a light microscope). In these early stages coherent processes dominate and the orientation, shape and subsequent evolution of the microstructures are determined by minimizing the strain energies involved.

The coarser perthites are incoherent intergrowths and it is often not possible to determine the intial exsolution mechanism. Coherent nucleation or spinodal decomposition will eventually coarsen, lose coherency and produce a texture indistinguishable from heterogeneous nucleation. Coarse perthites have a remarkable variety of textures and in some cases non-exsolution origins such as cation replacement have been suggested. Although exsolution is the favoured mechanism in most perthites, the extrapolation from cryptoperthites to the many coarse morphologies found in pegmatites requires considerable imagination, and it is likely that in at least some of these cases complex diffusion mechanisms involving sodium metasomatism may contribute to the final texture. Thus we must be cautious in using coarse perthites as indicators of thermal history.

In order to describe in general terms the sequence of processes taking place in the early stages of exsolution in alkali feldspars we will consider the evolution of a number of the microstructures which have been observed in cryptoperthites and relate them to relative cooling rates.

8.1.1 CRYPTOPERTHITES

(a) *The first stages*

In extremely rapidly cooled rocks, such as quenched material ejected during volcanic explosions, the alkali feldspars may remain homogeneous as monoclinic sanidine. Generally, however, the cooling is slower and most of the cryptoperthites which have been examined by electron microscopy appear to have undergone spinodal decomposition in the early stages. Heterogeneous nucleation has not taken place to any significant extent in these minerals. As we can see from Fig. 8.1 the relative likelihood of coherent nucleation or spinoda ldecomposition depends on the composition as well as the cooling rate. Fig. 8.2 is an experimentally determined TTT curve comparing the kinetics of nucleation and spinodal behaviour for a composition 37 wt % $KAlSi_3O_8$. The curve shows that a cooling rate faster than about 5°C per hour over the interval 600–400°C would be too fast for nucleation to take place.

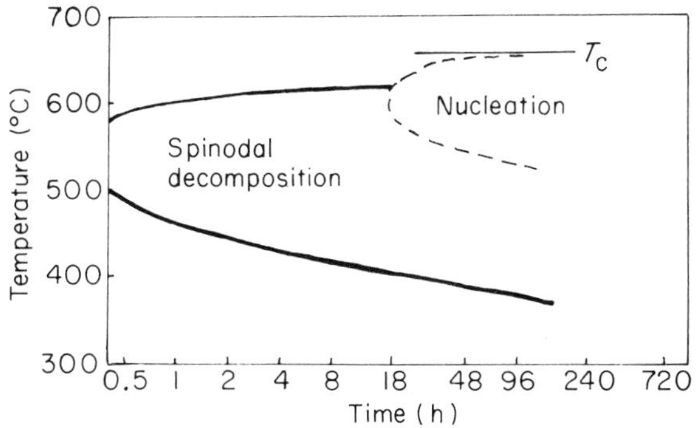

Fig. 8.2 Experimental TTT curve for the exsolution process in a high-temperature alkali feldspar of composition 37 wt % Or. (After Owen and McConnell, 1974.)

Observations of the wavelength of the spinodal modulation formed experimentally confirms the theory (section 5.3) that at temperatures near the spinodal temperature, the wavelengths will tend to infinity and the kinetics will accordingly be more sluggish. The extent of supercooling below the spinodal temperature determines the initial wavelength formed, although subsequent coarsening may take place on annealing. Therefore relatively coarse spinodal textures may arise directly by spinodal decomposition without significant coarsening, or by coarsening of an initially fine-scaled modulation, depending on the precise thermal history.

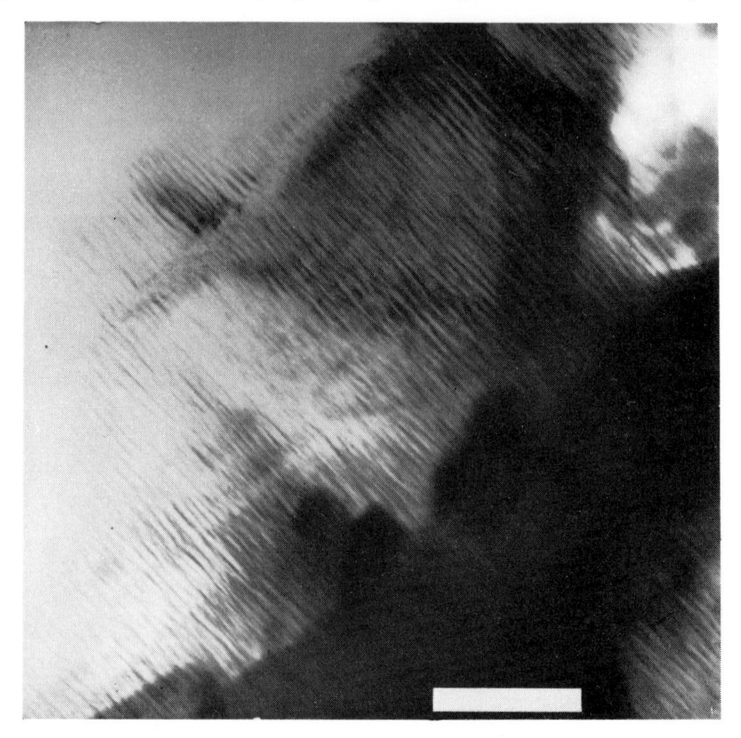

Fig. 8.3 Modulated microstructure consistent with a spinodal decomposition process in a natural sample of alkali feldspar. The length of the scale bar is 0.2 μm.

Fig. 8.3 shows a relatively fine spinodal microstructure in a natural specimen from a lava flow. At this stage both the Na- and K-rich components are monoclinic, as required for complete coherency. The albite-rich component therefore as yet shows no signs of transforming to the triclinic state, the structural collapse being prevented by the coherency strains. The orientation of the spinodal modulation is that involving the minimum strain energy for the coherent coexistence of two monoclinic feldspars. There is a very good agreement between the orientation calculated from the lattice parameters and elastic properties of the feldspars and the observed orientation.

(*b*) *The inversion in the Na-rich regions*

As the microstructure coarsens and the temperature falls, the coherency strains are no longer sufficient to maintain the monoclinic structure in the Na-rich regions. As the structural collapse to the triclinic form takes place, some loss of coherence is inevitable. To minimize the strain energy involved in sandwiching thin lamellae of a triclinic phase between monoclinic potassic regions, fine-scaled periodic twinning takes place in the sodic regions. Geometrically, there is an equal probability that the twins will be either albite or pericline laws, although albite twins tend to predominate when the degree of Al, Si-ordering is greater.

The twinning produces domains in which the distortion is in opposite senses (Fig. 7.1) so that the finer the twinning the more similar the 'average' structure is to the monoclinic state. This reduces the strain energy across the interface between the triclinic Na-rich and monoclinic K-rich regions. For the same reason the periodicity of the albite twinning depends on the thickness of the sodic regions—the thicker the lamellae, the coarser the twinning.

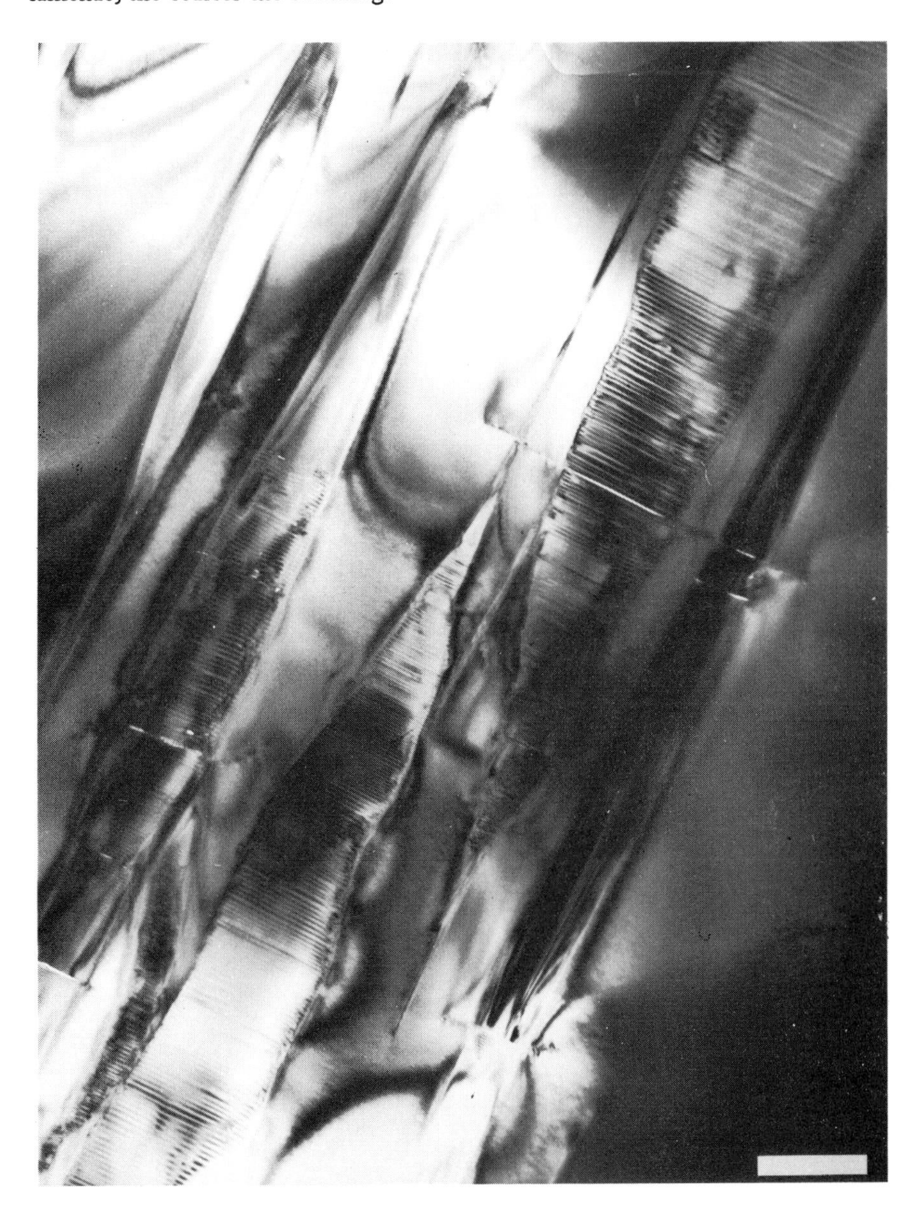

Fig. 8.4 Electron micrograph of a cryptoperthite of bulk composition Or_{72}. The wedge-shaped albite lamellae show fine-scaled albite twinning with a periodicity which depends on the thickness of the lamella. The microstructure suggests that the albite lamellae formed by a process of nucleation and growth. The length of the scale bar is 0.5 μm. (Photo courtesy of A. C. McLaren.)

Fig. 8.4 is an electron micrograph showing such fine-scaled twinning in the albite-rich regions of a cryptoperthite. In this case the bulk composition is more potassic (Or_{72}) so that the amount of albite phase is smaller and exsolution has taken place by a nucleation mechanism. The orientation of the albite lamellae is the same as that in spinodal decomposition, although they tend to have a lenticular shape and are irregularly distributed throughout the crystal.

(c) *Ordering in the K-rich regions*

Although there is still some uncertainty regarding the exact details, it appears that the microstructure associated with the ordering transformation in the potassic phase depends on the morphology of these potassic regions as well as on the other factors such as cooling rate. Within a spinodal texture Al, Si-ordering and the transformation to the triclinic state may be initially associated with the formation of a perturbed orthoclase structure but subsequently relatively coarse twins are formed via the 'intermediate microcline route' (see section 7.1.2). The twinning achieves the same purpose as that in the albite phase, to reduce the strain energy across the interfaces, although it is coarser in this case since the obliquity (deviation from the monoclinic cell) is lower.

As soon as the potassic phase becomes triclinic the strain involved tends to pull the interfaces around and they assume a wavy pattern with a wavelength associated with the scale of incipient twinning in the potassic phase. Fig. 8.5 is an electron micrograph of a cryptoperthite which appears to have exsolved by a spinodal

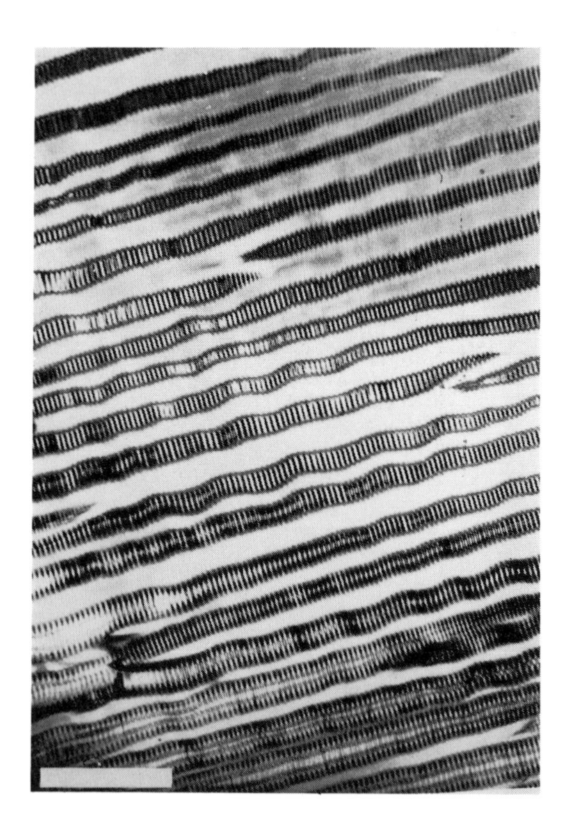

Fig. 8.5 Electron micrograph of a cryptoperthite showing periodic twinning in the lamellae of the albite-rich component. The lamellae have a wavy interface with the K-rich matrix. The length of the scale bar is 0.5 μm. (Photo courtesy of P. E. Champness.)

mechanism. The albite component is finely twinned and the orthoclase component is assumed to be partly ordered and have some degree of triclinicity. Thus the interface has a wavy pattern to reduce interfacial strain.

Fig. 8.6 Electron micrograph of a cryptoperthite showing the zig-zag interface between the finely twinned Na-rich feldspar and the K-rich feldspar matrix. The length of the scale bar is 0.2 μm. (Photo courtesy of C. Willaime.)

Fig. 8.7 Electron micrograph of a cryptoperthite showing the lozenge-shaped rafts of the twinned albite component in a matrix of the microcline phase. The length of the scale bar is 1.0 μm. (Photo courtesy of P. E. Champness.)

In samples which have had a longer thermal history this process has continued to a further extent, the boundaries readjusting themselves to give lower energy surfaces between the triclinic K-feldspar and the 'pseudo-monoclinic' (because it is finely twinned triclinic) Na-feldspar. This produces a zig-zag pattern in the interface (Fig. 8.6). The differences in the elastic properties of the phases eventually leads to 'rafting' or breaking up of the Na-rich regions, and results in lozenge-shaped islands of albite phase within a matrix of the microline phase (Fig. 8.7). These processes which occur in an initially spinodally decomposed alkali feldspar are dominated by the need to keep strain-energy terms to a minimum.

In more K-rich bulk compositions where the initial exsolution mechanism is one of nucleation (as shown in Fig. 8.4) the K-rich regions are large and relatively strain free. In this case the transformation to the triclinic state takes place with both twin laws present and eventually the typical cross-hatched microcline texture forms. Fig. 8.8 summarizes the evolution of the microstructures observed in cryptoperthites. The path taken and the stage at which the development stops depends on the composition and the cooling rate.

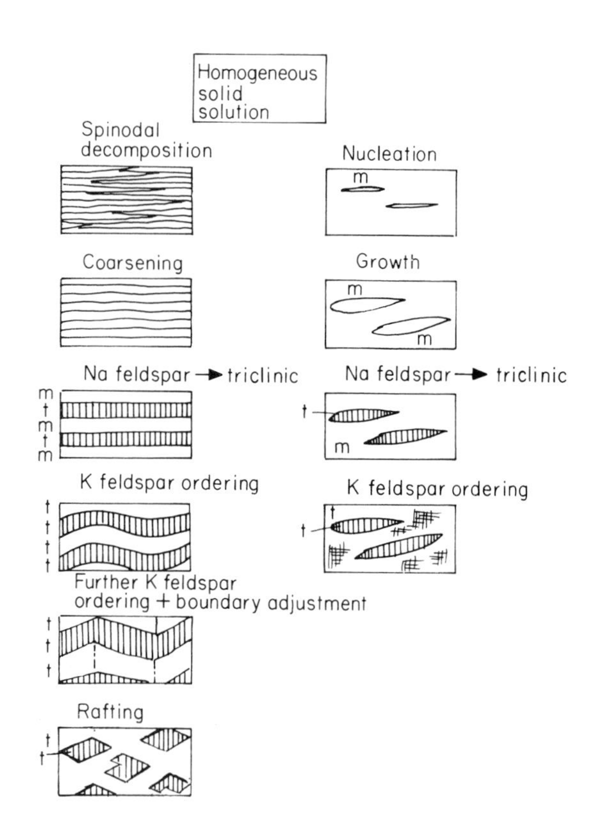

Fig. 8.8 Schematic diagram illustrating the development of the microstructures in cryptoperthites where the initial unmixing process is either one of spinodal decomposition or nucleation and growth. (Based on Lorimer and Champness, 1973 and Willaime *et al.*, 1976.)

8.2 Exsolution processes in the pyroxenes

Exsolution processes in the pyroxenes occur mainly as a result of the decrease in the extent of solid solution between Ca-rich and Ca-poor pyroxenes as the temperature decreases. The solubility of calcium is further decreased in the Ca-poor pyroxenes by the structural transformations from clino to orthopyroxene, or from high to low clinopyroxene. Both of these transformations further restrict the size of the M2 site (see sections 3.4.1 and 7.2).

The phases which result from these exsolution processes lie towards opposite sides of the pyroxene quadrilateral [Fig. 8.9(a)]. The most convenient phase diagram to illustrate the cooling behaviour is one across the middle of the quadrilateral from augite to pigeonite [Fig. 8.9(b)]. This shows the decreasing solid solution fields of pigeonite and augite as temperature falls. At the pigeonite end the high–low clinopyroxene transformation takes place after metastable cooling below the clino–orthopyroxene transformation.

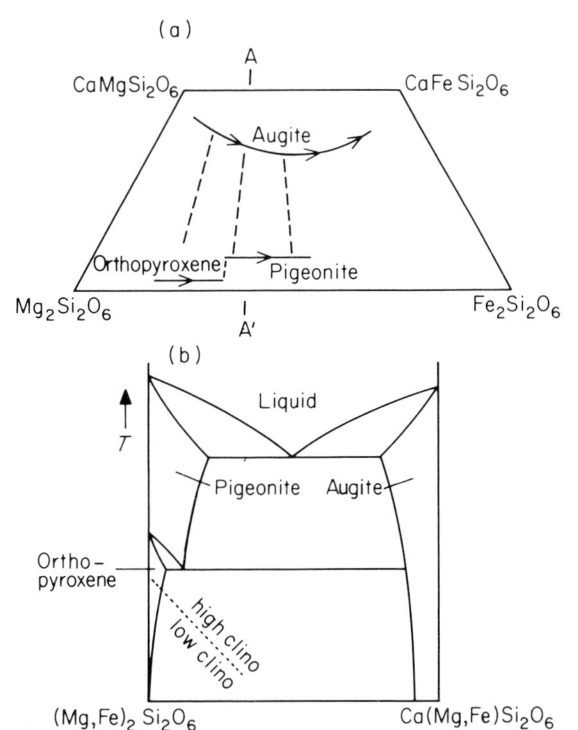

Fig. 8.9 (a) The pyroxene quadrilateral. (b) A schematic phase diagram across AA′ of the quadrilateral.

In the early stages of crystallization from a tholeitic magma, augite and orthopyroxene crystallize simultaneously. As the iron content increases during fractionation, pigeonite begins to crystallize instead of orthopyroxene. The dashed tie lines in Fig. 8.9(a) show coprecipitating phases during crystallization. Both the orthopyroxene and the pigeonite will crystallize with a higher Ca content than is stable at low temperatures, and exsolution of augite will tend to take place from both on cooling. The pigeonite is only retained (metastably) in relatively rapidly cooled basalts—in slowly cooled intrusions it is transformed to orthopyroxene.

In this section we will compare some of the features of the exsolution of augite from orthopyroxene and from pigeonite to illustrate some of the processes described in Chapter 5.

8.2.1 EXSOLUTION OF AUGITE FROM PIGEONITE

Augite may exsolve from pigeonite by a number of mechanisms depending on the cooling rate. Heterogeneous nucleation occurs at near-equilibrium temperatures while the structural similarities of the two phases enable spinodal decomposition to operate in rapidly cooled rocks. Homogeneous nucleation (which is generally difficult to distinguish from a coarsened spinodal texture) may also take place but apparently over a relatively restricted range of temperatures.

The schematic TTT curves in Fig. 8.10 illustrate these various modes of behaviour relative to the cooling rates. A wide variety of microstructures has been described from pigeonites in various geological environments and allows considerable discrimination of thermal histories. The four cooling curves superimposed on the TTT diagram may be used to account for the development of a number of the microstructures which have been described.

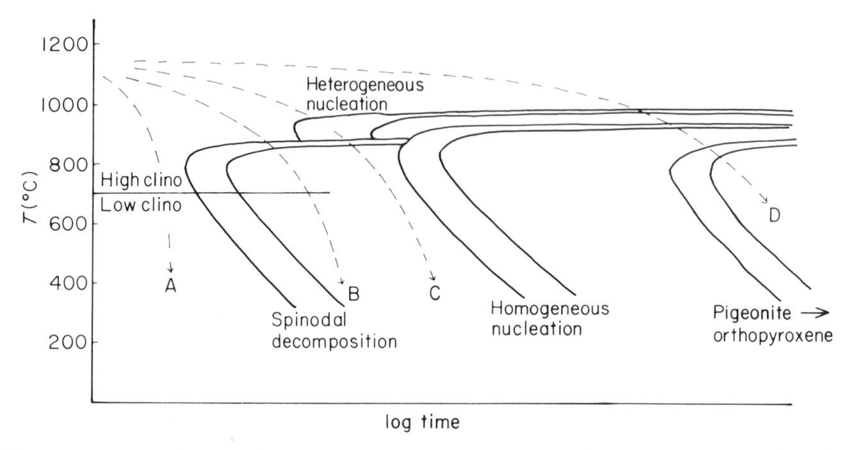

Fig. 8.10 Schematic TTT diagram illustrating the various modes of behaviour of pigeonite at different cooling rates.

Curve A shows the cooling curve for a pigeonite which misses the noses of all the decomposition curves and merely experiences the displacive transformation to low pigeonite. This transformation is shown as a straight line because it is virtually instantaneous. Fig. 8.11(a) is a pigeonite from an andesite rock from Weiselberg, Germany which shows only antiphase domains associated with this transformation.

Curve B is a slower cooling curve, although still too rapid to allow nucleation of augite. The solid solution decomposes by a spinodal mechanism, which in the early stages appears to form modulations in two directions simultaneously, resulting in a 'tweed' texture. As the spinodal coarsens only one of these modulations persists. Fig. 8.11(b) shows such a spinodal microstructure in a lunar pigeonite.

Slower cooling (curve C) leads to heterogeneous nucleation. The augite nucleates on grain boundaries and other defects and grows on (001) planes of the pigeonite, an

Fig. 8.11 A series of electron micrographs illustrating microstructures in pigeonites with different thermal histories. (a) Pigeonite from Weiselberg, W. Germany showing only antiphase domains associated with the high–low transformation. Consistent with cooling curve A of Fig. 8.10. The length of the scale bar is 0.2 μm (Photo courtesy of M. A. Carpenter.) (b) Spinodal microstructure developed in Apollo 12 lunar pigeonite. Consistent with cooling curve B in Fig. 8.10. The length of the scale bar is 0.25 μm. (Photo courtesy of P. E. Champness.) (c) Pigeonite from the Whin Sill, N. England showing exsolution lamellae of augite and curved antiphase domains in the pigeonite matrix (cf. Fig. 7.13). Second-stage lamellae of augite have exsolved in Ca-rich regions and appear to have nucleated on the antiphase boundaries. The length of the scale bar is 0.3 μm. (Photo courtesy of M. A. Carpenter.)

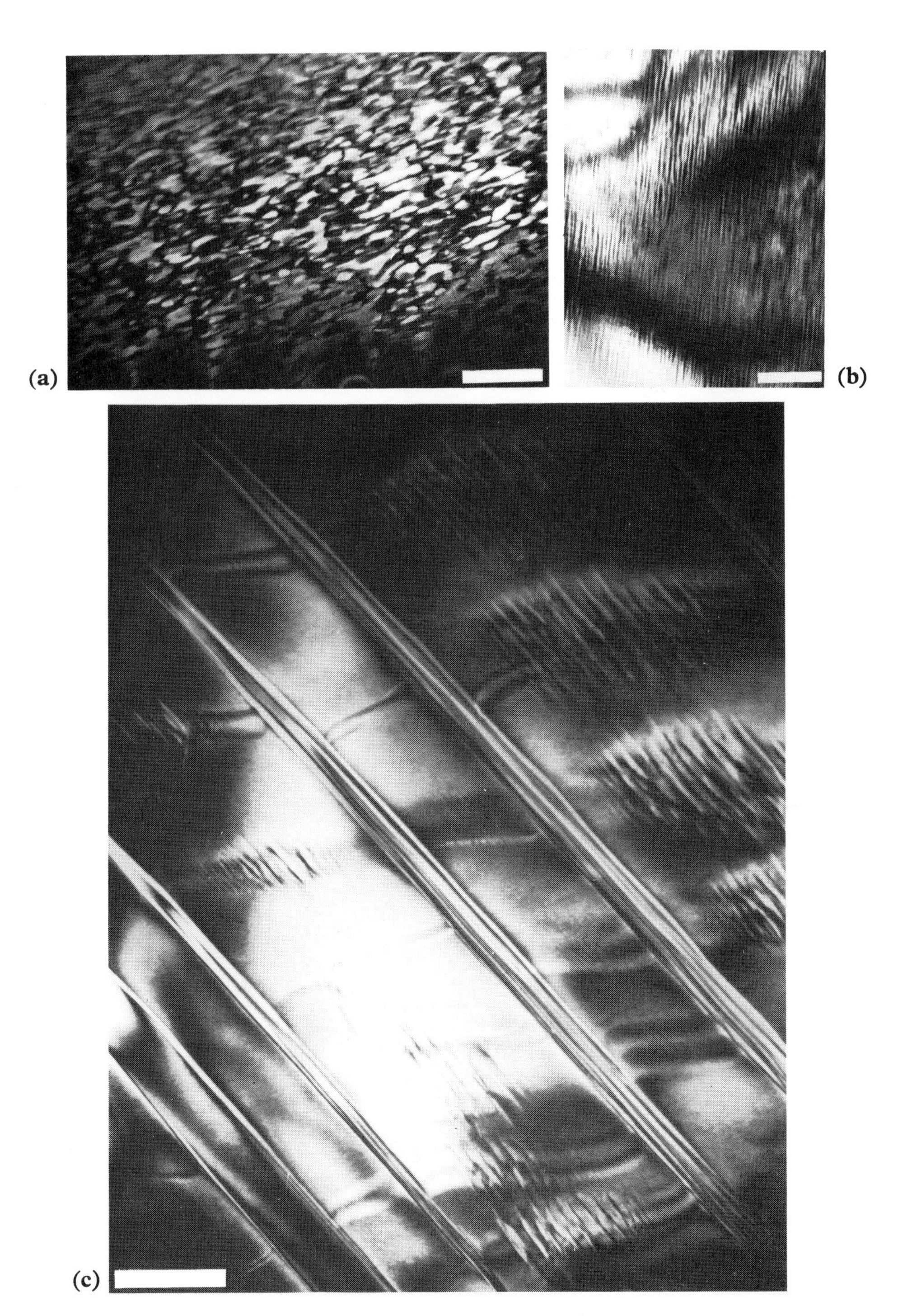
(a)
(b)
(c)

orientation in which the interfacial strain energy is a minimum. If the supersaturation is sufficiently great, i.e. the driving force for exsolution is high, other less favourable orientations of lamellae may also be formed. Thus although the interfacial energies associated with the exsolution of augite on (100) planes is about twice as high as that for (001) planes, this difference may be outweighed at high supersaturations. A similar explanation is likely for the two modulation directions of the spinodal texture. On coarsening, the energetically more favourable one is preferred.

The rate of calcium diffusion in pyroxene is relatively slow due to its large ionic radius, and it is not unusual to find stranded diffusion profiles adjacent to the augite lamellae. A zone immediately adjacent to the lamella is depleted in Ca relative to the central regions between lamellae. Thus at a lower temperature secondary exsolution may take place in these central regions, with a precipitate-free zone being formed around the augite. This secondary exsolution may be by heterogeneous nucleation or spinodal decomposition, depending on the cooling rate and the availability of nucleation sites in the crystal. Pigeonites from the Whin Sill, Northern England have cooled at a rate at which heterogeneous nucleation was able to occur, but calcium diffusion could not keep up with the cooling rate and secondary exsolution took place [Fig. 8.11(c)].

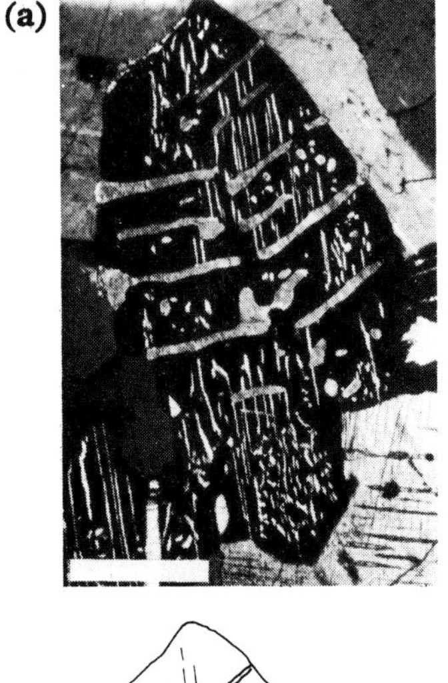

Fig. 8.12 (a) Optical micrograph of an inverted pigeonite from the Bushveld Complex. The sequence of exsolution and inversion transformations leading to this observed pyroxene intergrowth is illustrated in (b). The length of the scale bar is 0.3 mm. (b) (i) Growth of a twinned pigeonite crystal from the melt. (ii) Exsolution of augite from the pigeonite. (iii) Inversion of the pigeonite to orthopyroxene causing a loss in the regularity of the augite lamellae and the second stage exsolution of augite from the orthopyroxene.

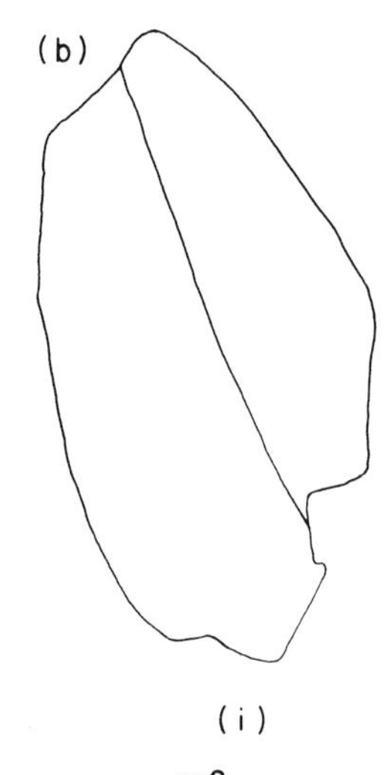

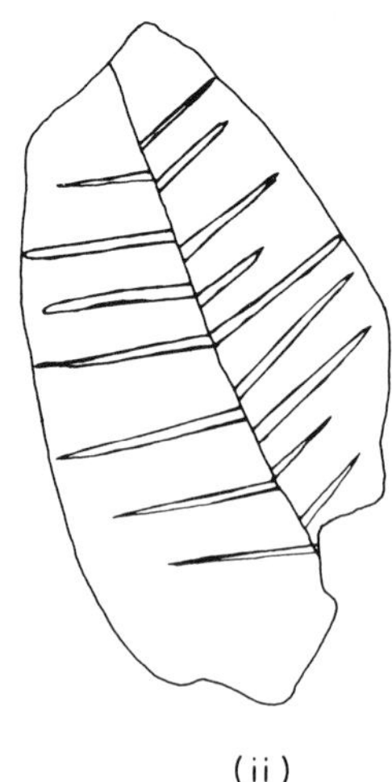

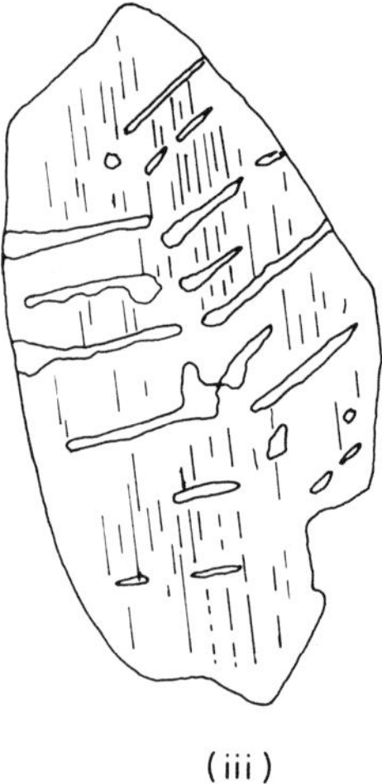

It is worth noting that secondary exsolution is not found in the alkali feldspars as the activation energy for the diffusion of Na and K in feldspar is lower than that for Ca in pyroxene. With more rapid diffusion rates stranded diffusion profiles are not as likely to form (see section 6.4.1).

Cooling curve D is the slow cooling extreme in Fig. 8.10 and in this case nucleation and growth of augite is complete before the inversion of the pigeonite to orthopyroxene. Subsequently, the orthopyroxene itself may exsolve augite as discussed in section 7.2.1. Fig. 8.12(a) is an optical micrograph of a pyroxene from the slowly cooled Bushveld Complex, and illustrates a number of episodes of exsolution and inversion. After initial crystallization from the magma the crystal was a twinned pigeonite, which subsequently exsolved augite on cooling. The augite lamellae form a 'herring-bone' pattern within the twinned pigeonite host. With further cooling, the whole pigeonite crystal inverted to a single crystal of orthopyroxene. The orthopyroxene, which has an even smaller tolerance for Ca substitution, exsolves a second set of augite lamellae. Fig. 8.12(b) illustrates this sequence diagrammatically.

The relationship between antiphase domains and exsolution lamellae

The formation of antiphase domains during the high–low transformation in pigeonite, and their relation to exsolution lamellae, provide additional information on thermal histories not available in most other two-phase systems.

If inversion in the pigeonite takes place before exsolution, antiphase boundaries will be apparently continuous on either side of the exsolution lamellae. If exsolution precedes inversion, antiphase domains will nucleate at the precipitate/matrix interface. As noted earlier (section 7.2.1) antiphase domains which have nucleated heterogeneously in this way will be larger than those which may have nucleated homogeneously within the pigeonite. The variation in Ca content in the diffusion profile may also affect domain size. It is usual to find that antiphase domains adjacent to augite lamellae are larger than those within the pigeonite.

If the high–low pigeonite inversion temperature, which is fixed for a given composition, is known then some idea of the temperature at which exsolution took place can be obtained.

Antiphase boundaries in pigeonite may also act as nucleation sites for the secondary exsolution of augite. In Fig. 8.11(c) the small platelets between the larger augite lamellae have nucleated at antiphase boundaries. Note that near the large lamellae the antiphase boundaries are 'clean', presumably due to the lower Ca content in these regions. In some orientations antiphase boundaries have a structure and composition which makes them ideal nucleation sites for augite (section 7.2.1).

At the present time there is insufficient data available to quantify the TTT curves and cooling rates shown in Fig. 8.10. To do this the following information is required:

1 the position of the solvus;
2 the position of the coherent spinodal;
3 pigeonite–orthopyroxene transformation temperatures;
4 high–low pigeonite transformation temperatures;
5 the diffusion coefficients for Ca which controls the coarsening kinetics.

Although a fair amount of semiquantitative data exists, diffusion coefficients are not known, nor is the effect of changing composition on these parameters. However when this situation is remdied pyroxene microstructures with their complexity and variation with geological environment will be ideal for quantitative thermal history determination.

8.2.2 THE EXSOLUTION OF AUGITE FROM ORTHOPYROXENE

The main point to be made here is that the structural differences between orthopyroxenes and augite do not allow the operation of a spinodal mechanism. As outlined in section 5.4, non-equilibrium behaviour under these circumstances may involve the formation of transitional phases—precipitates which have a structure more similar to the matrix than the equilibrium phase and hence a lower activation energy for nucleation. Their composition is enriched in the solute relative to the matrix.

The first of these transitional phases to form at rapid cooling rates are the GP zones which have the same structure as the matrix and are completely coherent with it. Other transitional phases may be formed at slower cooling rates and lower degrees of supercooling before the formation of the equilibrium phase. The relative kinetics of formation of these phases is shown in the TTT curves in Fig. 6.11.

Such transitional phases have been observed by electron microscopy in orthopyroxenes from the Stillwater complex. In this slowly cooled intrusion augite was able to nucleate heterogeneously and grow by the ledge mechanism (Fig. 5.19). Even at this cooling rate calcium diffusion in the orthopyroxene, in which the cation sites are relatively small, is very sluggish and unable to maintain equilibrium compositions between the lamellae and matrix over more than a micron. Thus the regions between augite lamellae metastably retain a relatively high calcium content down to tempera-

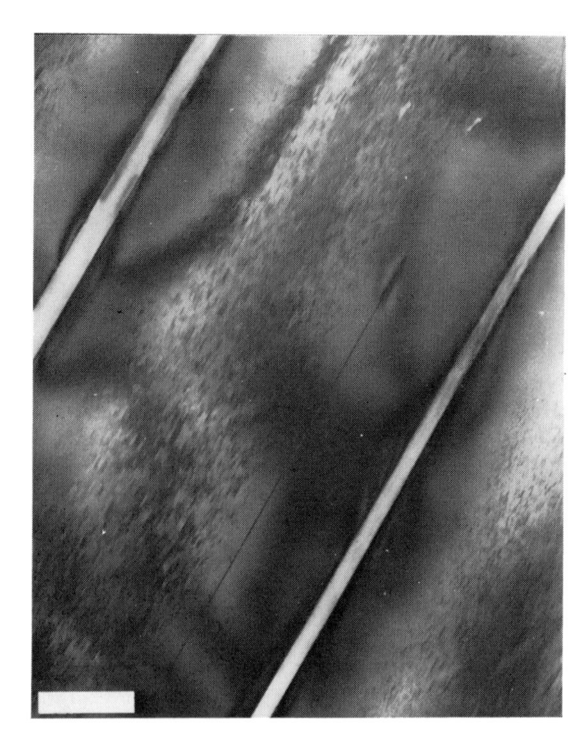

Fig. 8.13 Electron micrograph showing the microstructure of orthopyroxene from the Stillwater Complex. Lamellae of augite are flanked by precipitate-free zones, and secondary exsolution has taken place by the formation of Ca-rich GP zones in the central region. The length of the scale bar is 0.5 μm. (Photo courtesy of P. E. Champness.)

tures where secondary nucleation of the equilibrium precipitate is no longer possible. Under these non-equilibrium conditions transitional phases may form.

Fig. 8.13 is an electron micrograph of a Stillwater orthopyroxene. The augite lamellae are labelled A. Note the wide precipitate free zone around the lamellae and the fine GP zones in the central region. These have nucleated from the stranded Ca-rich regions at lower temperatures. The thin lamella labelled B is a second transitional phase.

8.3 The kinetics of pyrite exsolution from pyrrhotite solid solution

In the Fe–S system at relatively high temperatures, two iron sulphide phases exist. Pyrrhotite has a broad vacancy solid solution field, $Fe_{1-x}S$ (section 3.2.2) as shown in the high-temperature part of the phase diagram in Fig. 8.14. The second phase, pyrite, has a constant composition FeS_2 and is represented by a line on the phase diagram. A two-phase region exists between the two phases.

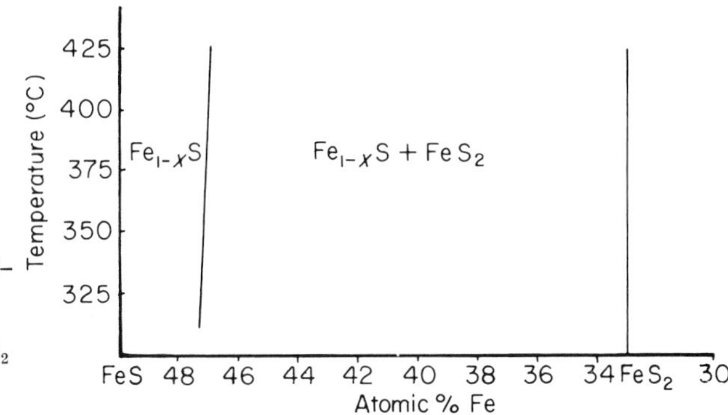

Fig. 8.14 Part of the phase diagram for the system FeS–FeS_2 at high temperatures. A two-phase region exists between stoichiometric FeS_2 and the pyrrhotite ($Fe_{1-x}S$) solid solution.

At 425°C pyrrhotite compositions more Fe-deficient than 46.9 atomic % Fe become unstable and exsolve pyrite. At 325°C this equilibrium composition is 47.3 atomic % Fe. The situation can be represented in terms of the schematic free energy vs composition curves for the two phases (Fig. 8.15). The common tangent defines the equilibrium compositions. As the temperature changes the relative positions of the two curves would change, thus changing the point of tangency on the pyrrhotite curve.

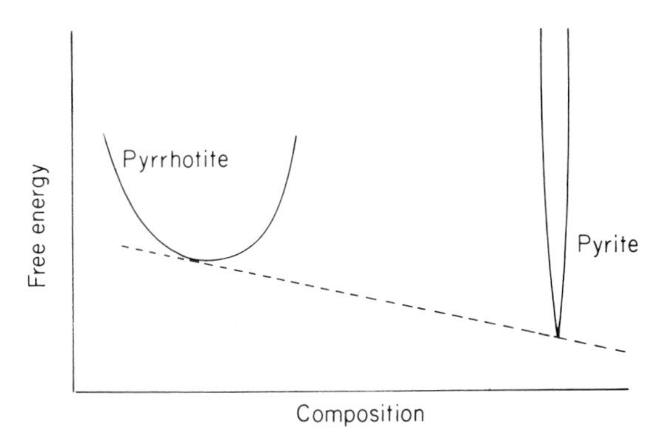

Fig. 8.15 Schematic free energy–composition curves for the pyrrhotite solid solution and pyrite.

Pyrrhotite has a structure based on hexagonally close-packed sulphur atoms with the cations disordered over the octahedral sites. Pyrite has a structure which can be compared to that of NaCl with the sulphur atoms forming dumbbell-shaped pairs. If the mid-points of the sulphur pairs are considered, these are in a cubic close-packed arrangement with Fe atoms in the octahedral sites. From the point of view of exsolution, however, the structures are quite unrelated. The mechanism of exsolution of pyrite from pyrrhotite is therefore by nucleation. No transitional phases are known. This lack of correspondence between the two structures leads to pyrite being exsolved as small spherical blebs.

Yund and Hall set out to examine experimentally some of the factors which affect the kinetics of such an exsolution process. Their results illustrate aspects of nucleation theory.

8.3.1 EFFECT OF SUPERSATURATION

At 325°C a pyrrhotite composition of 47.31 atomic % Fe is in equilibrium with pyrite. Samples of composition 46.29, 46.79, 46.85 and 47.13 atomic % Fe were prepared, corresponding to supersaturations of 1.02, 0.52,0.46 and 0.18, respectively. As the composition of the pyrrhotite solid solution becomes more Fe-deficient, the supersaturation increases and the free-energy difference ΔG between the solid solution and the exsolved phases becomes greater (Fig. 8.15).

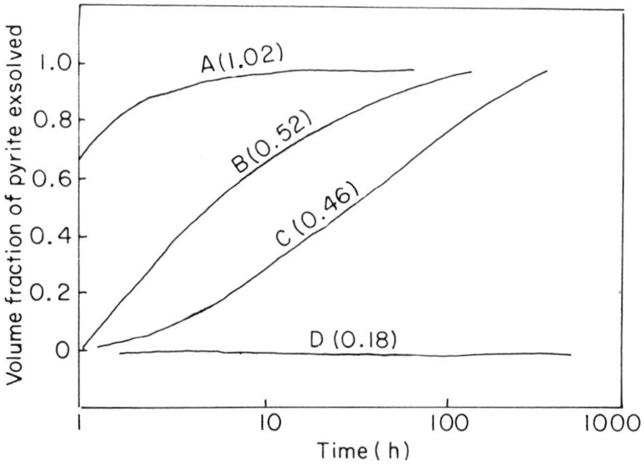

Fig. 8.16 Effect of supersaturation on the kinetics of exsolution of pyrite from the pyrrhotite solid solution at 325°C. The figures in parentheses are the supersaturation values in atomic % Fe. (After Yund and Hall, 1969.)

Annealing experiments were carried out at a constant temperature (325°C) for various periods of time, and the volume fraction of exsolved pyrite determined. The results are shown in Fig. 8.16. As would be expected, the effect of supersaturation is quite marked. Nucleation theory (see sections 5.2 and 6.2) predicts that the nucleation rate will be related exponentially to the volume free-energy change involved in forming a nucleus of the new phase. At near equilibrium conditions where the supersaturation is low, ΔG is small and hence the size of the critical nucleus is very large. The positive free-energy contributions of surface and strain energy outweigh the reduction in volume free energy. Fig. 8.16 shows that under these conditions no exsolution has taken place even after almost 500 hours (curve D).

With increasing supersaturation ΔG becomes larger, the negative contribution to the volume free energy becomes more dominant and the size of the critical nucleus decreases. The free energy of activation for nucleation $\left[\text{which is proportional to } \frac{1}{(\Delta G)^2}\right]$ also decreases and hence the nucleation rate increases.

8.3.2 EFFECT OF IMPURITIES

Some insight into the way a nucleus forms may be gained by the observation that very small amounts ($\approx$ 500 parts per million) of some impurity atoms may retard the nucleation rate by several orders of magnitude. The retardation is greatest when the impurity atoms are large, e.g. As, Sb and Bi. At higher supersaturation the effect of impurities on the nucleation rate is not as pronounced, although the growth of the pyrite grains seems to be inhibited.

Yund and Hall suggest the following explanation for their observations. The nucleation of pyrite involves considerable strain which could be reduced by the migration of vacancies to the nucleation site. This may be particularly important in cases where the structural misfit between the host and nucleus is large. It is known that impurity atoms have a high affinity for vacancies and other defects and there may be a large binding energy between them. If the impurity atoms are large and therefore very slowly diffusing, vacancy migration could be greatly reduced. Thus the strain associated with nucleation cannot be relieved and the nucleation rate is retarded. When the supersaturation is greater the strain energy is not as important and is outweighed by the volume free-energy reduction. Therefore the effect of impurities is not as pronounced.

The growth rate of pyrite grains may also be reduced by impurities segregating preferentially to the defect-rich pyrite/pyrrhotite interface. The slow diffusivity of these atoms may produce a drag on the interface and inhibit growth. At higher temperatures the greater mobility of the atoms greatly reduces the retarding effect of impurities on the exsolution rate.

The conclusion that can be drawn from this study may well apply in a more general way to nucleation processes associated with a large strain. The nucleation process itself is a random event and may be greatly affected by factors which have nothing to do with the pure mineral system. The precise atomic movements involved in forming a nucleus may be modified by the presence of even small amounts of impurities. In some cases, as we have seen here, these may reduce the nucleation rate; in other cases they may provide more suitable nucleation sites and accelerate it. As natural minerals invariably contain trace impurities these effects must be understood if laboratory studies of kinetics are to have any meaning in relation to natural processes.

8.4 Exsolution processes in spinels

The basic structure and variation in chemistry of spinels was described in section 3.2.5 where it was pointed out that at high temperatures a wide range of cation substitutions is possible, leading to the formation of solid solutions with complex compositions. These compositions generally reflect the bulk compositions of the rocks in which they occur. On cooling the extent of solid solution is usually reduced, particularly in cases where substituting cations have dissimilar ionic radii. Thus a variety of exsolution intergrowths have been observed in spinels. The exsolved spinel lamellae lie parallel to the {100} cube faces of the spinel structure.

Exsolution transformations involve the redistribution of cations within the cubic close-packed oxygen structure. The similarity in structure and cell dimensions between the end members of the spinel group allows the possibility for a certain amount of this redistribution to take place within a coherent oxygen framework, so that again we are dealing with a situation in which both spinodal decomposition and homogeneous nucleation may operate.

The relatively rapid diffusion rates of cations in the spinel structure results in a wide range of exsolution textures being formed depending on the composition and the cooling rate. Exsolution is on a scale from several microns to tens of Angstroms, and in some cases the cut-off temperatures for diffusion may be as low as 250°C, so that these oxide minerals may be particularly useful in evaluating thermal histories at temperatures where processes in other minerals have stopped. The ultimate aim is to have as many indicators as possible within the same rock, and the wide geological distribution of spinel group minerals makes them ideal for this purpose.

Work on the mechanisms and kinetics of exsolution processes in these minerals is only just beginning and there are still the usual problems of lack of the critical data required to quantify many of the observations which have been made. To illustrate some of the general features of the behaviour of spinels on cooling we will discuss briefly solid solutions between the phases ulvospinel Fe_2TiO_4, magnetite Fe_3O_4 and pleonaste (Fe, Mg)Al_2O_4. [The similarity of Mg and Fe^{2+} enables a complete solid solution to exist between spinel (sensu stricto) $MgAl_2O_4$ and hercynite $FeAl_2O_4$ down to room temperatures. Magnesium-rich members of this solid solution are called pleonaste, which we will treat as a single phase.]

8.4.1 EXSOLUTION IN THE SYSTEM MAGNETITE–ULVOSPINEL–PLEONASTE

The structures of the end members of the spinel group minerals are basically the same and no polymorphic phase transformations are known to occur. Exsolution in any binary (or pseudobinary) system can therefore be described in terms of a simple solvus. We shall assume here that in most cases the solvus will be symmetrical and concern ourselves only with the relative positions of the solvi on the temperature axis.

As a first approximation, which often holds quite well in spinels, we can estimate the relative extent of solid solutions between phases by comparing their lattice parameters. In the absence of any complicating structural differences the cell dimensions can give an idea of the ease of cation substitution between phases which is related to the enthalpies of mixing and hence the solvus temperature. The three phases with which we are concerned here have the following cell dimensions (*a* in Angstroms): pleonaste 8.11 (spinel 8.10 and hercynite 8.13), magnetite 8.39 and ulvospinel 8.53.

Simply on the basis of cell dimensions we might expect that the extent of solid solution between pleonaste and ulvospinel to be the most restricted, with a greater degree of solid solution possible between pleonaste and magnetite. The pair

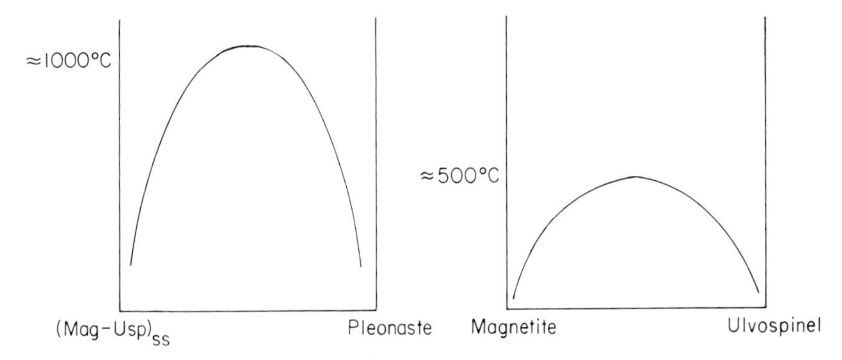

Fig. 8.17 The relative positions of the solvus temperatures for two spinel systems: (a) titanomagnetite–pleonaste and (b) magnetite–ulvospinel.

magnetite-ulvospinel are more similar than the previous two and hence a considerable degree of solid solution may be possible to quite low temperatures. The available experimental data confirm this. The pleonaste-ulvospinel solvus is cut off at about 1200°C by the liquidus. In the pleonaste–magnetite binary, complete solid solution is possible above about 800°C, while the solvus in the magnetite–ulvospinel system is at around 500°C.

The phase relations in such a system can be expressed by a pair of solvi; one between a homogeneous magnetite–ulvospinel solid solution and pleonaste, the other between magnetite and ulvospinel. The crest of the first solvus will depend strongly on the bulk titanium content, but at intermediate compositions it will be around 1000°C. The two solvi are schematically shown in Fig. 8.17.

As a result pleonaste will exsolve from the solid solution at relatively high temperatures leaving a matrix of magnetite–ulvospinel solid solution which will remain homogeneous until a lower temperature is reached. Fig. 8.18 shows a typical microstructure as seen in reflected light. The large lamellae are exsolved

Fig. 8.18 Reflected light micrograph showing the exsolution of pleonaste lamellae (black) from a matrix of magnetite–ulvospinel solid solution. The matrix has subsequently developed a fine 'cloth texture' (grey and white) due to exsolution on the magnetite–ulvospinel solvus. The length of the scale bar is 15 μm. (Photo courtesy of G. D. Price.)

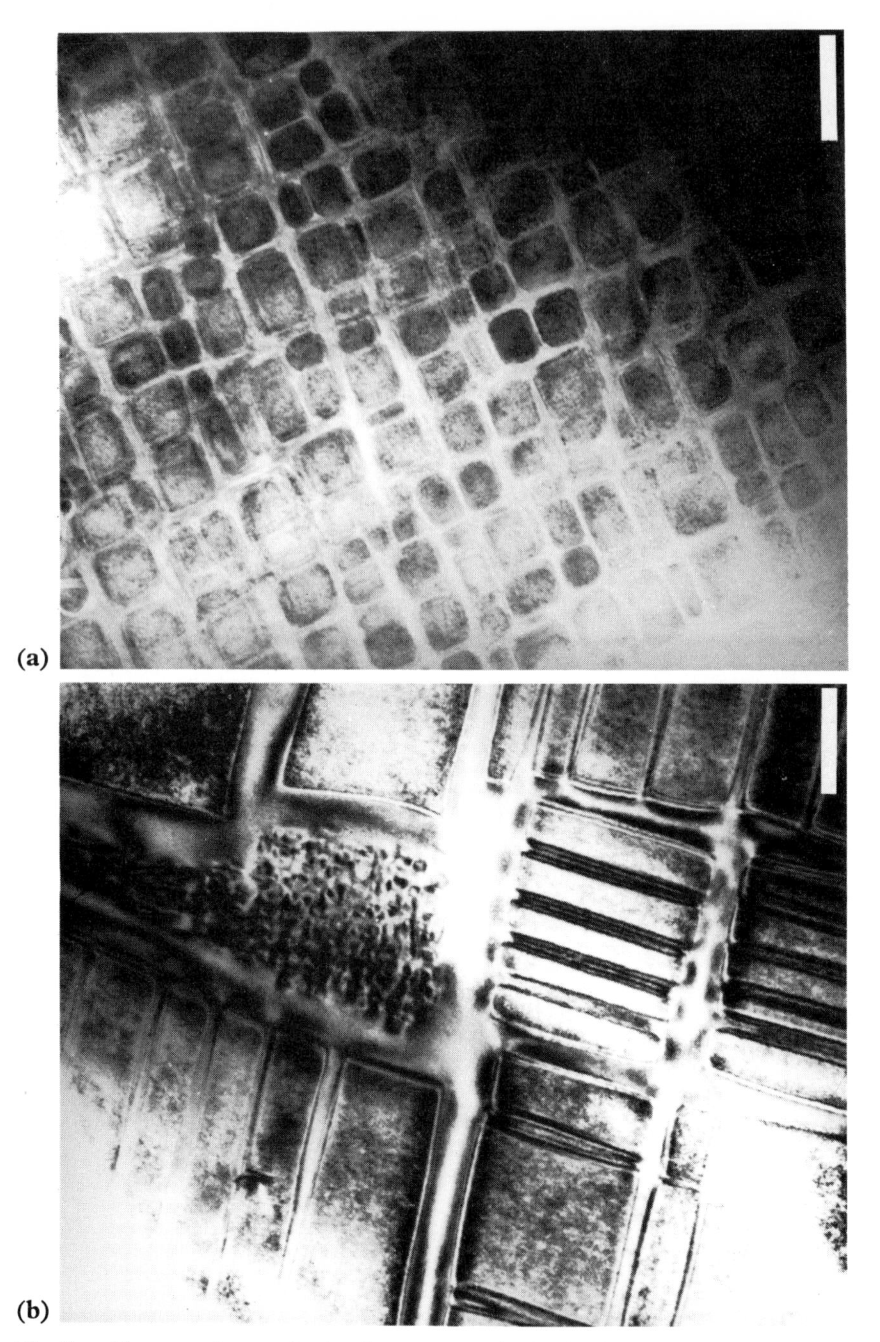

Fig. 8.19 Electron micrographs showing two stages in the development of the cloth texture in the magnetite–ulvospinel system. (a) Breakdown of the homogeneous solid solution to form magnetite-rich regions (blocks) in an ulvospinel-rich matrix, probably by a spinodal mechanism. From the Yamaska Intrusion, Canada. The length of the scale bar is 0.2 μm. (b) Coarsening of the microstructure in (a) results in larger blocks of magnetite within the lamellar matrix of ulvospinel. Secondary exsolution of magnetite-rich spinel has taken place in some of the coarser lamellae. Further coarsening would lead to an optically visible cloth texture as shown in Fig. 8.18. From the Taberg Intrusion, Sweden. The length of the scale bar is 0.2 μm. (Photos courtesy of G. D. Price.)

pleonaste, while the matrix shows a fine 'cloth texture' due to exsolution on the magnetite–ulvospinel solvus. In this case the microstructure associated with pleonaste exsolution is quite straightforward as the cooling rate is slow. Considering however that both homogeneous nucleation and spinodal decomposition mechanisms are possible, faster cooling rates may produce complex microstructures.

The much lower temperature of the ulvospinel–magnetite solvus means that exsolution textures will be very much finer and that even at moderately fast cooling rates the solid solution may be quenched metastably. In dyke rocks, for example, intermediate compositions of this solid solution appear to be homogeneous in reflected light. Calculations based on lattice parameters and estimated elastic constants suggest that the coherent spinodal may be depressed by as much as 200°C below the equilibrium solvus so that even spinodal decomposition may be a relatively sluggish process at these temperatures.

Figs 8.19(a) and (b) are two electron micrographs showing the stages in the development of the cloth texture shown in Fig. 8.18 Ulvospinel has exsolved apparently by a homogeneous mechanism to form lamellae on {100} planes. As the temperature falls, further growth leads to a grid of ulvospinel with 'blocks' of magnetite in between. Fig. 8.19(b) shows that in the centre of the larger ulvospinel lamella (which originally exsolved at a higher temperature and with a higher magnetite content than the finer lamellae) secondary exsolution of magnetite has taken place. The interfaces between magnetite and ulvospinel appear in this case to be coherent.

Interrelationships between different exsolving phases can provide a great deal of information on the sequence of processes taking place over a wide temperature range. The more complex the microstructure the more useful it is likely to be, once the relevant processes and their kinetics are understood. The complex chemistry of many spinels complicates this task and at the present time we must be content with qualitative comparisons of relative thermal histories.

8.4.2 ILMENITE 'EXSOLUTION' FROM IRON–TITANIUM SPINELS

Ulvospinel–magnetite solid solutions very often contain lamellae of ilmenite which appear to have exsolved on {111} planes of the spinel. The texture, shown in Fig. 8.20, resembles that of the Widmanstätten structure in iron meteorites and for many years was thought to imply that an extensive solid solution must exist between the

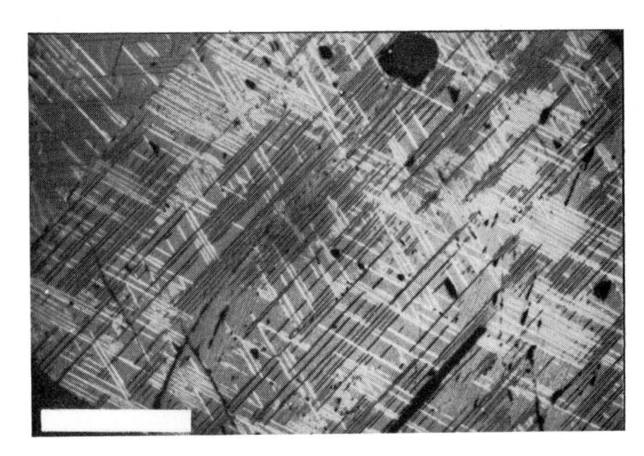

Fig. 8.20 Reflected light micrograph showing the Widmanstätten-like texture developed by the formation of ilmenite lamellae (black and white) on {111} planes of the ulvospinel matrix (grey). The length of the scale bar is 0.25 mm. (Photo courtesy of G. D. Price.)

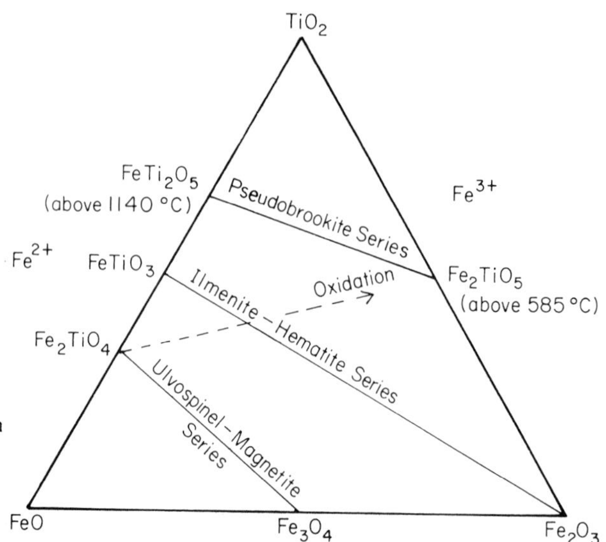

Fig. 8.21 The iron-titanium oxide system showing the major phases and solid solution series. The dashed line is the oxidation trend.

ulvospinel–magnetite series and the ilmenite–hematite series. The compositions of these phases are shown in Fig. 8.21.

The structures of the two solid solution series are quite different. The ulvospinel–magnetite solid solution is based on a cubic close-packed oxygen structure, whereas in the ilmenite–hematite solid solution the oxygen atoms are hexagonally close-packed. Very little solid solution could be expected to exist between such dissimilar structures. Another possibility is that at high temperatures the ilmenite structure transforms to a structure similar to the spinel and that a solid solution with this spinel structure exists between the phases. Transformation of the ilmenite end member to hexagonal close-packing on cooling would indeed produce ilmenite lamellae on {111} planes of the spinel, as on these planes the close-packed oxygen planes would be parallel and the best structural fit obtained.

Experimental work could not confirm this hypothesis and the degree of solid solution of the spinel phase towards the ilmenite composition was found to be much too small to account for the large proportion of ilmenite often observed. Attempts to homogenize the lamellae could not be carried out without a change in the bulk composition involving oxygen loss.

The work of Buddington and Lindsley (1964) confirmed that this type of 'exsolution' texture was not the result of cooling a one-phase solid solution into a temperature range where a solvus was stable. It occurs as a result of solid-state oxidation reactions where the oxygen fugacity (equal to the partial pressure of oxygen) is above the equilibrium value for the particular magnetite–ulvospinel composition.

In the previous section we considered the behaviour of spinels when the phases maintained their stoichiometry with a metal : oxygen ratio of 3 : 4 and hence were in equilibrium with respect to oxidation or reduction. The formation of ilmenite, however, involves a change in this stoichiometry. During oxidation, although the Fe : Ti ratio remains constant, the proportion of Fe^{3+} increases and the metal : oxygen ratio decreases. The dashed line in Fig. 8.21 shows such an oxidation trend. On this diagram, the oxygen fugacity varies increasing from the left to the right, i.e. as the phases increase their Fe^{3+} content and decrease Fe^{2+}. The stability of the phases with respect to oxidation is determined by the oxygen fugacity and for any oxide composition an equilibrium is established at a given oxygen fugacity and temperature. Above this oxygen fugacity the phase will tend to oxidize; below this oxygen fugacity it will tend to be reduced.

The reason oxidation exsolution occurs in many igneous titanomagnetites is explained by considering the changes in oxygen fugacity in a cooling igneous intrusion relative to the equilibrium oxygen fugacity for the titanomagnetite. Fig. 8.22(a) shows the equilibrium curves for two different ulvospinel–magnetite compositions. Note that at a given temperature, compositions with a higher magnetite content become stable as the oxygen fugacity increases. In Fig. 8.22(b) similar curves are drawn for two compositions of the ilmenite–hematite solid solution. By superimposing these two curves we define points of intersection which represent unique temperatures and oxygen fugacities at which the two solid-solution phases can coexist stably [Fig. 8.22(c)].

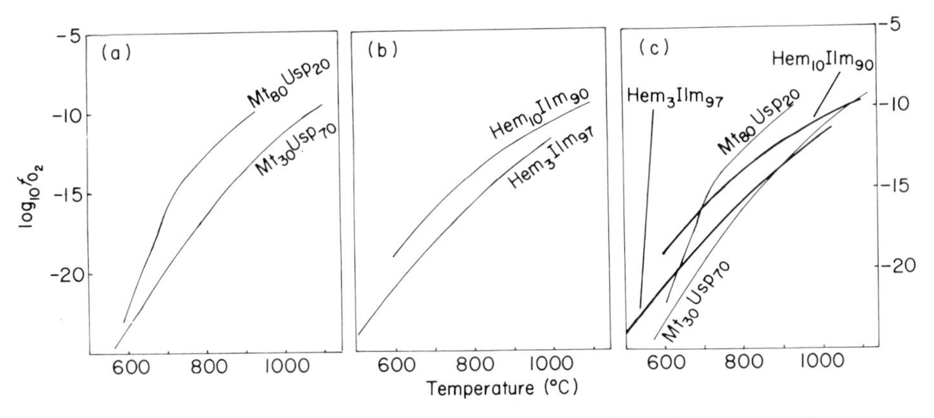

Fig. 8.22 (a) Equilibrium curves as a function of oxygen fugacity and temperature for two titanomagnetite compositions. (b) Equilibrium curves as a function of oxygen fugacity and temperature for two compositions of the hematite–ilmenite solid solution. (c) Curves in (a) and (b) superimposed. (After Buddington and Lindsley, 1964.)

In a cooling igneous rock the oxygen fugacity is buffered by the constituent solid and fluid phases present. The change in oxygen fugacity will follow the appropriate buffer curve and in general this may not be as steep as the equilibrium curves of Fig. 8.22. An example of a superimposed buffer curve is shown by the dashed line of Fig. 8.23. The fact that this cooling line crosses the equilibrium curves in a sequence leading to increasing magnetite contents indicates that oxidation will take place and that therefore an ilmenite–hematite solid solution of the appropriate composition will be 'exsolved'. In Fig. 8.23 we have shown only two equilibrium curves for each phase for simplicity. The whole family of curves at small compositional increments has been derived by Buddington and Lindsley and thus defines the equilibrium coexisting compositions of ulvospinel–magnetite solid solution and ilmenite–hematite solid solution for any temperature and oxygen fugacity. These curves form the basis of a geothermometer–oxygen barometer which can be applied to ilmenite–titanomagnetite intergrowths which have formed in this way, provided of course that equilibrium was closely maintained during oxidation. Note that not all ilmenite–titanomagnetite intergrowths form in this way. For some

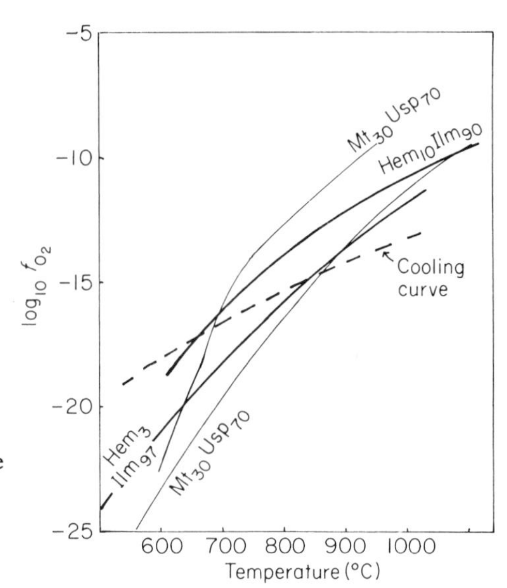

Fig. 8.23 A typical buffered cooling curve (dashed line) indicating the change in oxygen fugacity with temperature, superimposed on the set of equilibria of Fig. 8.22(c).

textures contemporaneous crystallization may be a more appropriate interpretation. The textural variations in such intergrowths have recently been reviewed extensively by Haggerty (1976).

The presence of microstructures due to oxidation phenomena in spinels adds a further dimension to the information which may be obtained by studying these minerals. Textural relationships between exsolved spinels and 'oxy-exsolved' ilmenite indicate the relative sequence of exsolution and oxidation so narrowing some of the uncertainties in determining the temperatures of these processes. Furthermore, in systems such as magnetite–ulvospinel–pleonaste, early oxidation at high temperatures tends to preferentially partition the Al into the surrounding host spinel. Within the Al diffusion distance the spinel rapidly approaches supersaturation and pleonaste solid solutions are 'forced' to exsolve often in intimate association with the ilmenite lamellae.

As well as indicating geological history, these processes have important consequences on the interpretation of magnetic properties of rocks. Palaeomagnetic studies are based on the premise that the Fe–Ti oxides present in rocks reflect the state of the magnetic field at the time they were formed. Exsolution and oxidation modify the magnetic properties of these minerals. Although the time scale for these processes is generally much shorter than that influencing the earth's magnetic field, minerals in different environments will undergo these changes to a greater or lesser extent. The evolution of the magnetic signature of rocks is therefore inseparable from the evolution of the microstructures.

8.5 Oxidation exsolution

Any oxide mineral containing transition metals that can exist in more than one oxidation state (e.g. Fe, Mn, Ti, Cr etc.) has a stability field which is not only defined by the temperature and pressure but also by the oxygen fugacity. The effect of a change in oxygen fugacity on the chemistry and structure of such a mineral can in general terms be treated in a similar way to transformations which occur as a result of changes in temperature or pressure. As the transformation involves a change in the bulk composition however, the mechanisms will be more complex, with diffusion processes taking place between the interior and exterior of the crystal.

Although the term oxidation exsolution usually refers to the formation of ilmenite lamellae in titanomagnetite (see section 8.4), lamellae may form in any oxidation process where the product has some structural similarity to the parent. The mechanisms of such processes are not well understood at present and in the case of ilmenite intergrowths in titanomagnetite, there is often some doubt as to whether a particular texture resulted from oxidation or contemporaneous crystallization, and hence whether the Buddington–Lindsley geothermometer can be applied.

Under near-equilibrium conditions the nucleation of ilmenite would occur preferentially at grain boundaries and cracks. If equilibrium is maintained the result may resemble an ilmenite overgrowth around a titanomagnetite grain, with very little of the exsolution-type texture. An oxidation overgrowth of this kind requires that the diffusion take place throughout the titanomagnetite. At lower temperature when diffusion distances become shorter the ilmenite will tend to grow into the crystal along {111} planes. This oriented or lamellar growth is energetically the next most favourable after grain-boundary growth is stopped by the decreasing diffusion distance within the crystal. The Ti diffusion to the ilmenite is coupled by the oxidation of Fe^{2+} to Fe^{3+} within the host titanomagnetite which becomes enriched in magnetite. This is evident by a change in the optical properties of the titanomagnetite in a zone surrounding each lamella. The width of this zone defines a diffusion distance.

From the point of view of textural development, grain-boundary intergrowths of ilmenite and titanomagnetite represent the earlier equilibrium stages of oxidation. Lamellar 'exsolution' intergrowths imply some deviation from equilibrium. However, as noted earlier and summarized by Haggerty, there are textures which may arise by processes other than oxidation and therefore some caution is necessary in their interpretation.

The thermodynamic principles which govern oxidation behaviour are the same as those which have already been applied to polymorphic transformations and phase separation. The free energy of a particular composition depends on the oxygen fugacity. When the oxygen fugacity changes a different composition or phase may have a lower free energy and the system will tend to change to achieve this composition. The kinetics of this change depend on the mechanism which operates and the temperature. The mechanism may in turn depend on the extent to which the oxygen fugacity differs from the equilibrium value—this is analogous to saying that in a temperature-dependent transformation the mechanism may depend on the extent of undercooling or superheating. We will illustrate some of these points by making some observations on the oxidation of olivine.

8.5.1 OXIDATION OF OLIVINE

The oxidation of olivine under equilibrium conditions is governed by a fairly complex set of reactions whose equilibria depend on the oxygen fugacity, the temperature and the iron content of the olivine. In order to understand the processes which may occur in natural olivines, we must first make reference to the equilibrium situation which is described by the set of lines in Fig. 8.24.

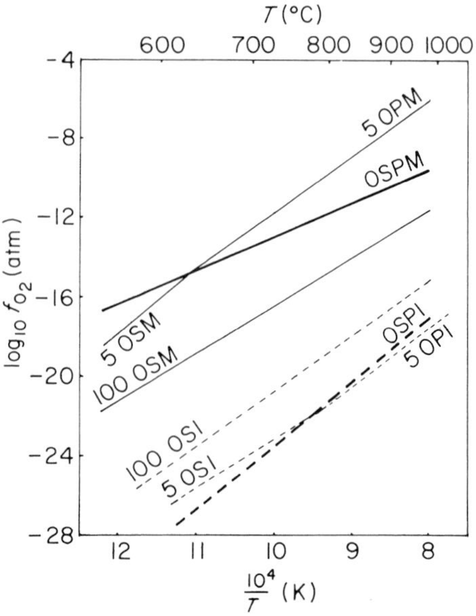

Fig. 8.24 Equilibrium lines for oxidation (full lines) and reduction (dashed lines) of olivine of two different fayalite contents Fa_{100} and Fa_5.
OSM is the oxidation reaction of olivine → silica + magnetite.
OPM is the oxidation reaction of olivine → pyroxene + magnetite.
OSI is the reduction reaction of olivine → silica + iron.
OPI is the reduction reaction of olivine → pyroxene + iron.
The line labelled OSPM separates the fields of OSM and OPM. (After Nitsan, 1974.)

In this diagram the positions of the equilibrium curves for both oxidation reactions (full lines) and reduction reactions (dashed lines) are shown for two different fayalite contents, Fa_{100} (denoted by 100 on the lines) and Fa_5 (denoted by 5 on the lines). The following reactions are shown:

Oxidation:

OSM	Olivine + O_2 → Silica + Magnetite
OPM	Olivine + O_2 → Pyroxene + Magnetite

Reduction:

OSI	Olivine → Silica + Iron + O_2
OPI	Olivine → Pyroxene + Iron + O_2

Here we will confine our attention mainly to the oxidation curves.

The equilibrium oxidation of olivine to silica + magnetite takes place at lower oxygen fugacities for pure fayalite (the line labelled 100 OSM) than for compositions with lower fayalite contents (e.g. 5 OSM). The oxidation to pyroxene + magnetite is a reaction which requires higher oxygen fugacities and higher temperatures and under equilibrium conditions occurs only in olivines with lower fayalite contents. Pure fayalite under equilibrium conditions does not cross into the field where olivine + pyroxene + magnetite are stable. Fa_5 is oxidized to silica + magnetite at lower oxygen fugacities and temperatures (5 OSM) and to pyroxene + magnetite at higher oxygen fugacities and temperatures (5 OPM). The thick line labelled OSPM merely separates the fields of OSM and OPM.

In the lower half of the diagram a similar set of lines exists for the reduction reactions which in pure fayalite produces iron and silica, while at lower fayalite contents and higher temperatures pyroxene and iron may be formed.

Between the oxidation and reduction equilibria the olivine is stable with respect to these reactions. Pure fayalite will therefore be stable between the lines 100 OSM and 100 OSI on Fig. 8.24. Although apparently stable over a range of oxygen fugacity, any iron bearing olivine can only have a stoichiometric metal : oxygen ratio when all of the iron is present as Fe^{2+}. This is only possible along a single curve on an f_{O_2}–T diagram (in this case, somewhere between the lines for OSM and OSI) termed the stoichiometric curve. Above the stoichiometric curve oxidation of Fe^{2+} to Fe^{3+} will take place. This results in a deviation from stoichiometry, with a higher oxygen : cation ratio and the formation of metal vacancies and Fe^{3+} centres. Below the curve reduction will take place, reducing the oxygen : cation ratio and forming defects associated with excess metal, for example oxygen vacancies and interstitial cations. The stability field of any olivine composition, lying between the appropriate oxidation and reduction curves is thus the extent of the solid solution between the most cation-deficient and cation-rich olivine possible before one of the reactions takes place. Olivine is thought to be able to contain up to about 0.5 atomic % Fe^{3+} before the oxidation reaction takes place. This type of oxidation and reduction of the Fe^{2+} within the olivine stability field is termed homogeneous.

Having described the equilibrium situation at some length we are now in a position to make some specific observations on the mechanisms of natural and experimental oxidation of olivine.

(*a*) *Equilibrium oxidation*

Under near equilibrium conditions the Fe^{3+} content of the olivine will increase until the appropriate oxidation reaction curve (OSM or OPM) is reached. As the equilibrium oxygen fugacity for the reaction is overstepped the reaction rate becomes significant, with nucleation of the new phases (say pyroxene–magnetite) taking place initially at grain boundaries and other high-energy sites. This results in a myrmekitic intergrowth of the two phases invading the host olivine from the outside of the crystal. The amount of overstepping required to drive the reaction will depend on the temperature.

(*b*) *Deviations from equilibrium*

The rate of the oxidation reaction is determined both by the oxygen fugacity and the temperature which define both the free-energy differences between the phases and the diffusion rates. If during cooling the deviation from equilibrium becomes larger and diffusion distances become shorter oxidation will tend to take place on a local scale, within the crystal.

A similar change has already been noted from grain-boundary oxidation to oxidation exsolution in titanomagnetite with falling temperature. In that case only one new phase was formed, whereas in the case of olivine oxidation two new phases must be simultaneously produced, and both are structurally quite different from the host olivine. This requires a substantial structural reorganization which may be possible at a grain boundary or dislocation structure but is unlikely to be able to take place within regions of 'good' olivine structure.

Is there another mechanism if the olivine is relatively free of dislocations? The curious dendritic intergrowths of magnetite within lamellar-shaped platelets which have often been observed in olivine (Fig. 8.25) may provide a clue. Although these were first described in olivines from the Rhum Intrusion almost 100 years ago, recent electron-microscope observations show that the texture is in fact a eutectoidal intergrowth of magnetite-pyroxene which has formed from a single phase which

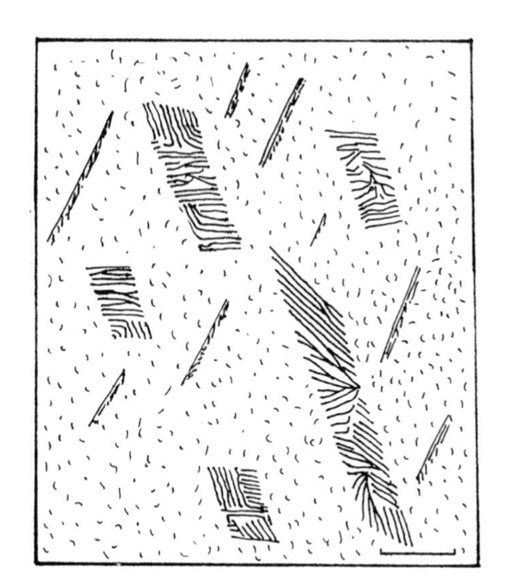

Fig. 8.25 Dendritic iron oxide within lamellar-shaped platelets in olivine from the Rhum layered intrusion. Drawn from an optical micrograph. The length of the scale bar is 5.0 μm.

previously occupied the lamella. Furthermore, similar lamellae have been found which have not broken down to pyroxene–magnetite and within these lamellae is a phase which is an olivine superstructure. The evidence suggests that in the early stages an oxidation exsolution of an 'oxy-olivine' phase has taken place. This is structurally similar to the host olivine and hence able to form well-oriented lamellae, presumably of an olivine-like phase enriched in Fe^{3+}. The superstructure may be formed by some ordering process. At a later stage these lamellae may break down to pyroxene–magnetite.

The principles involved in such a process are similar to those applied to the formation of transitional phases (section 5.4). When the equilibrium structure is not able to form due to structural and kinetic constraints, the formation of a transitional phase which is structurally more similar to the host may be more favourable kinetically. More detailed work must be done before these mechanisms are understood properly and in this context a few points about experimental oxidation must be made.

(c) *Oxidation experiments and non-equilibrium processes*

Reference to Fig. 8.24 shows that at 700°C, for example, pure fayalite will tend to oxidize to silica–magnetite at oxygen fugacities greater than about 10^{-17} atm. If fayalite is heated to this temperature in a good vacuum system, the oxygen fugacity will be around 10^{-6} to 10^{-8} atm, well within the field of olivine + pyroxene + magnetite. Under these conditions the fayalite will be highly metastable with respect to both OSM and OPM reactions, and the mechanism of oxidation may be quite different to that which operates at equilibrium. As we have stressed previously, within the non-equilibrium region the kinetics of available processes will dominate the transformations which take place.

In the last decade many experiments have been carried out on olivine oxidation. Often these have involved heating olivine in air (i.e. $f_{O_2} = p_{O_2} \approx 0.2$ atm). Typical results indicate a very complex situation with the formation of multiphase precipitates containing magnetite and/or hematite with pyroxene and/or various forms of silica. This is hardly surprising in view of the fact that the olivine is so highly metastable with respect to virtually any type of oxidation process that almost anything can happen depending on the many chance factors which may affect the relative kinetics of competing mechanisms. The results do not necessarily bear any relationship to processes which may operate under natural conditions.

In order to understand the mechanisms which operate under such metastable conditions it is necessary to study first the mechanisms under conditions as near to equilibrium as is experimentally feasible.

8.6 Exsolution in iron-bearing rutile

In most of the previous examples of exsolution phenomena we have been concerned with the cooling behaviour in a situation where a solid solution is stable at high temperatures, while at low temperatures phase separation leads to a more stable assemblage. It is not unusual however for metastable states to crystallize at low temperatures, either as disordered forms or as solid solutions. A metastable solid solution formed in this way may persist indefinitely if the temperature is below the kinetic cut-off for diffusion. If the solid solution was raised to a higher temperature however the kinetics of exsolution might be more favourable and phase separation may take place.

Rutile, TiO_2, which has crystallized hydrothermally at around 450°C, commonly contains up to 1 % Fe_2O_3 as one of its main impurities. The available phase equilibria data in this system suggest that at low temperatures very little Fe_2O_3 would be soluble in rutile, although the extent of miscibility is not known. The question arises as to whether the Fe_2O_3 in hydrothermal rutile may be present as a metastable solid solution. Recent experimental work suggests that this may be the case.

On the electron-microscope scale these iron-bearing rutiles are homogeneous. When annealed at higher temperatures exsolution takes place, implying that the original temperature of crystallization was below the kinetic cut-off for diffusion. The exsolution process involves a number of transitional stages culminating in the formation of hematite (or titanohematite) exsolution lamellae within the rutile. Here we will discuss three aspects of this process: the mechanism; the kinetics; the possible geological applications.

Before discussing the mechanism by which hematite can exsolve from rutile we must first compare their structures. For our purposes we can consider that the rutile

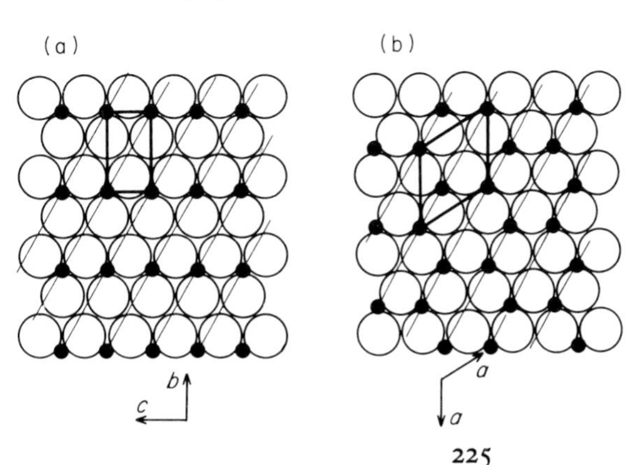

Fig. 8.26 (a) An idealized (100) layer of the rutile structure in which the puckered oxygen planes have been flattened so that the arrangement is hexagonally close-packed. The black circles are the cations and the large open circles are oxygen atoms. (b) The cation distribution in the hematite structure where the (0001) layers are very nearly hexagonally close-packed.

structure is derived from a distorted hexagonal close-packed arrangement of oxygen atoms with Ti in one-half of the octahedral sites. This distortion puckers the oxygen sheets and produces a tetragonal structure. It is commonly idealized in terms of the close-packing model as shown in Fig. 8.26(a). Hematite is very nearly hexagonally close-packed with cations in two-thirds of the octahedral sites [Fig. 8.26(b)]. There are consequently some similarities in lattice spacings between the two phases, but apart from this the structures are quite different in the way that the cations are ordered among the octahedral sites and no appreciable solid solution can exist between them. On equilibrium phase diagrams they would appear as line phases separated by a two-phase region.

8.6.1 THE EXSOLUTION MECHANISM

When exsolution takes place in a binary system where the two end members have dissimilar structures transitional phases may form under conditions when the direct nucleation of the equilibrium phase is kinetically impeded. The principles have been discussed in sections 5.4 and 6.3.2 and applied to the exsolution of augite from orthopyroxene (section 8.2.2). The mechanism of hematite exsolution from rutile is a further good illustration of these principles, with the formation of GP zones and a second transitional phase before hematite finally forms.

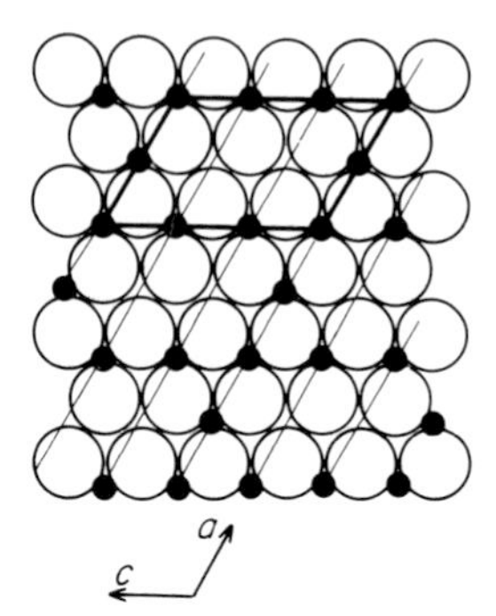

Fig. 8.27 The proposed distribution of cations in the intermediate (transitional) structure formed during the exsolution of hematite from rutile. (After Putnis, 1978.)

The formation of GP zones and their coarsening has already been illustrated in Figs 5.20 and 5.27. They form as two sets of very fine coherent platelets which share the rutile structure but locally distort it due to their compositional difference. The second transitional phase is a superstructure of the matrix rutile structure. Its unit cell is monoclinic and it is frequently twinned, presumably to reduce the interfacial strain (cf. albite twinning in the Na-rich component in perthites). The superstructure must be due to some ordered arrangement of the extra Fe^{3+} cations within the basic rutile framework. A suggested arrangement is shown in Fig. 8.27, which should be compared to the rutile and hematite structures in Fig. 8.26. This structure which is consistent with the electron-diffraction data also suggests that this phase has a cation : oxygen ratio of 2 : 3, i.e. $(Fe, Ti)_2O_3$. The equilibrium nucleation of hematite results in lamellae sharing certain lattice planes with the rutile so reducing the interfacial energy. The degree of lattice matching with the matrix decreases in a step-wise manner from GP zones → transitional phase → equilibrium phase.

8.6.2 THE KINETICS OF THE EXSOLUTION PROCESS

The relationship between the transitional phases may be better appreciated by considering the relative kinetics of their formation. Fig. 8.28 is an experimentally determined isothermal TTT diagram for the three phases. The curves represent the start of nucleation. The diagram illustrates the following points.

1 Over a range of temperature the transitional phases nucleate faster than the equilibrium phase. This reflects the relative nucleation barriers for the three phases.

2 The slope of the lower part of the curves is shallow, indicating a relatively high activation energy for diffusion. Using the method described in section 6.3 this activation energy was found to be 54.3 kcal mol^{-1} °C^{-1}. The immediate consequence of a low value for the energy barrier to nucleation and the high value for the activation energy for diffusion is that at temperatures approaching the solvus temperature the exsolution process will be rapid, but as the temperature decreases there will be a rather sharp kinetic cut-off. The slope of the TTT curve for GP zones indicates that at temperatures below about 450°C no significant exsolution will occur, even over very long periods.

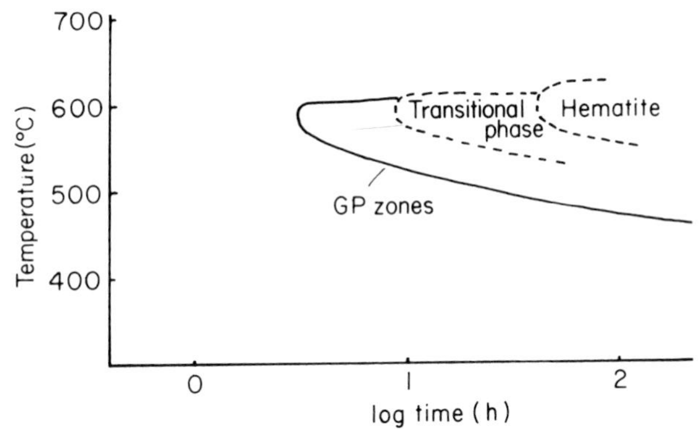

Fig. 8.28 Experimental TTT curve for the exsolution process in iron-bearing rutile. (After Putnis and Wilson, 1978.)

A comparison of the kinetics of nucleation of the three phases with the degree of lattice matching with the matrix leads directly to the most important general characteristic of such a precipitation sequence: the crystal structure of the transitional phases enable them to achieve good lattice matching with the matrix, thereby markedly reducing the interfacial free energy of the critical nucleus. This leads directly to the kinetic sequence observed.

8.6.3 APPLICATION TO THE PARAGENESIS OF THE RUTILES

The TTT diagram suggests that the natural rutile which was homogeneous before the annealing experiments must have formed below the kinetic cut-off temperature for the formation of GP zones, i.e. around 450°C. However, the application of the results to natural systems must be undertaken with some caution. We have already noted the dramatic effect of trace element impurities on the nucleation rate of pyrite from pyrrhotite (section 8.3). Furthermore, grain size can have an important effect on both the nucleation mechanism and the diffusion rates. Finally, in a natural hydrothermal system the presence of vapour phases may also affect the kinetics of the diffusion process. Thus while the TTT diagram is a first step to quantifying the rates of these processes, the effects of these other factors must be known before generalizations can be made.

References and additional reading

8.1

Lorimer, G.W. and Champness, P.E. (1973) The origin of the phase distribution in two perthitic alkali feldspars. *Phil. Mag.* **28**, 1391.

Owen, D.C. and McConnell, J.D.C. (1974) Spinodal unmixing in an alkali feldspar. In: MacKenzie, W.S. and Zussman, J. (Eds.) *The Feldspars*. Manchester University Press, Manchester.

Robin, Y.-P.F. (1974) Stress and strain in cryptoperthite lamellae and the coherent solvus of alkali feldspars. *Am. Mineral.* **59**, 1299.

Willaime, C., Brown, W.L. and Gandais, M. (1976) Physical aspects of exsolution in natural alkali feldspars. In: Wenk, H.U. (Ed.) *Electron microscopy in Mineralogy*. Springer-Verlag.

Yund, R.A. (1975) Microstructure, kinetics and mechanisms of alkali feldspar exsolution. In: Ribbe, P.H. (Ed.) *Feldspar Mineralogy*. Min. Soc. Am. Short Course Notes.

8.2.

Carpenter, M.A. (1978) Nucleation of augite at antiphase boundaries in pigeonite. *Phys. Chem. Minerals* **2**, 237.

Champness, P.E. and Lorimer, G.W. (1976) Exsolution in silicates. In: Wenk, H.U. (Ed.) *Electron microscopy in Mineralogy*. Springer-Verlag.

Lally, J.S., Heuer, A.H., Nord, G.L. and Christie, J.M. (1975) Subsolidus reactions in lunar pyroxenes: an electron petrographic study. *Contrib. Mineral. Petrology* **51**, 263.

8.3

Yund, R.A. and Hall, H.T. (1970) Kinetics and mechanism of pyrite exsolution from pyrrhotite. *J. Petrol.* **11**, 381.

8.4

Buddington, A.F. and Lindsley, D.H. (1964) Iron titanium oxide minerals and synthetic equivalents. *J. Petrol.* **5**, 310.

Haggerty, S.E. (1976) Opaque mineral oxides in terrestrial igneous rocks. In: Rumble, D. (Ed.) *Oxide Minerals*. Min. Soc. Am. Short Course Notes.

Lindsley, D.H. (1976) Experimental studies of oxide minerals. In: Rumble, D. (Ed.) *Oxide Minerals*. Min. Soc. Am. Short Course Notes

Price, G.D. (1979) Microstructures in titanomagnetites as guides to cooling rates of a Swedish intrusion. *Geol. Mag.* **116**, 313.

8.5

Nitsan, U. (1974) Stability field of olivine with respect to oxidation and reduction. *J. Geophys. Research* **79**, 706.

8.6

Putnis, A. (1978) The mechanism of exsolution of hematite from iron-bearing rutile. *Phys. Chem. Minerals* **3**, 183.

Putnis, A. and Wilson, M.M. (1978) A study of iron-bearing rutiles in the paragenesis TiO_2–Al_2O_3–P_2O_5–SiO_2. *Mineral. Mag.* **42**, 255.

9

More Complex Transformation Behaviour

In the previous two chapters we have discussed examples of mineral behaviour which, by and large, could be put into the categories of polymorphic transformations or exsolution. Deviations from equilibrium behaviour could be interpreted within this general context. When discussing cation ordering we pointed out that if the stable equilibrium state could not be achieved in the time available, some other partially ordered structure may form metastably. Similarly, if the exsolution of the stable end member in a binary system was impeded, other mechanisms of phase separation, requiring lower activation energies, were sometimes available. In this chapter we will examine briefly the behaviour of some minerals in which the distinctions between polymorphic processes and phase separation are not so easily defined. Under highly metastable conditions it is no longer possible to model the behaviour on simple nearest neighbour interaction models such as we described in Chapter 4 when we originally drew the thermodynamic distinction between ordering and exsolution.

We shall not discuss any alternative models which might describe such behaviour, except to restate one of the principal themes of this book: when a system is a long way from equilibrium the kinetics of available processes are more important than the thermodynamics of the stable states. Any reduction in free energy becomes acceptable and the most rapid mechanism is the one which operates.

The four examples discussed in this chapter all have one feature in common—the behaviour is dominated by the kinetics of available processes. The behaviour may be complex and not well understood, and so our interpretations are rather speculative. Our aim here is not necessarily to provide the answers but to pose the problems and suggest ways of looking at them. These problems are not merely isolated idiosyncrasies of certain minerals—the behaviour of many common minerals comes into this category, notable among them being plagioclase, perhaps the most widely studied mineral of all.

In general, the more complex the chemistry and structure of a solid solution, the more likely it is to form metastable phases, simply because of the greater number of atomic configurations possible. If the stable behaviour is known, it becomes easier to explain alternative behaviour and the appearance of complex phases. Often, however, there is no general agreement on which is the stable state of the system when equilibrium has not been reached over geological time. The problem of recognizing which configurations are possible and evaluating their relative stabilities is considerable. There are no direct methods for doing this at present, and experimental determinations of free energies of formation are made more difficult by the fact that metastable processes often result in complex intergrowths. One approach which we adopt is to study these minerals in different geological environments and evaluate the microstructures in terms of the extent to which re-equilibration has been possible. Another approach is to make an intuitive guess at the ideal equilibrium behaviour and then interpret the observed behaviour in terms of attempts at attaining this equilibrium.

9.1 The transformation behaviour of cubanite, $CuFe_2S_3$

We take cubanite as our first example because its behaviour is relatively simple, yet involves structural changes, ordering and exsolution. Furthermore, the low-temperature state is well known. The problem is that the transformation from the low-temperature form to the high-temperature form is not reversible experimentally. Instead of transforming back to low cubanite, high cubanite appears to exsolve chalcopyrite when it cools, an observation which has been made on many occasions. This rather curious behaviour can be interpreted in terms of the constraints on the transformations involved.

9.1.1 CUBANITE STRUCTURES

Above about 210°C cubanite $CuFe_2S_3$ falls within the large high-temperature solid-solution field at the centre of the Cu–Fe–S phase diagram (Fig. 3.12). Within this solid solution the structure consists of cubic close-packed sulphur atoms with variable amounts of Cu and Fe disordered over the tetrahedral sites. The structure is based on that of zinc blende which is described and illustrated in section 3.2.3. High cubanite is one specific composition within this solid solution.

In nature, cubanite occurs in a low temperature form, often as exsolution lamellae within chalcopyrite. Natural low-temperature cubanite however, has a structure based on hexagonal close-packing of sulphur atoms with Cu and Fe arranged in an ordered way over the tetrahedral sites. The structure is based on that of wurtzite, with the cation ordering producing an orthorhombic superstructure of the wurtzite subcell. The low cubanite structure has been described and illustrated in section 3.2.1.

This orthorhombic phase undergoes a sequence of transformations on heating. Firstly, the cations disorder within the hexagonally close-packed structure. We will use the terms ordered low cubanite and disordered low cubanite to distinguish these forms. At a slightly higher temperature (≈210°C) the low cubanite undergoes a major transformation which involves the change in sulphur packing. The result is the disordered cubic close-packed structure referred to as high cubanite.

While the order–disorder transformation between the two forms of low cubanite is readily reversible, the major structural change between low and high cubanite is not. Although the change in sulphur packing from h.c.p. to c.c.p. is relatively rapid on heating, it is very sluggish on cooling, and the transformation has never been reversed experimentally. Thus we have the situation of high cubanite failing to transform back to its low-energy state and looking for some other way to reduce its free energy.

9.1.2 THE METASTABLE BEHAVIOUR OF HIGH CUBANITE

The first point to make is that any behaviour other than the transformation back to low cubanite is alternative metastable behaviour. Next we must ask what are these possible alternatives. One of the features of sulphide solid solutions is that some form of cation or vacancy ordering will tend to take place on cooling, even at rapid cooling rates. We have already seen this in the behaviour of metal-rich chalcopyrites (section 7.4). Ordering is certainly a much more rapid process in sulphides than a major structural change and so we might expect that some form of cation ordering could occur in high cubanite.

This brings us to the next point: not all compositions can form an ordered structure within a given distribution of cation sites. With a Cu : Fe ratio of 1 : 2 an ordered structure is simply not geometrically possible within the cubic close-packed sulphur structure without the formation of a large cumbersome hexagonal superstructure. Without going through all of the details (which are in the original source reference) we will conclude that the symmetry of the cubic sub-lattice does not allow the formation of simple superstructures due to ordering at this composition. Ordering in high cubanite is therefore not a suitable alternative transformation.

A composition which can form a simple superstructure by ordering cations is $CuFeS_2$. The chalcopyrite structure simply doubles the unit cell (Fig. 2.4). Experiments have confirmed that a phase with the chalcopyrite structure is exsolved from high cubanite as it cools. Fig. 9.1 shows a sequence of electron micrographs of this process. Fig. 9.1(a) is the disordered high-cubanite phase. The linear features are dislocation structures. In short-term cooling experiments (≈ 5 min) nucleation of the chalcopyrite firstly takes place heterogeneously on these defects [Fig. 9.1(b)], followed at a lower temperature by homogeneous nucleation throughout the crystal [Fig. 9.1(c)]. If cooling is considerably slower (≈ 5 h), the chalcopyrite lamellae are able to form a coarser microstructure [Fig. 9.1(d)].

Presumably there is some degree of disorder possible within this chalcopyrite so that the composition may deviate slightly from $CuFeS_2$. Nevertheless, its observed exsolution from high cubanite indicates that the driving force for this behaviour is the reduction in free energy associated with cation ordering. The fact that this is not possible at the $CuFe_2S_3$ composition, inevitably leads to the exsolution of a different ordered composition. The matrix phase remains in the disordered state.

The behaviour of cubanite can be summarized by the schematic free energy–composition diagrams for ordered chalcopyrite, ordered low cubanite, and the disordered cubic solid solution at low temperatures (Fig. 9.2). The greatest reduction in free energy of the high cubanite (C_0 on the figure) would be ΔG_1, on transforming to low cubanite. As this is too sluggish kinetically, the exsolution of chalcopyrite is an alternative mode of behaviour which although involving a much smaller overall reduction in free energy (ΔG_2) is a relatively rapid process.

This description of the behaviour of cubanite, as well as our previous example of ordering in metal rich chalcopyrites, suggests that in the experimental time scale at least metastable processes play a large part in the behaviour of this copper iron sulphide solid solution.

9.2 Ordering and exsolution in omphacite

Omphacite is a pyroxene which is commonly formed in rocks metamorphosed under the high-pressure, low-temperature conditions of the blueschist facies where the maximum crystallization temperatures are around 450–550°C. Omphacite also

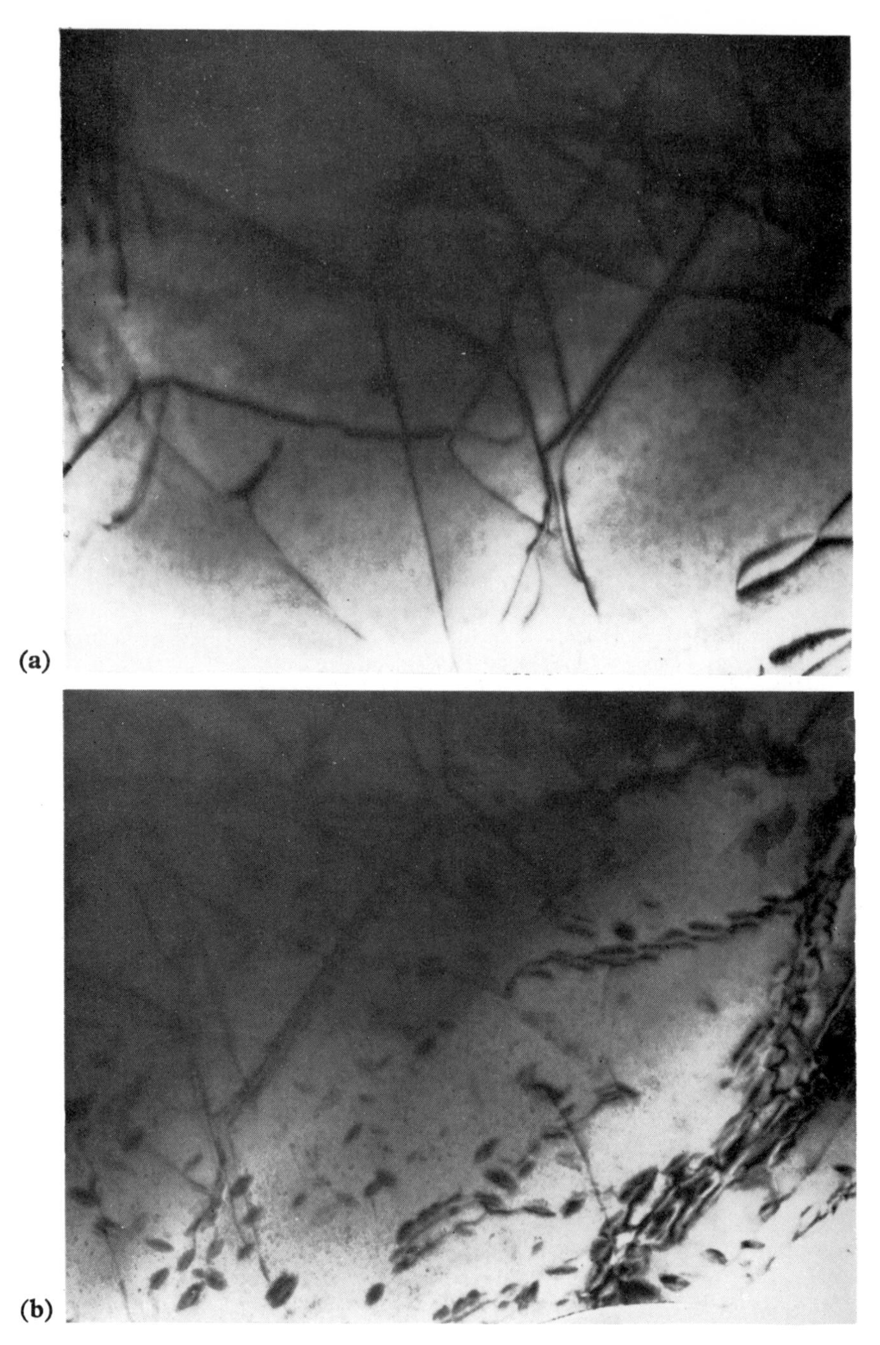

Fig. 9.1 A series of electron micrographs illustrating the behaviour of high cubanite on cooling. All of the micrographs are taken from the same area of the crystal fragment.
(a) Disordered high cubanite. The linear features are dislocation structures. (b) Heterogeneous nucleation of chalcopyrite lamellae on the defects, followed by (c) homogeneous nucleation throughout the crystal. The sequence (a) to (c) takes place with relatively rapid cooling.
(d) With slower cooling, the chalcopyrite lamellae form a coarser microstructure. The length of the scale bar is 0.2 μm.

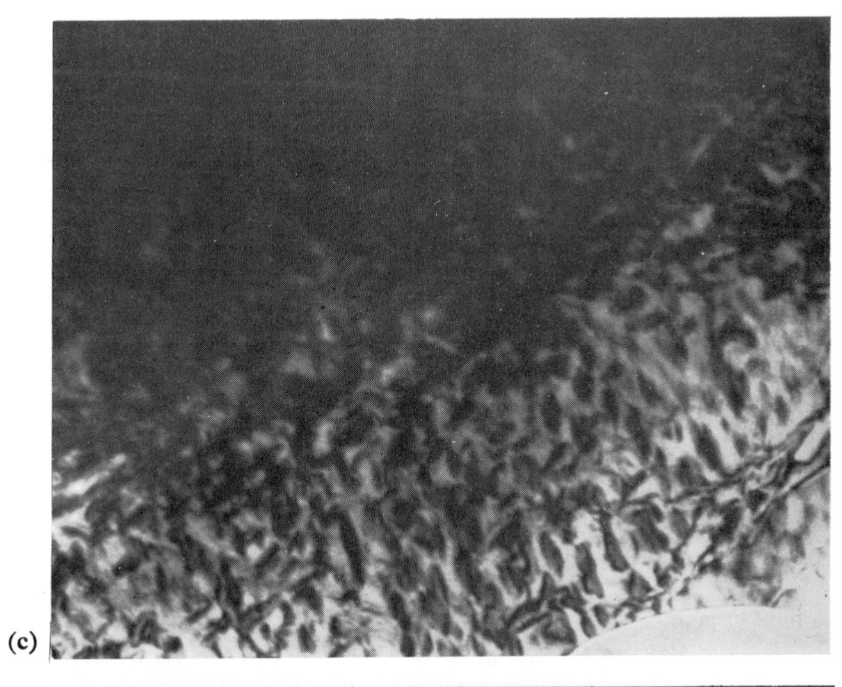
(c)

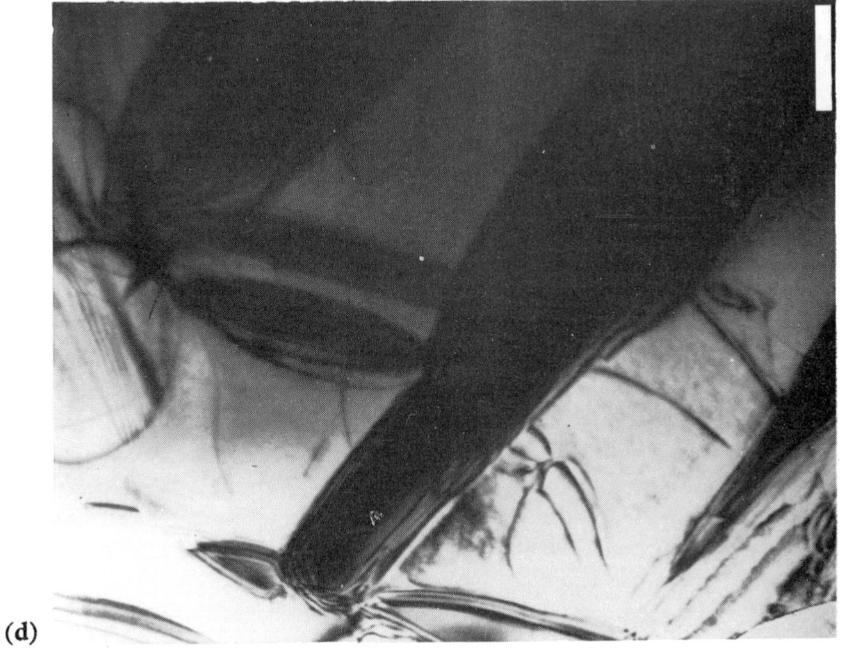
(d)

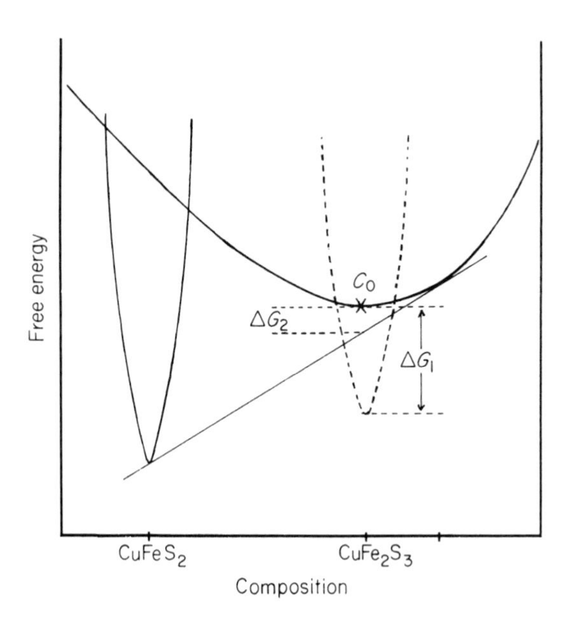

Fig. 9.2 A schematic free energy–composition diagram for ordered chalcopyrite, low cubanite (dashed line) and the disordered solid solution at low temperature. The tangent from the chalcopyrite curve to the solid-solution curve shows the free-energy reduction ΔG_2 associated with the exsolution of chalcopyrite, compared with ΔG_1 for the transformation to stable low cubanite.

occurs in higher temperature eclogite rocks. Minerals which crystallize at relatively low temperatures, below their equilibrium ordering temperature, may incorporate a considerable amount of disorder metastably, and if ordering or exsolution transformations are sluggish at these temperatures, equilibrium may never be reached as the mineral cools. Furthermore, the processes which do operate under such highly metastable conditions are controlled kinetically rather than thermodynamically.

The omphacites form a compositional field roughly half-way between the end members diopside $CaMgSi_2O_6$ and jadeite $NaAlSi_2O_6$. The Ca and Na ions occupy the larger M2 sites, while Mg and Al are in the M1 sites. A coupled substitution of cations

$$Na^+ + Al^{3+} \leftrightharpoons Ca^{2+} + Mg^{2+}$$

as well as some replacement of Al^{3+} by Fe^{3+} and Mg^{2+} by Fe^{2+} in the octahedral sites produces the omphacite compositional field (Fig. 3.23). When the cations are completely disordered, omphacite has a structure similar to that of diopside.

9.2.1 OMPHACITE ORDERING SCHEMES

The presence of antiphase domain structures in natural omphacites from a number of blueschist regions suggests that they crystallized in a more or less disordered state and that subsequently cation ordering transformations took place, thereby reducing the symmetry. In this compositional field however there are a number of possible ordering schemes which could operate and the complexity of the microstructures of omphacites indicates that more than one transformation pathway can be followed even within a single crystal. This complexity makes it extremely difficult to determine the cation configuration in ordered omphacite as quantitative methods of structure analysis require a single homogeneous crystal of the phase.

We can however devise a number of possible configurations which fit the observed electron-diffraction data.

If we consider omphacite as $Na_{0.5}Ca_{0.5}Mg_{0.5}Al_{0.5}Si_2O_6$ we have the following possibilities.

1 Note firstly that problems of charge balance do not allow the formation of a fully-ordered structure in which specified sites are alternately occupied by specified cations. The only way this is possible is to alternate layers of jadeite composition and diopside composition. If taken a step further this begins to look like an exsolution process with the omphacite solid solution breaking down into end-member phases at low temperatures. Whether an intimate mixture of jadeite and diopside layers is more stable at low temperatures than complete phase separation is not known but this is a problem which arises time and time again in minerals and will be considered further below.

2 If Al^{3+} and Mg^{2+} occupy alternate M1 sites, the alternate M2 sites must be statistically occupied by $\frac{1}{3}Na\frac{2}{3}Ca$ and $\frac{2}{3}Na\frac{1}{3}Ca$ to maintain charge balance. On the atomic level therefore Na, Ca cations cannot be fully ordered and we might expect that there may be some preference for Na^+ to occupy sites adjacent to the Al^{3+} sites. Similarly, if the Na^+ and Ca^{2+} were perfectly ordered on M2 sites the M1 sites would be statistically occupied by $\frac{1}{3}Al\frac{2}{3}Mg$ and $\frac{2}{3}Al\frac{1}{3}Mg$.

Electron-diffraction data suggest that within single crystals of omphacite a number of different ordering schemes may operate in different regions coexisting on an extremely fine scale.

In a situation where a number of ordered or partially ordered phases with very similar free energies exist, and where ordering is taking place under metastable conditions, the kinetics of each ordering process will determine the transformation pathway followed. Furthermore, as noted earlier, natural omphacites have a range of composition and this factor, as well as the initial state of order of the omphacite could become very important parameters which might influence the nature of the final product.

9.2.2 ORDERING AND EXSOLUTION

Any transformation which takes place in omphacite after it has crystallized will involve some diffusion of Al (+Fe), Mg, Na, Ca. Transformations which involve longer range diffusion of these cations will be extremely sluggish at these low-temperature conditions, and ordering transformations might be expected to be more rapid than those involving phase separation.

Omphacite microstructures frequently show the presence of omphacite exsolution lamellae within an omphacite matrix, where one or both phases may be ordered. The relationship between antiphase boundaries and exsolution lamellae shows that in many cases ordering throughout the crystal has proceeded first and then has been followed by exsolution processes (Fig. 9.3).

Any ordering process which specifies even to some extent the cation content of certain sites, must by its nature restrict the composition of the phase formed. The degree of order which can be achieved by a polymorphic ordering transformation will depend on the bulk composition of the omphacite as well as its geological history. At some compositions a single partially ordered phase may be formed. At a different composition the degree of order which can be achieved may be less, and

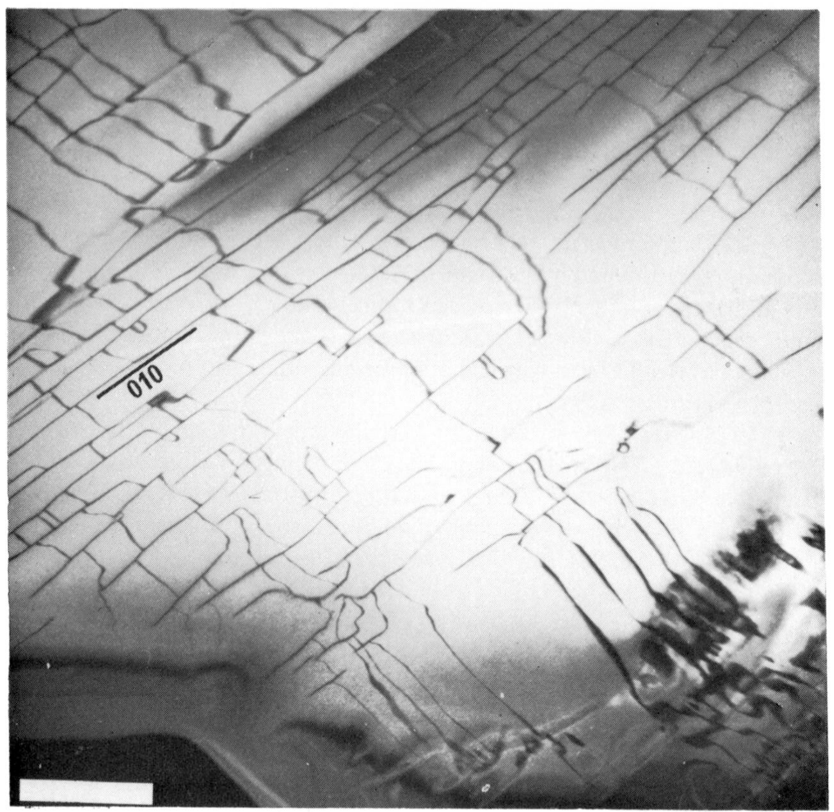

Fig. 9.3 Electron micrograph of omphacite illustrating the relationship between the fine exsolution lamellae [lying parallel to (010)] and the antiphase domain boundaries (which lie approximately normal to them). The microstructure suggests that the formation of the antiphase domains (due to ordering) preceded the formation of the lamellae. The length of the scale bar is 0.5 μm. (Photo courtesy of M. A. Carpenter.)

under suitable geological conditions, the first-formed partially ordered phase may decompose by phase separation into regions with compositions which can achieve more ordered structures. In some cases only one of these phases may be ordered, while the other containing the 'rejected' cations may remain disordered. In other cases both phases may be partially ordered.

This is an example where the free energy of a disordered phase can be reduced by both ordering and exsolution transformations. Ordering usually precedes exsolution because of the more favourable kinetics, while exsolution processes may subsequently reorganize the partially ordered regions into compositions which can achieve a better degree of order. Within this general scheme a large number of mechanisms seem to be able to operate and equilibrium is probably never reached.

9.2.3 IS OMPHACITE STABLE AT LOW TEMPERATURES?

The stable equilibrium state of omphacite is not known under atmospheric temperature conditions. This brings up the question of the stability relationships between the partially ordered omphacite structure and the end members jadeite + diopside. Is omphacite stable relative to this pair under these conditions? The answer is not known, but the problem is similar to those encountered in a number of mineral systems. It is worthwhile therefore to outline the general nature of the problem in this case.

At the high pressures of blueschist metamorphism, omphacite with Al^{3+} in octahedral co-ordination is a stable phase relative to feldspar bearing assemblages in which Al^{3+} is in tetrahedral co-ordination. The composition of the omphacite which crystallizes depends on the P, T-environment of the metamorphism and the bulk composition of the rock in which it forms.

If at low temperatures the omphacite were no longer stable relative to jadeite+diopside, the necessary transformation would involve exsolution transformations with long-range diffusion, which at the low temperatures involved would be extremely sluggish even over geological time. Thus ordering and local exsolution within the omphacite composition region could all be interpreted as metastable behaviour relative to the formation of jadeite+diopside. Alternatively, the partially ordered omphacite structures which are geometrically possible at the midpoint composition between these end members may represent free-energy minima which lie below that for the jadeite+diopside assemblage.

Whichever is the correct interpretation, it is most unlikely that the very wide range of microstructures which have been observed in natural omphacites represents a stable equilibrium state. Any relationship between omphacite and other minerals must also therefore be considered in terms of metastable processes.

9.2.4 OMPHACITE MICROSTRUCTURES AS TIME–TEMPERATURE INDICATORS OF METAMORPHISM

The detailed observations of omphacite microstructures which have been made by Carpenter have shown that, despite the complexity of the antiphase and exsolution textures which are often present, omphacites from different geological environments have quite distinct microstructural features which may be used as indicators of thermal history. In many crystals regular antiphase domain structures exist which are consistent in size and appear to be characteristic of particular environments.

As the ordering process is taking place under conditions of substantial undercooling, the initial size of the antiphase domains is probably very small and virtually independent of the crystallization temperature or the composition. The coarsening of these domains reduces the free energy associated with domain boundaries and will be a function of temperature and time. The ordering process under these conditions is rapid compared to domain coarsening which involves longer range diffusion. Long annealing times at higher temperatures should produce coarser antiphase domains than shorter times and/or lower temperatures.

Although the kinetics of antiphase domain coarsening in omphacite is not yet known, domain coarsening is a diffusion controlled process and if the crystallization temperature can be independently determined a comparison of the antiphase domain sizes can be used to compare the relative thermal histories of the samples. In general, for ideal isothermal growth,

$$(\text{Domain size})^2 \propto e^{-Q/RT}.\ \text{time}$$

where Q is the activation energy for the motion of antiphase boundaries.

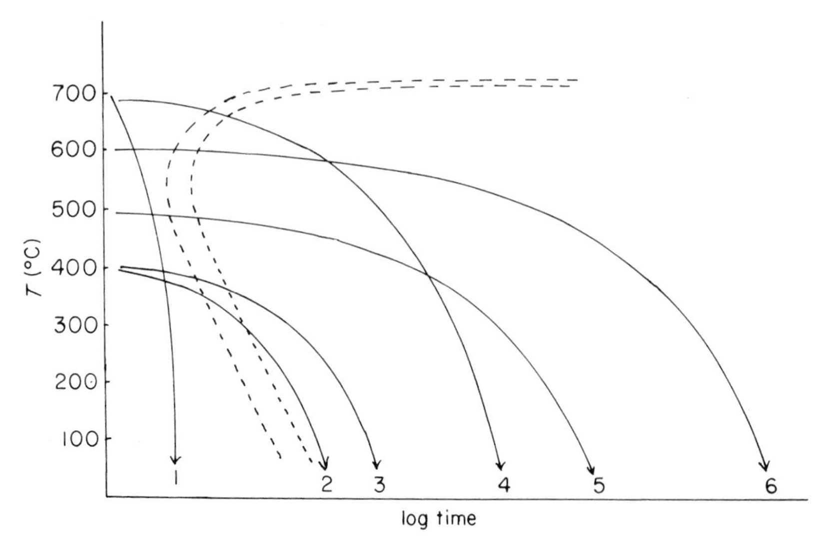

Fig. 9.4 Schematic TTT plot showing thermal histories of omphacite bearing rocks deduced from antiphase domain sizes, and, for comparative purposes, plotted as simple cooling curves from peak metamorphic temperatures. The dashed C-curves represent the start and finish of ordering. (After M. A. Carpenter.)

In Fig. 9.4 a number of cooling curves are drawn on a schematic TTT plot of ordering in omphacite to illustrate this approach. A description of the specimens to which the cooling curves relate and the data on which they are based is given in Table 9.1.

Table 9.1

	Locality	Estimated peak P, T conditions	Size of antiphase domains
1	South Africa (eclogite in kimberlite pipe)	$P \sim 30$ kbar $T \sim 1000°C$	Disordered (no domains)
2	Blueschist from Russian River	$P \sim 6$–7 kbar $T \sim 400°C$ (?)	50 Å
3	Zermatt area West Alps (blueschist)	$P < 10$ kbar $T \sim 580$–600°C	≤ 100 Å (rims) Cores of crystals disordered
4	Red Wine Complex Canada (Blue melanocratic gneiss)	$P \sim 6$–8 kbar $T \sim 600$–700°C	300–400 Å
5	Syros, Greece (blueschists)	$P \sim 13$ kbar $T \sim 450$–500°C	300–500 Å
6	Tauern Area, East Alps (eclogite)	$P \sim 18$–20 kbar $T \sim 580$–600°C	500 Å–0.2 μm

Ref.: M. A. Carpenter (pers. comm.)

9.3 The pyrrhotite problem

At relatively high temperatures, the Fe–S system consists of only two sulphide phases, pyrrhotite $Fe_{1-x}S$ and pyrite FeS_2 (see sections 3.2.2 and 8.3). Pyrrhotite has a structure with hexagonally close-packed sulphur atoms and cations in the octahedral sites. The solid solution towards Fe-poor compositions is achieved by vacancies in these octahedral positions, and at high temperatures these vacancies are disordered. This disordered state however is not stable at low temperatures, and the pyrrhotite problem is essentially one of determining what is stable at low temperatures and to what extent kinetics plays a role in the nature of the low-temperature phases which form on cooling.

The low-temperature situation is very complex and a large number of pyrrhotite-superstructure phases have been found both in nature and experiment. The superstructures are formed when the vacancies become ordered in some way, and in general we can assume that a minimum electrostatic energy will be achieved if the vacancies are arranged with maximum possible separation. One fact which emerges from experimental work is that virtually every pyrrhotite composition is able to achieve some form of vacancy ordering and that this ordering transformation is very rapid. As a result a large number of possible ordered (or partially ordered) phases exist.

We begin by considering the two end-member compositions of the pyrrhotite series. The Fe-rich end member is FeS which has no vacant sites (apart from random defects) and the most Fe-deficient composition is Fe_7S_8.

9.3.1 FeS AND Fe_7S_8

Below about 140°C, FeS (troilite) undergoes a simple displacive transformation and forms a superstructure in which the *c* axis of the high-temperature form is doubled, and is hence referred to as the 2*c* superstructure. This end member is effectively stoichiometric and below the transformation a miscibility gap exists between the 2*c* structure and the other Fe-deficient pyrrhotites.

Fe_7S_8 at low temperatures is able to achieve a highly ordered vacancy arrangement (Fig. 3.10). The vacancies are confined to alternate layers and ordered within these layers so that they are approximately equidistant and at their maximum separation. In this context Fe_7S_8 is a unique composition in that the iron : sulphur ratio makes such an ordered vacancy arrangement possible. The superstructure formed has four times the *c* axis of the disordered cell and is referred to as the 4*c* superstructure.

Although a fully ordered structure is possible at this composition, very rapid cooling of disordered Fe_7S_8 may lead to a large number of mistakes in the stacking arrangement of full and vacancy-containing layers. Such mistakes may lead to different superstructures being formed and since we are effectively dealing with a one-dimensional situation (i.e. the way the layers are stacked) the structure is very flexible with regard to accommodating different possible arrangements. A further effect of rapid cooling is that the formation of a fine antiphase domain structure may result in a large number of non-conservative boundaries (Fig. 5.35) and hence in a deviation from the ideal Fe_7S_8 composition for the 4*c* superstructure. The point we are making is that even at this 'favourable' composition a fully ordered, compositionally restricted low-temperature phase may only be achieved by relatively slow cooling. The problems of antiphase domain structures and mistakes or stacking faults of various kinds becomes much more acute in the intermediate compositions between FeS and Fe_7S_8.

Geometrically it is possible to construct ordered structures for the intermediate pyrrhotites by adding more cation layers to the Fe_7S_8 structure. This maintains the ordered vacancy arrangement while increasing the iron content. For example, the 4*c* superstructure can be denoted by the layer sequence

$$\mathrm{F\,D_a\,F\,D_b\,F\,D_c\,F\,D_d\,F\,D_a} \quad \ldots\ldots$$

where F is a full layer and D_a, D_b, D_c, D_d are the four different positions for the vacancy layers as shown in Fig. 9.5(a).

By inserting extra full layers in a systematic way different superstructures and different compositions can be produced. For example, the layer sequence

$$\mathrm{F\,D_a F\,F\,D_b\,F\,D_c\,F\,F\,D_d\,F\,D_a} \quad \ldots\ldots$$

results in a 5*c* superstructure and a composition Fe_9S_{10} [Fig. 9.5(b)]. The layer sequence

$$\mathrm{F\,F\,D_a\,F\,F\,D_b\,F\,F\,D_c\,F\,F\,D_d\,F\,F\,D_a} \quad \ldots\ldots$$

results in a 6*c* superstructure and a composition $Fe_{11}S_{12}$ [Fig. 9.5(c)].

Clearly, at some compositions a relatively well-ordered structure of this kind is possible although if we attempt to derive an ordering scheme for say, $Fe_{10}S_{11}$ the superstructure becomes more complex (with an 11*c* repeat) and rather like a mixture of Fe_9S_{10} and $Fe_{11}S_{12}$. In fact, any pyrrhotite composition may achieve an ordered state of some kind although the resulting superstructures become very

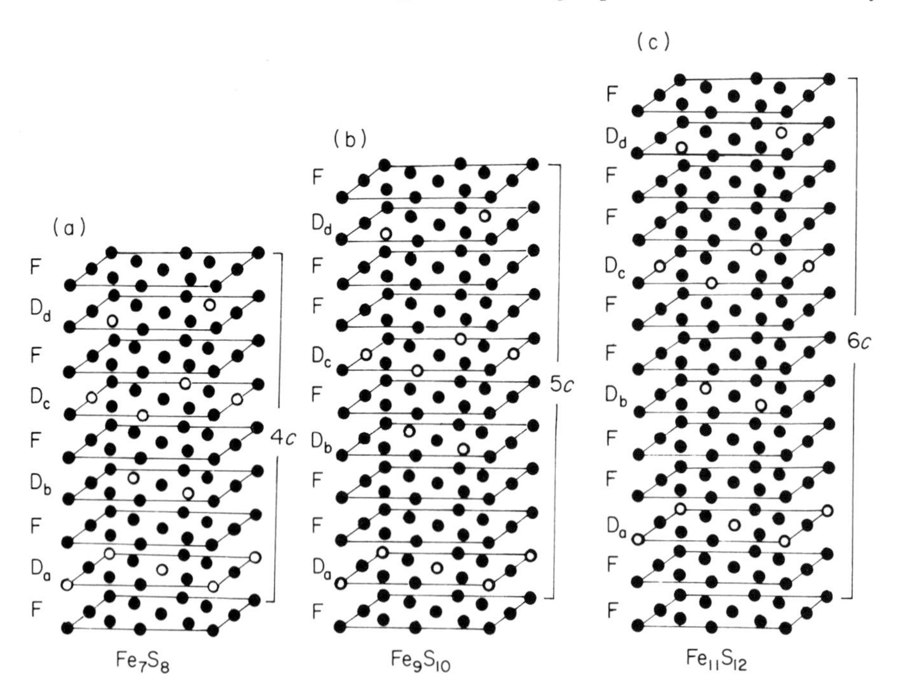

Fig. 9.5 Diagrams illustrating the ordered distribution of vacancies in (a) the 4*c* structure of Fe_7S_8; (b) The 5*c* structures of Fe_9S_{10}; (c) the 6*c* structure of $Fe_{11}S_{12}$. The filled circles are Fe atoms and the open circles are vacant sites. Only the cation layers are shown.

cumbersome and may involve non-integral repeats of the *c* axis. As the structures become more complex the number of different types of antiphase domains and stacking arrangements becomes much greater. For a composition such as $Fe_{10}S_{11}$ a number of different structures are possible, each with a certain degree of order and it is likely that under these circumstances the structure which is produced will depend on the cooling rate, the greatest degree of organization being achieved at the slowest cooling rates.

In natural intermediate pyrrhotites many different superstructures have been reported and it is not unusual to find that a small crystal is an intimate coherent intergrowth of a large number of these phases. In other words equilibrium has not been achieved even over a geological time scale. It is not surprising therefore that low-temperature stability relations remain controversial.

9.3.3. THE COOLING BEHAVIOUR OF PYRRHOTITE SOLID SOLUTIONS

We shall begin with the hypothesis that the assemblage $FeS + Fe_7S_8$ (i.e. $2c+4c$ structures) has a lower free energy than the intermediate superstructures at low temperatures and examine how we might expect pyrrhotite solid solutions to behave if this were the case. This hypothesis is based on the fact that both FeS and Fe_7S_8 are well-organized structures which would be associated with free-energy minima. Intermediate compositions which are not able to achieve the same degree of organization might be less stable. The existence of intermediate superstructures could then be attributed to their ease of formation relative to phase separation to the more stable $FeS + Fe_7S_8$ assemblage.

In accordance with this hypothesis we can draw the free energy–composition diagram at low temperatures to illustrate the possible cooling behaviour (Fig. 9.6). For convenience, free-energy curves for only a few structures are shown, although as we have pointed out, some form of superstructure may be possible at every composition. At higher temperatures the relative positions of the curves would

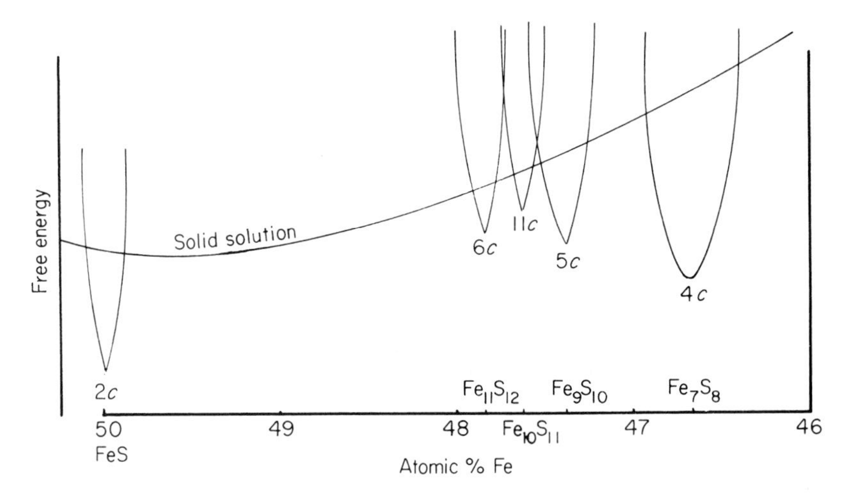

Fig. 9.6 Hypothetical free energy–composition curves for some of the ordered pyrrhotite structures, assuming that the assemblage $2c + 4c$ (i.e. $FeS + Fe_7S_8$) is more stable than the ordered intermediate structures.

change until a temperature is reached where the curve for the solid solution lies below that for any of the ordered phases.

In the situation we have described there are many apparent possibilities for both phase separation and ordering and the kinetics will determine the sequence of transformations. The resultant microstructures are likely to be very complex indeed, as is found in practice. The fact that the free energies of the various ordered phases will be similar, due to their chemical and structural similarities adds a further element of unpredictability and suggests that ordering and exsolution may be virtually contemporaneous under non-equilibrium conditions.

If we examine the likely cooling behaviour of a solid solution of composition around $Fe_{10}S_{11}$, the following possibilities emerge.

1 As the temperature falls, the more Fe-deficient solid solutions will become unstable with respect to the exsolution of the 4*c* phase. Ideally therefore exsolution should proceed as the temperature continues to fall, with the matrix becoming more Fe-rich. At some lower temperature, this matrix phase should break down to the stable 2*c*+4*c* assemblage. Thus the resultant texture would consist of lamellae of the 4*c* phase within a matrix of finely intergrown 2*c*+4*c*. The problem is that the temperature below which 2*c*+4*c* is the stable assemblage is around 100°C and at this low temperature very little phase separation would be possible kinetically. In practice the strong tendency for the vacancies to order at higher temperatures results in complex non-integral superstructures within the matrix phase, which become metastably stranded at low temperatures. The resultant assemblage is 4*c* + intermediate pyrrhotite superstructures.

2 If the cooling rate was such that the original exsolution of the 4*c* phase from the $Fe_{10}S_{11}$ solid solution did not take place, a temperature will be reached where the $Fe_{10}S_{11}$ itself will tend to order. In the first instance a superstructure may be formed at the $Fe_{10}S_{11}$ composition. With slightly slower cooling rates or longer annealing times a further lowering of free energy may take place with some phase separation to compositions which can achieve a better ordering scheme, in this case 5*c*+6*c*. In other words, any intermediate pyrrhotite may be able to form a superstructure on rapid cooling, but may subsequently exsolve to form intergrowths of phases with simpler superstructures.

At the Fe-rich end of the system the exsolution of 2*c* phase from intermediate pyrrhotite solid solutions would reduce the overall free energy at low temperatures. Again the matrix phase should ultimately break down to 2*c*+4*c*, but is more likely in practice to retain the complex superstructures.

Three different types of pyrrhotite assemblages would form under the conditions described:

1 4*c*+Intermediate pyrrhotite.

2 Intergrowths of intermediate pyrrhotite with complex microstructures consisting of various superstructures.

3 2*c*+Intermediate pyrrhotite.

9.3.4 NATURAL PYRRHOTITES

Although natural pyrrhotites are extremely complex, especially on an electron-microscope scale, there is sufficient evidence to support the hypothesis suggested in the previous section for it to be used as a general framework from which to interpret experimental results and observations on natural material. This evidence can be outlined as follows.

1 Natural pyrrhotites tend to fall in three narrow compositional ranges: near FeS, near Fe_7S_8 and intermediate pyrrhotites around the 5*c*- and 6*c*-composition field. Intergrowths between them are of the three types summarized in the previous section. The question remains whether the intermediate pyrrhotite superstructures are stable at low temperatures relative to $2c+4c$.

2 This question is partly answered by the fact that complex superstructures are known to give way to simpler superstructures after long annealing times. During thermal metamorphism of pyrrhotites complex non-integral superstructures change their composition and form intergrowths of the simpler superstructures. This is in accord with the suggestion that the simpler superstructures which have a greater degree of order are more stable at low temperatures and would be formed during slower cooling or thermal annealing. Furthermore, the 5*c* and 6*c* types tend to be the most common intermediate pyrrhotite superstructures, suggesting that they are more stable than the complex superstructures. The next point is to ascertain whether 5*c* and 6*c* phases are stable relative to the $2c+4c$ assemblage.

3 Electron-microscope observations of a natural pyrrhotite from Norway show that intimate integrowths of $2c+4c$ do occur on a very fine scale. This is regarded as strong evidence that equilibrium has been achieved in very small regions producing the stable $2c+4c$ assemblage. The low temperatures at which the breakdown of intermediate pyrrhotites to $2c+4c$ occurs inevitably leads to very fine intergrowths which would not be observable other than by electron microscopy. It is not yet known how common such an assemblage may be, although it is worth noting that the free energy difference between ordered 5*c* or 6*c* structures is not likely to be very different from that of the stable assemblage. The driving force for any transformation is therefore low, and at the temperatures involved the kinetics of the exsolution process must be extremely sluggish.

9.3.5 CONCLUSIONS

We conclude that at low temperatures the assemblage FeS (2*c* structure)+Fe_7S_8 (4*c* structure) is more stable than the intermediate pyrrhotites. Transformations within the pyrrhotites are governed by the kinetically competing processes of vacancy ordering and exsolution to more stable ordered phases. Ordering, which requires only short-range diffusion is more favourable kinetically and may be an alternative mode of behaviour to exsolution which requires longer range diffusion processes. Much of the complexity in observed behaviour arises because of the large number of superstructures possible and the similarity in free energy of various structures and intergrowths. As a result natural pyrrhotites are, on the whole, metastable assemblages.

9.4 The intermediate plagioclase feldspars

No book on mineral behaviour would be complete without some attempt to outline the complex behaviour of the intermediate plagioclase feldspars, and in this final section we will describe some of the important features of this system. It is important to note from the outset that despite the enmorous amount of research which has been carried out on plagioclase feldspars, the details of their behaviour are still not well understood. Indeed, the rather straightforward treatment we will give this system here belies some of the complexities, and it is still controversial in parts. Our approach therefore will be to question the possible modes of stable and metastable behaviour and suggest how the observed microstructures etc. could be interpreted within the overall framework of the general principles.

The features of feldspar chemistry and structure which govern the behaviour of the plagioclases are fairly well understood and have already been outlined in the previous sections on the structure of feldspars (section 3.6.2) and on the cooling behaviour of the albite and anorthite end members (section 7.1). There it was stressed that the way in which each end member achieved its low-temperature ordered form is quite different and that the ordered arrangement of Si and Al in each phase is unique. Because of the charge linkage between Ca and Al on the one hand, and Na and Si on the other it is not possible to form a homogeneous ordered plagioclase between the end member compositions of around An_{0-2} and An_{95-100}. In other words the addition of some anorthite component to ordered albite necessarily introduces some Al, Si-disorder into the low-albite ordering scheme; similarly the addition of albite component to ordered anorthite must also introduce some disorder. The only type of ordered structure in intermediate compositions would be one in which albite-type and anorthite-type regions were separated from each other by some boundary.

For this reason our discussion of intermediate plagioclases is based on the interpretation that at low temperatures the most stable assemblage is low albite and low anorthite, and that under equilibrium conditions all intermediate compositions would eventually exsolve into intergrowths of these end members.

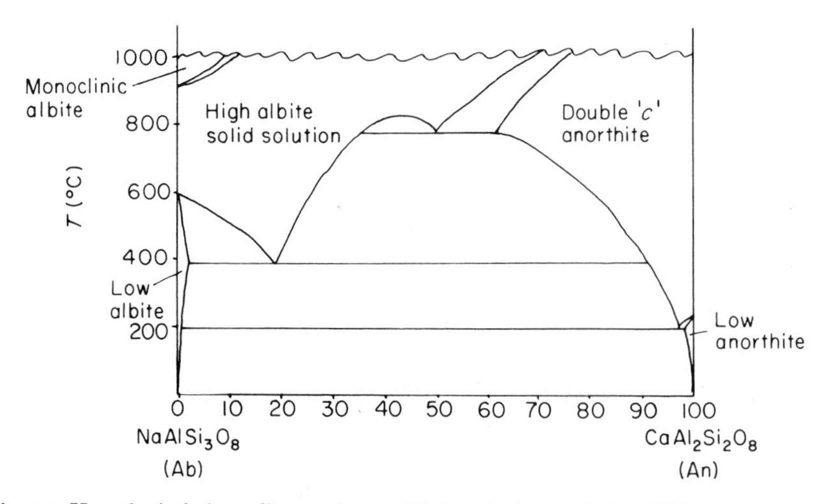

Fig. 9.7 Hypothetical phase diagram for equilibrium in the plagioclase feldspars.

We can summarize this interpretation by a schematic equilibrium phase diagram (Fig. 9.7) in which we illustrate the polymorphic transformations of the end members with the inversion intervals intersecting a simple solvus. Such a solvus would be necessarily assymetric due to the much higher ordering temperatures at the anorthite end of the high-albite solid solution. Furthermore, the substitution of albite component into ordered anorthite, while introducing some Al, Si-disorder, is possible to a much greater extent than the substitution of anorthite component into ordered albite. The latter substitution would involve the formation of Al—O—Al bonds, thus contravening the aluminium avoidance principle (section 3.6.2). The albite ordering scheme is therefore restricted to nearly pure-albite compositions while the anorthite ordering scheme allows a much greater latitude in Al : Si ratio.

We will assume here that if given sufficient time the behaviour of all plagioclases could be interpreted in terms of an equilibrium diagram similar to Fig. 9.7. However we know the plagioclases rarely, if ever, behave in this way. The problems arise because solid-state processes in the system are extremely sluggish due to the high free energy of activation associated with the diffusion of Si and Al. Even at geologically slow cooling rates equilibrium is not maintained and instead of forming intergrowths of albite and anorthite, intermediate compositions are retained metastably, and on cooling develop characteristic microstructures which depend on both composition and cooling rate. In other words, the processes which take place in these plagioclases are metastable alternatives to the true equilibrium behaviour. Under these conditions it is important to realize that the behaviour will not necessarily be related to an equilibrium phase diagram, but will be controlled kinetically.

In the absence of equilibrium exsolution behaviour we must look for other alternatives by which an intermediate composition might reduce its free energy. Before discussing the various types of intergrowths observed in plagioclases we will briefly take up a point first raised in section 3.6.2, that is, what might be the result of any attempt at cation ordering in intermediate compositions?

9.4.1. ORDER-MODULATED STRUCTURES IN INTERMEDIATE PLAGIOCLASE

Cation ordering in the intermediate plagioclases necessarily involves some compromise arrangement in the distribution of Al, Si and Ca, Na due to the charge linkage already mentioned. Any attempt at Al, Si-ordering must be based on an anorthite-type ordering scheme as the albite ordering scheme does not allow large deviations from the Al : Si ratio of 1 : 3 and is therefore irrelevant in these intermediate compositions. Diffraction evidence also suggests that this is the case.

In any local ordering process based on the anorthite scheme we must assume that there is an equal probability that regions will form which are in either of the two antiphase configurations (see Fig. 7.9).

However, we must also consider the tendency for Na, Ca-ordering taking place. For intermediate compositions it is not possible to form regions in which both Al, Si and Ca, Na are ordered. To maintain local charge balance in an Al, Si-ordered region, the Ca, Na cations must remain disordered, and conversely if Ca, Na are ordered Al, Si must be locally disordered. The problem of finding a compromise solution is solved very neatly by the intermediate plagioclases. Although the details have yet to be worked out, it appears that a unique free-energy minimum is achieved within an order-modulated structure such as shown schematically in Fig. 9.8(a). The peaks and troughs in the diagram represent maximum degrees of Al, Si-order, with + and − representing the two antiphase possibilities inherent in the anorthite ordering scheme. The boundary between these antiphase domains is necessarily disordered in relation to Al, Si. Within these disordered boundary regions Na, Ca may order thus reducing the overall free energy.

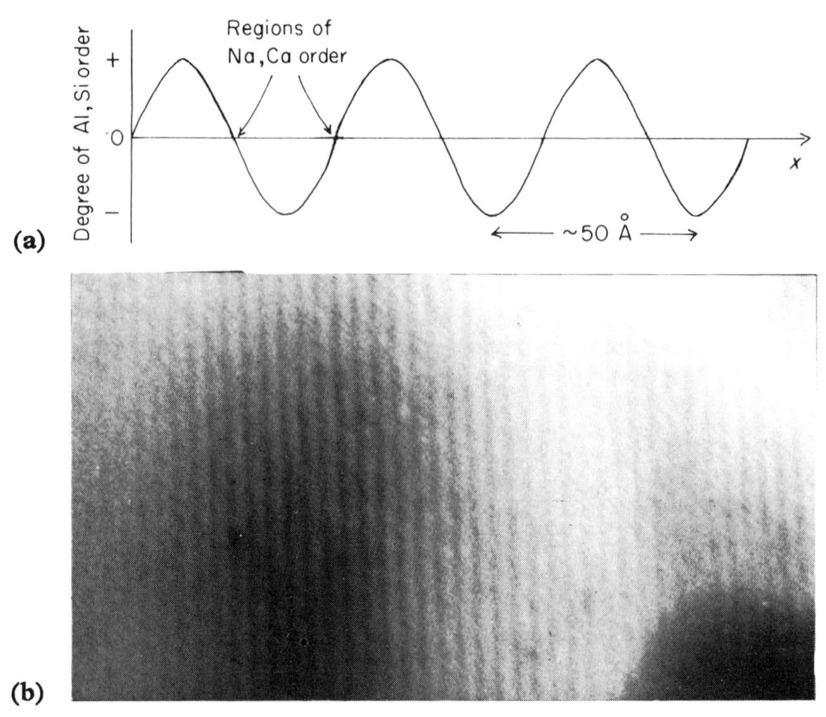

Fig. 9.8 (a) Model for the order modulated structure of intermediate plagioclase. The peaks and troughs are regions of maximum Al, Si-order in antiphase. Between these regions Al, Si are disordered and some Na, Ca-ordering can take place. (b) Electron micrograph of the order-modulated structure in intermediate plagioclase. The wavelength of the modulation is about 50 Å.

This interaction between Al, Si- and Ca, Na-ordering stabilizes the order-modulated structure whose wavelength and orientation are dependent on the composition. The scale of this modulation is extremely fine with a wavelength around 50 Å as shown in the electron micrograph in Fig. 9.8(b). In the rest of this section we will refer to these fine scaled microstructures as the *order-modulated structures* representing states of partial Al, Si and Ca, Na-order.

As well as these fine structures, many natural plagioclases show coarser compositional intergrowths due to phase separation. These intergrowths are of three principal types and are found in different compositional regions.

1 In the central region between about An_{45} and An_{60}—the Bøggild intergrowth.

2 Near the albite end of the system between An_0 and An_{16}—the Peristerite intergrowth.

3 Near the anorthite end of the system between about An_{65} and An_{90}—the Huttenlocher intergrowth.

These intergrowths are generally lamellar and may occur on a scale up to the resolution limit of an optical microscope.

9.4.2 THE BØGGILD INTERGROWTH

The Bøggild intergrowth is a coherent lamellar intergrowth of two intermediate plagioclase components and occurs in the central part of the system between the compositions An_{45} and An_{60}. The lamellae differ in composition by about 10% An. The microstructure [Fig. 9.9(a)] is consistent with a spinodal-decomposition process

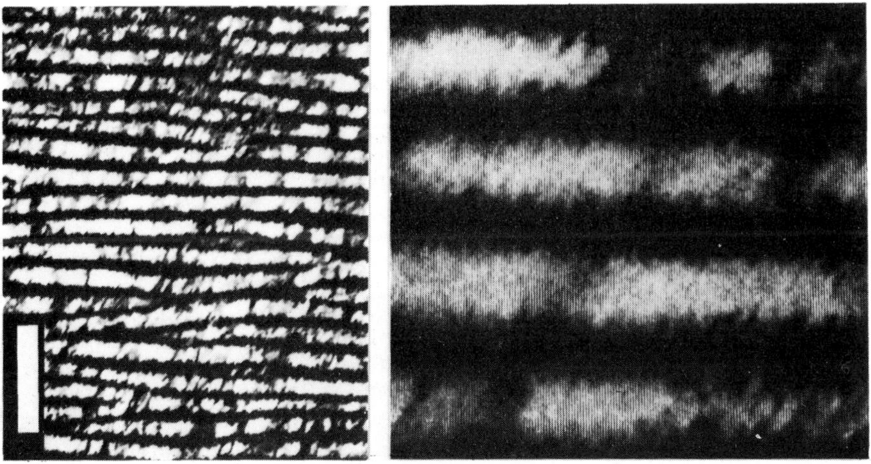

Fig. 9.9 (a) Dark-field electron micrograph showing the lamellar Bøggild intergrowth in a labradorite of composition An_{52}. The length of the scale bar is 0.5 μm. (b) High-resolution micrograph of the same specimen showing the 30-Å order-modulated structure in one of the components. (Photos courtesy of A. C. McLaren.)

which could take place below the crest of a solvus such as shown in Fig. 9.7. The variation in composition across the lamellae also indicates a sinusoidal variation as would be expected from a spinodal process. The overall symmetry remains that of the high-albite solid solution, with coherency and elastic strains between the lamellae determining the amplitude, wavelength and orientation of the intergrowth.

It has also been observed that one or both of the sets of lamellae show the order-modulated fine structure. The spinodal decomposition may have preceded the order modulation indicating that both exsolution and ordering processes are operating to reduce the free energy of the system. The onset of the order modulation takes place at a higher temperature in the more calcic lamellae than in the sodic lamellae. Thus, at more rapid cooling rates the sodic lamellae may retain the disordered high-temperature structure, while under slower cooling the sodic lamellae may also develop the order modulated structure at lower temperatures.

At the present time the relationship between the exsolution process and the onset of ordering is not understood, and some lines of evidence suggest that ordering preceded exsolution. The Bøggild intergrowth is the least understood of the plagioclase microstructures.

Fig. 9.9(b) is a high-resolution dark-field electron micrograph showing the coarser lamellae of the Bøggild intergrowth within which the very fine order-modulated structure can be clearly resolved in one of the components.

It is apparent that while the initial spinodal decomposition process may be interpreted with reference to an equilibrium solvus, the subsequent ordering processes bear no relationship whatsoever to any behaviour which could be inferred from an equilibrium phase diagram.

9.4.3 PERISTERITE INTERGROWTHS

Albite-rich plagioclases with bulk compositions between about An_2 and An_{16} generally occur as intergrowths of pure albite (An_0) and a more calcic plagioclase. The intergrowths are coherent lamellae which may initially occur in two orientations,

Fig. 9.10 Transmission electron micrograph of the peristerite intergrowth in a specimen of composition $An_{7.6}$. The lamellae are of albite and a more calcic plagioclase. The length of the scale bar is 0.5 μm. (Photo courtesy of A. C. McLaren.)

forming a 'tweed' structure similar to that observed in pyroxenes (section 8.2). On coarsening, only the more stable of these orientations is developed and most coarser peristerites show only a single set of lamellae (Fig. 9.10). Further coarsening may result in some loss of coherency.

The calcic component of the intergrowth generally has the order-modulated structure described in the previous section, and it is assumed that the ordering process took place at some lower temperature after the initial phase separation.

The principles governing this type of phase separation are not necessarily related to the possible existence of a two-phase region between ordered and disordered albite as shown by the inversion interval in Fig. 9.7. The coherency of the lamellae suggests some degree of undercooling below the equilibrium temperatures for phase separation. Under these conditions the exsolution of the ordered albite component from the solid solution will result in a decrease in free energy even when the remaining calcic component becomes metastably stranded. This concept has already been discussed in relation to the intermediate pyrrhotites (section 9.3) where exsolution of the more stable FeS or Fe_7S_8 phases occurs over a range of bulk compositions. In the present case the calcic phase behaves as a 'passive' component and plays no direct role in the free-energy reduction associated with the phase separation. It is simply the phase which is left when the compositionally restricted ordered albite phase is formed. At some lower temperature the calcic phase attempts to reduce its free energy by cation ordering resulting in the formation of the order-modulated structure.

The development of the peristerite intergrowth is a kinetically controlled process and the best examples are found in rocks which have had a prolonged annealing at relatively low temperatures. Although the extent of this development depends on the thermal history, an insufficient number of peristerites from various environments have as yet been studied to enable any more than qualitative comparisons to be made.

9.4.4 THE HUTTENLOCHER INTERGROWTH

The intergrowths at the anorthite-rich end of the plagioclase system (between about An_{65} and An_{90}) have a number of similarities to the peristerites at the albite end. They consist of an ordered end member component (in this case anorthite) intergrown with an intermediate plagioclase which has the order-modulated structure. Again the intermediate plagioclase behaves as the passive component and the exsolution process is driven by the reduction in free energy of the anorthite component as it transforms to the fully ordered structure. The anorthite component may contain a considerable amount of albite in solid solution at high temperatures and its composition will generally depend on the bulk composition and the cooling rate.

We can interpret the Huttenlocher intergrowth by the following sequence of transformations:

1 the crystallization of the high-albite solid solution;

2 compositional fluctuations on cooling may form calcic and sodic regions in a lamellar intergrowth;

3 the possibility of a greater degree of Al, Si-order in the calcic component provides the driving force for the phase separation;

Fig. 9.11 Electron micrograph showing the microstructure of the Huttenlocher intergrowth in a calcic plagioclase (about An_{85-88}). The darker lamellar regions are of the order modulated intermediate plagioclase structure in which the periodic antiphase boundaries are well resolved. The light areas are more Ca rich and have the anorthite structure. The length of the scale bar is 0.1 μm. (Photo courtesy of T. L. Grove.)

4 the formation of the order-modulated structure in the sodic component may further stabilize the intergrowth;

5 the anorthite-rich component eventually forms the stable low anorthite structure due to positional ordering of the Ca atoms.

One of the important features of this type of exsolution behaviour which distinguishes it from the 'normal' exsolution phenomena discussed in Chapter 8 is that the transformations are not governed by the existence of discrete stable end-member compounds but by the kinetics of the processes. In other words the compositions of the lamellae in a Huttenlocher intergrowth do not define any special points on an equilibrium phase diagram, but merely indicate the extent to which the phase separation has been able to proceed. The phase separation itself results from the free-energy reduction due to ordering of one of the components rather than to segregation alone.

The cooling history is the controlling factor which determines the ultimate microstructure of a given plagioclase composition. The relatively coarse Huttenlocher intergrowths such as shown in Fig. 9.11 are formed in plagioclases which have had a long thermal history at fairly high temperatures, e.g. in layered intrusions or in granulite or amphibolite facies metamorphic terrains. The rapidly cooled plagioclases from volcanic rocks show no evidence of exsolution at all within the compositional range An_{75}–An_{100}. In the more calcic compositions the ordering transformation to the anorthite structure forms antiphase domains whose morphology also depends on the cooling rate. At more rapid cooling the domain boundaries are smooth and curving, while, at slower cooling zig-zag orientations commonly develop as the boundaries migrate to lower energy structural planes.

When the cooling is sufficiently slow some of the more sodic bulk compositions will develop very fine exsolution textures which form a tweed structure similar to that found in the early stages of the formation of the peristerite intergrowth. The slower the cooling rate the more calcic the bulk compositions in which exsolution is found. The slowest cooling rates result in the greatest compositional differences between the sodic and calcic lamellae, about An_{66} and An_{85}. At this stage the exsolution texture is also visible optically. Note however that the sodic component has the order-modulated structure and in no way represents a stable equilibrium phase, or the end member of a simple solvus as is sometimes suggested.

The most comprehensive study of the effect of thermal history on the microstructure of calcic plagioclases has been made by Grove (1977) who concluded that a number of features of Huttenlocher plagioclases are useful indicators of thermal history. The fact that most natural plagioclases are chemically zoned provides a variable bulk composition which for a given cooling history will produce a variation in microstructure across a single crystal. Thus a careful study of one such crystal may provide a great deal of information on the geological history of the host rock.

9.4.5 SUMMARY

Fig. 9.12 illustrates the bulk compositions over which the three common intergrowths may form and shows the lamella compositions in plagioclases which have had long annealing times at the appropriate temperatures. Nevertheless, even these represent only the nearest approach to equilibrium and are in fact formed by metastable processes in which both ordering and exsolution play a role. More rapid cool-

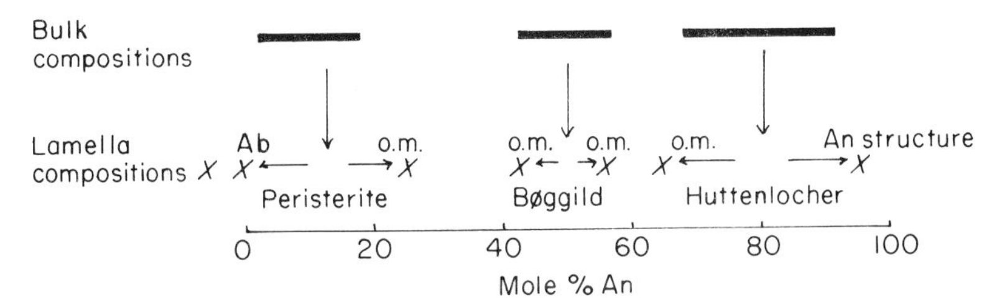

Fig. 9.12 Summary of plagioclase intergrowths showing the bulk compositions over which the three common intergrowths may form, and the approximate lamella compositions which may be achieved. o.m. Denotes order-modulated structure. (After Smith, 1975.)

ing rates result in further deviations from equilibrium and, given the combination of a variety of cooling rates and compositions, very complex intergrowths may be formed. In this section we have avoided many of these complications in an attempt to give an overall simplified picture. The interpretation of the more complex microstructures however will generally differ only in detail and the principles which we have used to describe the main intergrowths will still apply.

References and additional reading

9.1

Cabri, L.J., Hall, S.R., Szymański, J.T. and Stewart, J.M. (1973) On the transformation behaviour of cubanite. *Can. Mineral.* **12**, 33.

Putnis, A. (1977) Electron microscope study of phase transformations in cubanite. *Phys. Chem. Minerals* **1**, 335.

9.2

Carpenter, M.A. (1978) Kinetic control of ordering and exsolution in omphacite. *Contrib. Mineral. Petrol.* **67**, 17.

Carpenter, M.A. (1978) Omphacite microstructures as Time–Temperature indicators of metamorphism. (Abstr.) *Phys. Chem. Minerals* **3**, 61.

9.3

Morimoto, N., Gyobu, A., Mukaijama, H. and Izawa, E. (1975) Crystallography and stability of pyrrhotites. *Econ. Geol.* **70**, 824.

Pierce, L. and Buseck, P.R. (1974) Electron imaging of pyrrhotite superstructures. *Science* **186**, 1209.

Putnis, A. (1975) Observations on coexisting pyrrhotite phases by transmission electron microscopy. *Contrib. Mineral. Petrol.* **52**, 307.

9.4

Grove, T.L. (1977) Structural characterization of labradorite—bytownite plagioclase from volcanic, plutonic and metamorphic environments. *Contrib. Mineral. Petrol.* **64**, 273.

McConnell, J.D.C. (1978) The intermediate plagioclase feldspars: an example of a structural resonance. *Zeit. Kristallography* **147**, 45.

McLaren, A.C. (1974) Transmission electron microscopy of the feldspars. In MacKenzie, W.S. and Zussman, J. (Eds) *The Feldspars*. Manchester University Press, Manchester.

McLaren, A.C. and Marshall, D.B. (1974) Transmission electron microscope study of the domain structures associated with the *b*-, *c*-, *d*-, *e*- and *f*-reflections in plagioclase feldspars. *Contrib. Mineral. Petrol.* **44**, 237.

Smith, J.V. (1975) Phase equilibria of plagioclase. In: Ribbe, P.H. (Ed.) *Feldspar Mineralogy*. Min. Soc. Am. Short Course Notes.

Index